ABOUT ISLAND PRESS

Island Press is the only nonprofit organization in the United States whose principal purpose is the publication of books on environmental issues and natural resource management. We provide solutions-oriented information to professionals, public officials, business and community leaders, and concerned citizens who are shaping responses to environmental problems.

In 1994, Island Press celebrated its tenth anniversary as the leading provider of timely and practical books that take a multidisciplinary approach to critical environmental concerns. Our growing list of titles reflects our commitment to bringing the best of an expanding body of literature to the environmental community throughout North America and the world.

Support for Island Press is provided by Apple Computer, Inc., The Bullitt Foundation, The Geraldine R. Dodge Foundation, The Energy Foundation, The Ford Foundation, The W. Alton Jones Foundation, The Lyndhurst Foundation, The John D. and Catherine T. MacArthur Foundation, The Andrew W. Mellon Foundation, The Joyce Mertz-Gilmore Foundation, The National Fish and Wildlife Foundation, The Pew Charitable Trusts, The Pew Global Stewardship Initiative, The Rockefeller Philanthropic Collaborative, Inc., and individual donors.

HOLDING OUR GROUND

Holding Our Ground

PROTECTING AMERICA'S FARMS AND FARMLAND

TOM DANIELS
AND
DEBORAH BOWERS

ISLAND PRESS

WASHINGTON, D.C. / COVELO, CALIFORNIA

© 1997 Island Press

All rights reserved under International and Pan-American Copyright Conventions. No part of this book may be reproduced in any form or by any means without permission in writing from the publisher: Island Press, Suite 300, 1718 Connecticut Ave., NW, Washington, DC 20009.

Daniels, Tom.
    Holding our ground : protecting America's farms and farmland / Tom Daniels and Deborah Bowers.
       p. cm.
    Includes bibliographical references and index.
    ISBN 1-55963-482-0 (pbk.)
    1. Land use, Rural—United States.   2. Regional planning—United States.   3. Agriculture—Economic aspects—United States. 4. Agriculture and state—United States.   I. Bowers, Deborah. II. Title.
HD256.D36   1997
333.76′17′0973—dc21   96-52665
                                                                                                                 CIP

Printed on recycled, acid-free paper

Manufactured in the United States of America
10 9 8 7 6 5 4 3 2 1

To the men and women who own and operate America's farms and ranches, and to existing and future public-private partnerships that will protect the nation's outstanding land resources.

Contents

List of Figures   xi / List of Tables   xii
List of Photographs   xiii / Preface and Acknowledgments   xv

ONE   The Importance of Farmland   1
TWO   Making the Case for Farmland Protection   15
THREE   Managing Community Growth: What You Can and Cannot Do   31
FOUR   The Business of Farming in America   59
FIVE   Farmland Protection and the Federal Government   75
SIX   State Farmland Protection Programs   87
SEVEN   Agricultural Zoning   105
EIGHT   Controlling Sprawl: Urban Growth Boundaries and Urban Service Areas   133
NINE   The Purchase of Development Rights   145
TEN   The Transfer of Development Rights   171
ELEVEN   Land Trusts: Private-Sector Land Conservation   193
TWELVE   Transferring the Farm and Estate Planning   217
THIRTEEN   Creating a Farmland Protection Package   235
FOURTEEN   Making the Connection: Land Protection and the Big Picture   251

APPENDICES   Sample Forms and Agreements
   A   Model Agricultural Zoning Ordinance   263
   B   Agricultural Use Notice / Nuisance Disclaimer   270
   C   Model Agricultural Conservation Easement   272
   D   Sample Governor's Executive Order   279
   E   Sample Testamentary Easement   282
   F   Agreement Creating an Urban Growth Boundary Agreement   284
   G   Sample Easement Sale Application Ranking System and Application   289
   H   Cooperative Agreement for Public-Private Partnership in Farmland Protection   293
   I   Sample Reimbursable Conservation Easement Acquisition Agreement   295
   J   State Assignment of Conservation Easement   297
   K   State Agricultural District Laws   299
   L   IRS Form 8283 for Easement and Property Donations and Bargain Sales   300

Notes   303 / Glossary   313 / Bibliography   319 / Contacts   323
Index   325 / About the Authors   335

List of Figures

1.1 The Cycle of Farmland Conversion
1.2 Metropolitan Areas of the United States, 1990
3.1 GIS Map of Farms under Easements and Easement Sale Applications in Northwestern Lancaster County, Pennsylvania
3.2 The Zoning Map
3.3 Subdivision: The Creation of New Lots
3.4 Comparing the Fiscal Impacts of Farmland and Residential Development
4.1 The Food and Fiber Industry
7.1 Agricultural Zone (A) on the Community Zoning Map
7.2 Types of Agricultural Zoning
7.3 Agricultural Zones and Permitted Development
8.1 Urban Growth Boundary Example, Lancaster County, Pennsylvania
8.2 Using the Purchase of Development Rights to Create a Growth Boundary
9.1 Farmland Preservation Versus Farmland Conversion in Maryland
10.1 Transferring Development Rights from One Property to Another
10.2 Montgomery County's Agricultural Reserve
10.3 Montgomery County Preserved Farmland
10.4 Application to Sell TDRs to the Pinelands TDR Bank
11.1 Land Trusts throughout the United States, 1992
11.2 Limited Development by Dutchess County Conservancy, Dutchess County, New York
13.1 Farmland Protection in Carroll County, Maryland
13.2 Farmland Protection in Marin County, California

List of Tables

1.1 America's Land
3.1 The Cost-of-Community-Services Studies
4.1 U.S. Farm Size and Farm Product Sales, 1992
4.2 Farms by Value of Products Sold, 1982–92
4.3 U.S. Farm Size, 1982–92
4.4 Farms by Size and Share of Farm Ouput and Government Payments
5.1 Land Evaluation Scores Based on Soil Productivity, Adapted from DeKalb County, Illinois
5.2 Sample LESA System Site Assessment, Adapted from McHenry County, Illinois
5.3 Land Capability Ratings
6.1 State Farmland Protection Programs
6.2 Factors to Consider in Creating an Agricultural District
6.3 Provisions of State Agricultural District Laws
7.1 Sample of County Agricultural Zones Using Minimum Lot Sizes
7.2 Sliding-Scale Agricultural Zoning, Clarke County, Virginia
9.1 Purchase of Development Rights by State to August 1996
9.2 Leading Locally Supported PDR Programs as of February 1997
9.3 Points-Based Appraisal for the Purchase of Development Rights, Montgomery County, Maryland
9.4 Securitized Installment Purchase Agreement Programs
10.1 Examples of Active TDR Programs, July 1996
11.1 Farmland Preserved by Major Land Trusts
11.2 The Strengths and Weaknesses of Public-Private Partnerships
12.1 Federal Estate Tax Rates and Taxes Due, 1996
13.1 Advantages and Disadvantages of Farmland Protection Tools

List of Photographs

1.1 Land-consuming sprawl uses up more space than necessary and makes the landscape look like Anywhere, USA.
2.1 Compact development conserves farmland.
3.1 Farmland slated for development.
3.2 House lots subdivided off a farm.
4.1 Modern farming requires a large investment in land and machinery.
6.1 Farming next to houses is not easy for farmers or residents.
7.1 Agricultural zoning provides protection for a farming neighborhood.
8.1 An urban growth boundary in Oregon runs along the road and separates farmland from development.
9.1 The 221-acre Harold Trimble farm, Drumore, Pennsylvania, preserved through the sale of development rights.
11.1 Gene Garber points to the farms that he and his father, Henry, preserved through donating conservation easements to the Lancaster Farmland Trust.
13.1 Part of a 1,300-acre contiguous block of preserved farmland in Lancaster County, Pennsylvania.
13.2 Farmland overlooking Tomales Bay in Marin County, California.

Preface and Acknowledgments

In recent years, dozens of concerned citizens, farmers, planners, and local elected officials have contacted us for information and advice on how to protect farms and farmland. Although the issue of protecting farmland and the open space it provides is not new, the intensity of the challenge has increased. Farmers are harder pressed to make a living in farming, and communities are struggling to accommodate increasing populations and the development that goes with population growth.

Over the last twenty years many articles and reports have been written about farmland protection. These publications have tended to focus on a particular technique, such as agricultural zoning, or on the loss of farmland in a single county. Even the handful of books on farmland protection have concentrated on specific techniques or programs in certain counties or states.

We believe that citizens, elected officials, planners, and landowners should have a solid understanding of the issues behind farmland protection before specific techniques and programs are devised and large amounts of money spent. All too often, well-intended individuals, communities, and states have come up with programs that do little more than throw money at the problem of disappearing farmland and open space. We hope to help landowners and communities form *strategies* to protect farmland and implement those strategies to achieve long-term farmland protection goals.

How to Use This Book

There are several things you should keep in mind while reading this book. First, you should be aware of the political realities of putting together, monitoring, and revising land protection programs. Remember that supporters of farmland protection cover a wide range of political beliefs. Many conservatives believe in protecting farmland, as do many liberals and moderates. The more supporters you can include and organize, the more effective your efforts are likely to be.

Second, you should look for possibilities of creating public-private partnerships between government and private groups; this may include negotiating agreements with the development community. Third, you should have a solid knowledge of the variety of land protection techniques and understand how to use them as part of a land protection strategy. And fourth, you should understand the needs and motivations of farmers and owners of open spaces and work with these landowners. If the landowners in your community are opposed to land protection, then your strategy must begin with landowner education, focusing on how landowners can benefit from a land protection strategy.

In chapters 1 and 2, we discuss the reasons for protecting farmland and how to explain those reasons to others. In chapter 3, we present the legal basis for managing community growth, the land-use planning process, and the importance of forming a community vision: how the community wants to grow and the role of farmland and open space in that vision. Chapter 4 describes the business of farming, federal farm programs, and the role of land in farmers' decisions. Chapter 5 evaluates the efforts of the federal government in protecting farmland. Chapter 6 analyzes the state-level farmland protection techniques. Chapter 7 explains the use of zoning to protect agricultural land. Chapter 8 describes how local governments and the state of Oregon are using urban growth boundaries to create more compact development and discourage sprawl.

In chapter 9, we explain how a purchase-of-development-rights program works, along with suggestions for funding a program and how landowners can best make use of existing tax laws. Chapter 10 discusses the *transfer* of development rights, where that has succeeded, and what factors are necessary for a successful program.

Chapter 11 looks at the increasingly active role private land trusts play in protecting farmland. In chapter 12, we examine the financial, tax, and estate planning issues involved in passing the farm to the next generation. Chapter 13 reviews the strengths and weaknesses of the many farmland protection tools and emphasizes the importance of putting together several tools in a farmland protection package. Chapter 14 summarizes the "big picture" issues and discusses the link between farmland loss and urban decline.

Acknowledgments

We wish to thank Katherine Daniels, Stephen Grossi, Sam Goodley Jr., Robert H. Daniels, Mark Lapping, Pat O'Connell, John Bernstein, Bob Wagner, David Skjaerlund, Jeff Winegard, June Mengel, Kathleen Bridgehouse, Darlene Petersen, Jean Hocker, David Schuyler, Jordan Henk, Phil Rainey Jr., Jerry Greiner, Kevin Kasowski, Joan Rosen, and our editor at Island Press, Heather Boyer.

CHAPTER I

The Importance of Farmland

The well-being of a people is like a tree.
Agriculture is its root,
Manufacture and Commerce are its branches
and its leaves.
If the root is injured,
the leaves fall, the branches break away,
and the tree dies.
—*Chinese Proverb*

Land is our most precious heritage.
—Robert West Howard
The Vanishing Land

Land and how it is used affects everyone. Lately, this simple fact has dawned on more and more Americans. We have become a spread-out suburban society, dependent on our cars for shopping, commuting to work, and recreation. And the suburban juggernaut of tract housing, office parks, commercial strips, and shopping malls continues to push farther into the countryside. The results have not been good. Across the nation, sprawling suburban development has depleted downtowns, fostered racial segregation, increased the cost of government, threatened air and water quality, eaten up plant and wildlife habitats, and paved over some of America's best farmland.

Throughout the United States people are feeling a growing sense of urgency about land and their communities. While many people believe that growth is inevitable, and even desirable, they recognize that auto-dependent sprawl is neither a pleasing nor a sustainable pattern of development. They fear that if a historic landscape, a pristine natural area, or prime farmland is not protected today, it may be gone tomorrow. And the burden of today's growth will ultimately fall on our children and our children's children.

Here are a few examples. In Minnesota, 24,000 acres of farmland go out of production each year, a rate of 1 acre every twenty minutes. Michigan, with the nation's second most diverse agriculture after California, loses about 10 acres of farmland to other uses every hour. In Colorado, an estimated 90,000 acres of ranch land are being lost to development each year, or nearly 300 acres a day. Farmland loss in Florida is estimated at 150,000 acres a year, the fastest rate in the nation.

And nationally, over 1 million acres of farmland—an area nearly the size of

PHOTO 1.1
Land-consuming sprawl uses up more space than necessary and makes the landscape look like Anywhere, USA.

the state of Delaware—succumb to suburban sprawl each year. That adds up to *more than 2,700 acres taken from farming every day.*

California, the nation's leading agricultural state and the source of most of the fruits and vegetables we eat, loses about 100,000 acres of farmland each year. University of California-Davis Professor Alvin Sokolow says, "Farmland losses appear more significant, however, when measured by the quality of the land resource affected, location, and impact on local economies and environments. The Central Valley is California's only remaining region with the land and water to sustain large-scale farming."[1]

And now even the fertile Central Valley is threatened. The huge valley stretches more than four hundred miles long and a hundred miles wide and is home to six of the nation's top ten farming counties. Fresno County, the number one farm county, produces more food than twenty-four states! But a 1995 study sponsored by the American Farmland Trust estimates that the population of the Central Valley will triple by the year 2040, and sprawling development patterns will consume another 1 million acres of the most productive farmland in America.[2] Conflicts over precious water supplies are certain to intensify, since farmers currently irrigate 6.7 million acres of valley farmland.

Similarly, *Beyond Sprawl*, a 1995 report sponsored in part by the San Francisco-based Bank of America, concluded that "unchecked sprawl has shifted from an engine of California growth to a force that now threatens to inhibit growth and degrade the quality of our life."[3]

The advance of suburbs into the countryside has made farmers less certain about the future of their farms. Here's one example. John Meikle and his family operate a 300-acre, 200-head dairy farm in Cache County, Utah, about five miles from the city of Logan. Each year, Logan and its suburbs creep closer. In 1995, ten acres that the Meikles had rented for cropland was sold for $200,000. The Meikles are facing the choice of what to do about the future. Can they make enough money in milking cows to earn a decent living? Should they continue to farm and hope that their sons will have both the interest and the financial ability to take over some day? Should they sell their farm to another farmer at an affordable "farm price"? Should they sell their land for development and buy a farm somewhere else? Or should they sell the farm for the highest dollar and go into another line of work?

Farmers throughout America face the same choices as the Meikles. And the choice of one farmer will be influenced by the decisions of other farmers, landowners, and the local government. While farmers in the densely populated Northeast feel the most pressure from sprawl, farmers near big cities in many parts of the country are being squeezed for space. Even near rural communities in the Midwest, farmland commands a premium. Iowa dairy farmer Steve Hopkins explained that when he was looking to buy farmland, "It put us in competition with people wanting to own an acreage rather than a place to farm. Any decent county seat town has a strong acreage demand."[4]

Farmers recognize the threat of increased operating costs, rising land taxes, and general headaches from nonfarm neighbors when urban sprawl invades the countryside. Many serious, successful farmers want to continue to farm where they are. Some nonfarming landowners and less successful farmers look to cash in from selling their land at urban values. As a result, there is controversy among the farm community about regulating how much farmland can be developed and how fast.

Farmers and rural landowners who rent to farmers are generally very independent people. Yet the more they cooperate, the more likely they are to stay independent and in farming. Nothing hurts a farming neighborhood more deeply or rapidly than the sight of the first farmer who sold the farm for development driving around in a fancy new car. Then every farmer starts to think about selling out. By contrast, when farmers see themselves as farming together, some remarkable things can occur. In Whitman County, Washington, farmers asked to have more stringent zoning placed on their property to limit their ability to sell land for nonfarm development. The Oregon Farm Bureau has stood up in favor of land-use planning and some of the strongest agricultural zoning in the nation because these measures protect farmland, the source of farmers' livelihoods. And in fourteen states, farmers have voluntarily sold the right to develop their land. They have taken a stand to say that their land should remain in farming.

Farmland Protection: A Wake-Up Call

Despite the facts that farmers make up only 2.4 percent of the nation's population and only one in sixteen rural dwellers lives on a farm, farmers own or rent most of the private land in America. As a result, they hold the key not only to the nation's food supply, but also to managing community growth, maintaining an attractive landscape, and protecting air, water, and wildlife resources.

Most of America's population growth and land-use changes are happening in rural-urban fringe regions, ten to forty miles (and sometimes farther) outside of urban centers. In the rural-urban fringe, the sharply defined boundaries between cities and countryside are being blurred by two types of development. The first is the continued wave of large residential and commercial projects as population centers expand. The second, referred to by geographer Peirce Lewis as "the Galactic City," features scattered homes and commercial strips held together by highways.[5] In between the houses and stores, there are often large open spaces of farmland, forest, and idle land. This dispersed development has greatly increased the confrontation between farmers and nonfarm neighbors.

Both types of development have created well-known problems:

1. Developers bid up land prices beyond what farmers can afford and tempt farmers to sell their land for development.
2. The greater number of people living in or next to the countryside heightens the risk of confrontation between farmers and nonfarmers.
3. Complaints increase from nonfarm neighbors about manure smells, chemical sprays, noise, dust, and slow-moving farm machinery on commuter roads.
4. Farmers suffer crop and livestock loss from trespass, vandalism, and dog attacks. Stormwater runoff from housing developments washes across farmland, causing erosion, and competition for water supplies increases.
5. As farmers become more of a minority in their communities, nuisance ordinances may be passed, restricting farming practices and in effect making farming too difficult to continue.

6. As farms are developed, farm support businesses are pushed out. Remaining farmers stop investing in their farms as they expect to sell their land for development in the near future.
7. Open space becomes harder to find, the local economy changes, and rural character fades.

At the same time, newcomers to the countryside value farmland for:

- the open space and scenic vistas
- protecting air and water quality
- wildlife habitat
- the sense of rural character

Ironically, newcomers can destroy the farms and farmland that they value. And farmers have often sowed the seeds of their own decline by selling off road frontage for house lots to urban refugees. Most of these newcomers still work and shop in the cities and suburbs, some are retired, and others may commute to work through their computers. But they tend to see rural land as an amenity and a place to live, not as productive farmland.

What the Protection of Farms and Farmland Can Achieve

We do not advocate trying to protect or preserve every acre that is currently in farm use. Population growth, economic growth, and real estate development will continue to consume farmland, and some of this growth may be desirable. We do encourage creative, cooperative, and strategic efforts to slow down *the rate* of farmland conversion. Farmland protection efforts should attempt to protect the best-quality farmland that has the greatest chance of staying in farm use. Undoubtedly, this means that some good farms will be lost because of their location close to intense development. But communities that can protect a viable amount of farmland and support the farm operators will reap economic, fiscal, environmental, and aesthetic benefits for years to come.

There is no simple or single solution for protecting farms and farmland. Government regulations and money cannot do the job without the commitment of landowners to farm and their ability to earn a living on the land. In fact, a good price for farm products is probably the most important factor in the long run. But good farm management, effective and efficient local land-use planning, and affordable property taxes are also important.

Most states and the federal government have done little to limit the flood of low-density residential and commercial sprawl into the countryside. All states offer farmers use-value assessment of farms in an attempt to hold down farm property taxes. But use-value assessment has had limited success in the face of urban and suburban growth (see figure 1.1).

Some states and counties have undertaken innovative land-use planning programs; others have spent millions of dollars in an attempt to protect agricultural land from development. Farmland protection is mainly a local and

FIGURE 1.1
The Cycle of Farmland Conversion

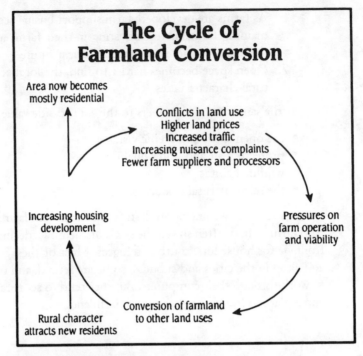

Impact on Remaining Farms

As the density increases in previously predominantly agricultural areas, the impact upon agricultural operations can exponentially increase. The greatest impact of increasing residential development is not just the potential loss of farmland, but the impact on existing farm operations. An increase in the number of non-farm residences in rural areas can often place greater pressures upon farm operations, making it more difficult for them to continue or expand.

regional issue. Decisions on the part of landowners, elected officials, and the general public determine what land will be developed and what land will be protected for future generations.

While communities try to achieve that elusive balance between growth and a pleasant environment, powerful growth forces are at work. Robert Heuer, who has written extensively about farmland loss in the greater Chicago area, says that the conversion of farmland is the result of a "civic-industrial complex." This complex is made up of the real estate and development industries, financial institutions, and the media—especially newspapers, with their dependence on real estate advertising. The complex feeds on the continual outward expansion of cities.

Heuer goes on to say, "the civic-industrial complex acts on what we have always been and remain today: a people blessed with an abundance of land and cursed by an abiding commitment to carve up every square foot of it. I asked one southern suburban [Chicago] official about growth patterns not creating wealth so much as shifting resources from the city and older suburbs. Her response was that we just want what the western and northern suburbs have gotten."[6]

The American Dream of the post–World War II era has for many been a large house on an acre or two in the suburbs or countryside. Today, this pattern of housing is chewing up farmland at an alarming rate. Changing the American Dream to encourage people to live in higher-density housing—say, six homes per acre, will be an enormous challenge. Americans will have to decide whether they want to save their most productive land for agriculture or let it be used for housing, pavement, offices, and shopping malls. While there is much debate about the merits of protecting farmland, it does seem that most Americans would like to see their country grow most of its own food.

What's So Special About Land?

Land is a special resource. It is very difficult, if not impossible, to move, and as Will Rogers observed, "They ain't makin' any more of it." Land is also a remarkably flexible resource. It can be used as a building site for homes, businesses, factories, or institutions or paved over for roads. It may contain valuable mineral resources. Land can be farmed or forested for hundreds of years, if done carefully; or land can be left as a natural area for wildlife habitat, watershed management, recreation, and education.

Land stirs emotions, personal values, and debate. Landowners may see land as:

- security and the foundation for a family
- a legacy to pass on to their children and their children's children, and so on down the generations
- an investment asset and a store of wealth (Americans have more of their wealth tied up in real estate—about 20 percent—than any other country except Japan)
- a commodity to be bought and sold
- a renewable resource that produces food and fiber
- beautiful open space and a way to keep in touch with nature

Elected officials often view land as:

- the property tax base or wealth of the community
- a site for new growth and development

The general public usually looks at land as:

- providing space for housing, work, and shopping
- providing open space for scenic vistas, wildlife habitat, recreation, air and water quality, and farming

These different attitudes toward land often clash in the community's effort to manage growth. American-style land development occurs within a land market based on private property rights and influenced by government spending programs and local ordinances. Private property is a basic and long-standing institution in America, and most of the nation's land is held by private landowners. But a landowner cannot use land in ways that threaten the health and safety of neighbors. Both government infrastructure programs and local land-use ordinances are intended to produce safe, healthy development patterns.

Land also has "public good" aspects. For example, the public may derive enjoyment from views of open farmland, yet the farmer is not able to charge a price for the views. This tension between the private ownership of land and the public interest in land use is a fundamental and continual issue in community efforts to manage growth.

America's Land Resources

The United States covers about 2.3 billion acres, of which local, state, and federal governments own almost 40 percent. Most of the privately held land in the United States, about 940 million acres, is owned by farmers. Around 530 million acres of farmland are classified as grassland or rangeland, mainly in the western states. There are an estimated 360 million acres of cropland and another 50 million acres of potential cropland. The remaining private land consists of 400 million acres of forest, much of it poorly managed; about 50 million acres of urban development; and another 27 million acres of scattered settlements.[7] (See table 1.1.)

Although these figures do not suggest that America is running out of land, the *quality of land* is a different issue. Prime farmland—designated as Class I and II soils by the Natural Resources Conservation Service—is the most productive and easiest to farm. There are only 43 million acres of flat, fertile, Class I farmland, which land-use expert Peter Wolf calls "the gem of the nation."[8] About 200 million acres are considered Class II soils. So, overall, less than one-third of America's farmland rates as prime for production.

Because prime farmland is level to gently sloping and is well drained, it is also the cheapest land to develop for houses, offices, and factories. As more

TABLE 1.1
America's Land

All Land		
	Publicly owned	839 million acres
	Privately owned	1,417 million acres
	Total	2,256 million acres
Privately Owned		
	Cropland	410 million acres
	Grassland and Rangeland	530 million acres
	Total Farmland	940 million acres
	Forest Land	400 million acres
	Developed Land	77 million acres
	Total Private Land	1,417 million acres

Note: An acre is 43,560 square feet, or a parcel of land 200 feet by 217.8 feet.
Source: 1992 Census of Agriculture and *1992 Summary Report, Natural Resources Inventory.*

prime farmland is developed, production must shift to lower-quality soils that are more prone to erosion, produce lower crop yields, and are more difficult to farm profitably.

Farmland: The Land That Feeds Us

When you are in the supermarket, think about where all the foods you are choosing came from. Unless it's summertime and your supermarket makes a special effort to carry local produce, the fruits and vegetables probably came from California, the beef was most likely raised on a feedlot in the Great Plains, the bread and cereal were created from grains grown in the Midwest, the orange juice from Florida orange groves (if not imported), peanut butter from Georgia, cheese from Wisconsin; the milk and eggs may be from your own or an adjacent state.

Farmland is a strategic resource, fundamental to our nation's security. Fertile soils, temperate climate, and available water have made the United States the most productive agricultural nation in the world. America has 7 percent of the world's tillable land but produces 13 percent of the world's food. Each American farmer, on average, feeds over 120 people, and American consumers pay the smallest percentage of their income for food (about 12 percent) of any country.

Most of us take for granted the food we eat. While we fill our shopping baskets, we don't consider the land, the water, or the people who grew, packaged, and transported the food for our convenience. Most states produce only a small fraction of their food needs. If a truckers' strike lasted more than two weeks, most supermarket shelves would be bare. But barring drought, flood, insects, and diseases, there is another, longer-term threat: the paving over of America's best farmland.

There are three main farming regions in America: California, the South, and the Midwest. The Central Valley dominates California agriculture and produces most of the nation's fruits and vegetables. Most of the South's agriculture (not including Texas) is found in the Mississippi Delta, central and southern Florida, and on the inner coastal plain of the Carolinas, Georgia, and Alabama. The great grain and livestock belt of the Heartland stretches from central Ohio northwest to the Dakotas and southwest through Illinois, Missouri, Kansas, and Texas.

But the strength of American agriculture is the variety of its farms. Every state has farms. And agriculture is important to the economy and appearance of many states, counties, and municipalities. Agriculture is the leading industry in Pennsylvania, for example, a state most people associate with steel mills and Philadelphia lawyers! Farms provide numerous jobs in farm implement, feed, seed, and hardware businesses as well as in firms that process, transport, and market farm products. In fact, roughly one out of every ten American jobs is related to agriculture.

The Loss of Farmland: How Serious?

Most American cities began in places with rich farmland and ample water supplies; and even today some of the nation's most fertile land is within or close to metropolitan areas. In a 1993 study, the American Farmland Trust reported that over half of the nation's farm product sales came from counties that are either part of a metropolitan area (30 percent) or adjacent to one (26 percent) and feeling the pressure of suburban and ex-urban growth.[9]

America's farmland resources have come under increasingly intense pressures since World War II. The period 1945 to 1990 was highlighted by a sharp decline in the number of farms and the sprouting of new suburbs around cities both large and small. Although the amount of farmland decreased by 17 percent, advances in technology nearly doubled crop yields and livestock production. The remaining farms grew larger to become more efficient.

Meanwhile, the federal government aided suburban sprawl through home financing and the mortgage interest deduction, as well as lavish spending on highways, schools, and sewer and water facilities. Interstate highways and improved communications and utilities made more distant land accessible for residential and commercial development. Gas stations, fast-food stands, and other franchises began to spread outward along major roads, with subdivisions of tract homes filling in between. Often parcels of farmland and open space became trapped between subdivisions and commercial strips as builders and developers leapfrogged around large landowners who were not yet willing to sell for development.

An alarming trend has been the increase in the amount of land used per person, indicating larger house lots and a more dispersed population. By 1990, urban and suburban settlements consumed almost one-third more land per person than in 1970. With the rise of "edge cities," large shopping and employment centers on the outskirts of metropolitan regions, housing subdivisions have been pushing ever farther into the countryside.

Over the past thirty years, the competition for land has accelerated. Land prices, even accounting for inflation, have increased, indicating a greater scarcity of land to meet demands. Between 1970 and 1990, the number of metropolitan counties (counties that contain a city of at least fifty thousand people) has grown from 446 to 599 of the nation's 3,137 counties. (see figure 1.2). These metropolitan counties include 16 percent of the land area of the United States and nearly three-quarters of the nation's population. About one-fifth of America's prime farmland is located within metropolitan counties. When counties adjacent to metropolitan counties are included, these expanded regions include over one-third of the nation's prime farmland.

Estimates of the loss of farmland to other land uses vary from about 1 million acres to 1.5 million acres a year.[10] At that rate, the amount of land in urban use is projected to double in fifty years. But the impact of urban development on farmland and open space cannot be measured just in terms of acres converted or higher land prices. Often the knowledge that development is likely will cause farmers to reduce investment in buildings and equipment, as they

FIGURE 1.2
Metropolitan Areas of the United States, 1990.

foresee the eventual sale of the farm. Another important factor is the compatibility, or lack of it, between new houses, offices, stores, and factories and nearby farms. Once farmland becomes interspersed with houses and other development, farm operations change to small vegetable farms, horticulture greenhouses, and horse farms or the land is sold to grow more buildings.

At the same time, the structure of agriculture has changed dramatically. Today, the top 20 percent of the nation's 1.925 million farms produce 90 percent of all farm output. In 1992, the average age of an American farmer was nearly fifty-four years old. And 40 percent of all farmland was rented.

These conditions suggest that the next ten to fifteen years could be crucial to the future of America's farmland. As older farmers retire, what will they do with their land? Pass it on to other farmers? Sell it for development? Will owners of rental land keep renting to farmers? And will the number of smaller and medium-size family farms continue to shrink?

The Search for Solutions

Since Maryland first enacted a property tax break for farmland in 1956, programs to protect farmland and open space have been implemented in every state. Some states have gone much further and established programs that actually save farmland from development, thereby assuring future farming opportunities. One of the aims of this book is to examine how well the many farmland and open-space protection techniques have performed. But a more important purpose is to compel state and local governments, farmers, landowners, and concerned citizens to reexamine their protection efforts and explore new ways to achieve protection goals.

There are several ways that farmland protection efforts can work to overcome patterns of sprawl and help promote local agriculture. To achieve these goals, the general public and elected representatives must recognize that protecting farmland makes good sense for their communities. Farmers and agricultural businesses benefit from a stable land base. The loss of farmland often means not only the loss of agriculture-related jobs but an imbalance in local finances as farmland is converted into houses and property taxes rise to pay for new schools and other public services. Nonfarmers benefit from the open space, watershed protection, and rural character that farmland provides.

If farmland is to be protected, it must first be profitable to operate a farm. One of the most overlooked aspects of farmland protection is helping farmers to stay in business. Farmers must have the protection and freedom to expand or change their operations in order to remain competitive.

Private citizens and public officials throughout America have searched for solutions to slow the pace of growth or channel growth into more compact development that maximizes the use of existing services.[11] Many places have undertaken innovative efforts to stem the tide of suburban sprawl and protect valuable farmland and open spaces. By blending financial incentives, voluntary action on the part of landowners, and government land-use regulations, these places have tried to maintain what is special about their communities: the appearance, the economy, the character, and the social cohesion.

Protecting land resources should have the same high priority as protecting clean air and clean water. It is not too late to stem the loss of valuable lands; but citizens, organizations, and governments need to be energized and empowered to take action, and landowners and the development community must be made part of the debate about how to manage the growth of communities and regions. Protecting land resources requires money, and an important element in land protection is educating the public because tax dollars and political support are essential for effective land protection efforts. At stake is the future of our communities and our quality of life for generations to come.

Protecting Farmland or Protecting Open Space?

Most nonfarmers think of farmland as open space rather than as part of a farming business and the local farming industry. The emphasis of this book is on protecting farmland as an economic asset that also happens to be pleasing to look at.

The protection of nonfarm open space often involves smaller properties, a wide variety of landowners, and issues of parks, trails, greenways, public recreation, and wildlife habitats. Open-space protection is mostly a concern of growing suburbs where the landscape is filling up with development and not much farming remains. Even so, many of the tools and programs discussed in this book can be used to protect open space that is not farmed.

Farmland protection may be thought of as open-space protection without public access to the property. Just because land is zoned for agriculture or under a conservation easement, the public has no right to go walking on it. The land is still private property. But the fact that protecting farmland also protects open space is important for gathering public support for starting and maintaining farmland protection programs. The protection of open space is really one more reason—in addition to benefits to local public finances and economic development—why protecting farmland makes good sense for a community.

CHAPTER 2

Making the Case for Farmland Protection

If we don't properly manage this growth, the nation will lose some critical agricultural production capacity. Indeed, we are already losing it. As we lose unique farmland, we must rely more on foreign countries for fruits and vegetables. And as we lose more prime farmland, found disproportionately near cities, agriculture is forced onto marginal land that is more expensive to farm and leads to greater environmental impacts.[1]

—Ed Thompson, Jr.
American Farmland Trust

If we decide we want farmland, we must build the political framework.[2]

—Volker Eisele, vintner
Napa County, California

If you want to protect farmland from development in your community, you first need to understand why farmland needs to be protected. Aside from the aesthetic reasons that are plain to see, you need a good grasp of the fiscal and economic benefits of retaining farmland. Once you understand why protection is necessary, you need to tell others. Organize a group of interested citizens who are willing to inform the public and elected officials. Work to build a community consensus for land-use goals, tools to protect land resources, and incentives for farming. Most important, work closely with farmers and farm groups, such as the Farm Bureau, and include them in your organization.

Protecting farmland is good fiscal policy. While every community is different, a growing number of studies conducted by the American Farmland Trust, Rutgers University, and others show that farmland provides fiscal benefits by generating more in local taxes than it demands in local services. Farmland and open space require few public services, unlike residential development, which puts a heavy burden on local governments to build schools, provide public safety through police and fire services, and develop infrastructure such as sewer and water facilities and roads.

Cost-of-community-services studies, conducted by the American Farmland Trust (AFT) in several communities nationwide, have consistently shown

>**Farmland protection advocates must demonstrate that protecting farms and farmland will benefit both the taxpayers and the farmers.**

that farmland pays more in taxes than it demands in public services. Meanwhile, because they demand more in public services than they produce in tax revenue, houses and apartments end up costing the community.

In a 1994 study, for example, AFT found that in three Minnesota cities, residences cost more than twice as much as farmland in public services. For every tax dollar paid by residences, the towns spent an average of $1.04 in public services, including education, fire and police protection, and roads. But for farmland, the cities spent an average of fifty cents in services for every dollar raised.

The Center for Urban Policy Studies at Rutgers University conducted a fiscal impact study in 1992, during development of the New Jersey State Plan. The study estimated that the state would realize a savings of $1.3 *billion* over twenty years if development were kept adjacent to cities rather than continually being allowed to leapfrog into rural areas. If the state plan were adopted, the study said, 78,000 acres of farmland would be consumed by residential and commercial development by the year 2010; without the plan, 108,000 acres would be developed.

A 1993 study in San Joaquin County, California, estimated the cost to the county of an acre of farmland converted to housing at $182,000.[3]

Some studies have even indicated that industrial and commercial growth, while seeming to boost the tax base, can have effects that are not fiscally beneficial. For instance, a study in Augusta County, Virginia, concluded that surplus revenues generated by industrial growth in 1991 were not large enough to offset the deficit created by residential development that accompanied the industrial growth. The lesson for citizens is to insist that economic development officials and politicians check their assumptions about what is good for the local economy.

The community's choice is between paying for land protection or paying to service new development.

Another way to look at paying for land protection versus paying to service new development is this: Both cost money, but money spent on protection is more likely to produce the outcome that local residents want: a healthier, less congested, and prettier environment. And protected land tends to increase the value of nearby property.

There have been cases in which elected officials have been so burdened with the costs of development that preserving land, even by buying it at fair-market value, was a better deal than having it developed into suburban density housing. In Washington Township in Morris County, New Jersey, township officials conducted a fiscal impact analysis to determine the cost of providing services to the three hundred or more homes that were planned for a 850-acre tract. Not only would all the new homes require the construction of a new school and provision of other services, but the tract was located in the heart of an area the township was looking to preserve through the state's farmland preservation program. In 1994, township officials decided it would make more sense to purchase the property outright, through an $11 million bond issue, than to see it

PHOTO 2.1
Compact development conserves farmland.

developed. The property was purchased, and gradually the development rights were sold to the state farmland preservation program.

Several townships in Pennsylvania's Bucks County came to the same conclusion in 1995, committing millions of dollars to purchase development rights or to purchase land outright to keep it from being developed. Because education expenses make up the majority of local government budgets, the primary issue was the cost of schools—overcrowded conditions, soaring construction costs, rising school taxes. To stop imminent development in the only sure legal way, the townships set out to buy the development rights or buy the land.

Protecting farms and farmland is good economic development policy. Most local officials seem to think that economic development means bringing businesses into the community. All too often, the retention of existing businesses is overlooked. A farm is a business, and it is connected to other farms and farm support businesses to form a local agricultural industry. Officials can be made to see that promoting local agriculture is a form of economic development: jobs, investment, income, and tax base.

Any argument demonstrating the negative fiscal impacts of farmland loss should be balanced with a detailed examination of the positive economic aspects of protecting farmland and farming in a locality. The state department of agriculture and local university extension can provide statistics on the economic value of a county's farming operations. But the statistics must reflect all segments of the agricultural economy, including produce markets, nurseries, orchards, horses, farm equipment dealers, and farm supply businesses. All of these operations are interconnected and provide services or products to each other.

Protecting local farming promotes a diverse local economy. Farming provides jobs not just on the farm but also in the transportation, processing, and marketing of farm products and in farm support businesses—the feed, seed, hardware, and machinery dealerships. In several states, notably Vermont, Pennsylvania, Wisconsin, Oregon, and Virginia, farmland and open space are the foundation of an important tourist industry. Farms also provide valuable hunting and fishing sites.

If a county or state does not work to protect its farmland, and farmers believe that development is inevitable, they will not make long-term investments in buildings and equipment and will eventually sell for development. This impact is felt by farm-support businesses that are forced to close down. In Pennsylvania, for example, the number of farm equipment and supply dealers in the eastern part of the state, dramatically affected by sprawl, decreased from 600 in 1982 to 275 in 1992.

Besides benefiting the community, protecting farmland benefits farmers.

Most serious and successful farmers recognize that suburban sprawl raises property taxes, even with use-value taxation for farmland. More people means more public services and more children to educate. Rising property taxes can be a powerful reason to move or leave farming altogether. Putting new development in places where it can be serviced efficiently and where it will not conflict with farm operations can benefit both farmers and the entire community.

According to Oregon farmer Ken Buelt, "The Oregon Farm Bureau Federation is a strong supporter of Oregon's land-use system as a whole and of restrictions on farmland in particular. . . . Under Oregon's system, areas not needed for urban growth are protected from sprawl. The result is that the price of land—agriculture's biggest capital input—reflects agricultural values. Farmers can thus afford to reduce their unit costs by expanding their operations at prices that reflect farm values. . . . Farmers are not forced to pay residential values for agricultural land."[4]

While many farmers in metropolitan areas may oppose agricultural zoning to protect the high land values that result from sewer lines, highways, and nearby development, this does not necessarily mean they want to sell their land for development. But they may not be able to afford to farm or keep land as open space. Incentives, such as property tax breaks or purchase of development rights, can help landowners keep their property and pass it along intact to the next generation.

To a farmer, the best way to protect farmland is to make farming more profitable. Food-processing plants create a demand for and add value to locally produced crops and livestock. Locally organized farmers' markets and community-supported agriculture projects offer direct marketing of farm products to consumers. Niche marketing of specialty crops, animals—from goats to emus—and farm-grown fish (aquaculture) is of increasing importance. Local regulations that allow farm-based businesses, such as welding, woodworking, or bed-and-breakfast inns, can help farmers supplement their farm income.

But as growth comes closer, farmers' choices about what to do with their land do become limited by what other farmers do, by whether farm support businesses stay open, by whether land is still available to rent, and by many other factors that affect the profitability of farming in a particular locale.

Few farmers will resist a realistic alternative to development, and many farmers will provide at least some vocal support for a purchase-of-development-rights program, which gives farmers cash compensation for restrictions placed on their land.

The most obvious and immediate effect of any level of farmland protection for farmers is that it acknowledges their importance to the community when they may have been wondering whether farming had a future there at all. A "right-to-farm" ordinance, for example, tells farmers that the law supports their way of life and work. A purchase-of-development-rights program gives farmers an opportunity to stay in farming by getting cash out of the land, reducing debt, and keeping the land affordable for the next generation of farmers. Among the more obvious benefits to the farmer in a purchase-of-development-rights program is the money the farmer receives. But the success of the program for the locality as well as for the individual farmer depends on a great number of farmers participating in the immediate community. A certain number of farms and a certain number of acres preserved, creating enough farming activity to sustain the farm service businesses and to provide enough farmland for rent, are essential. Once this is accomplished, the benefit to the individual farmer is substantial and permanent.

A farmer benefits in other ways from selling development rights: property taxes may be reduced; better estate planning is often undertaken; land surveys are conducted; and boundary uncertainties or disputes are resolved. These are things that are usually put off, because they create a great deal of stress. The farmland preservation process helps the farmer get on track in accomplishing these tasks.

Farmland protection will minimize conflicts with nonfarm neighbors. Farmland protection through zoning and the purchase or transfer of development rights can mean fewer houses built in farming areas. This will likely result in fewer conflicts between the farmer's activities and the way of life of nonfarming residents.

Over the last twenty-five years, many Americans have moved to the countryside just beyond the suburbs of major cities. This type of country living can offer the best of urban and rural lifestyles. There is access to urban jobs, shopping, and cultural events; yet the countryside is peaceful, safe, pretty to look at, and blessed with clean air and water. Often people move in next to farms because of the scenery. But farming as practiced today is essentially an industrial process that can include heavy machinery, chemical sprays, and intensive livestock operations. Farms often generate noise, dust, and odors that can spill over onto neighboring properties. And slow-moving farm machinery can tie up traffic on narrow country roads. Also, the busier traffic makes it hard for farmers to move farm equipment along roads; and as the land base becomes more fragmented, farmers must travel farther to farm rented parcels.

Pesticide spray drift, the grumble of farm machinery early in the morning or late at night, and the smell of manure do not fit with nonfarm neighbors' expectations of a pristine rural environment. On the other hand, nonfarm neighbors often cause problems for farmers. Vandalism, trash, theft of crops, and dogs harassing livestock are common.

Farmers also become a political minority. In some cases, communities have enacted nuisance ordinances that restrict the hours when farmers can operate machinery. In one Illinois suburb, a farmer was actually arrested for operating machinery after curfew!

Right-to-farm laws in every state provide protection for farmers from nuisance suits involving standard farming practices. But such legal protection does not always prevent nonfarm neighbors from calling up to complain about farming activities or even taking farmers to court.

Protecting farmland promotes good land-use planning.

We can only guess how future farming technology may affect the cost and production of food, or what value society will place on open land, or how local farms could be even more vital to their communities than they are now. Once farmland and open space are covered with houses, stores, offices, and factories, it is unlikely that these lands will return to farming or natural areas any time soon. Development should be placed where it will not interfere with farming operations and will not gobble up more open space than necessary. A community needs to assess its land resources and decide how much development it wants and can support over the long run in order to maintain job opportunities, a variety of services, and an overall good quality of life for its residents now and in the future. A farmland protection program that has strong public support usually results in additional public policy choices that conserve land and tax dollars. Gradual, phased, and compact development on the edge of existing cities and towns means not only that farmland and open spaces can be conserved, but also that public service costs can be held to a minimum.

Keeping local farming means keeping a local supply of fresh food.

It's hard to believe, but many communities in the United States haven't given this a thought over the last four decades as their farms have been covered with asphalt and concrete. This means those residents are entirely dependent on food grown elsewhere, and on cheap and plentiful supplies of energy to produce, package, and transport food to them.

A local food supply conserves on energy and can offer fresh fruit, vegetables, milk, meat, and grains. In the past twenty years, local farmers' markets have enjoyed a resurgence. Not only do these markets carry a variety of foods, they also create a wonderful social gathering. Pick-your-own operations and roadside stands are both popular and a way to get consumers out into the countryside to see where their food is grown.

Another increasingly popular food source is the more than one thousand community-supported agriculture projects in which consumers buy shares in a farm's output directly from the farm. Avoiding the middle man enables farmers to earn higher profit margins.

Protecting farmland protects environmental quality.

Farmland and open space provide many environmental benefits, including wildlife habitat, water recharge areas, scenic views, and the conservation of historic landscapes. Farmland protection programs have the added benefit to society of protecting these valuable assets.

For example, in 1995 New York City began a multimillion dollar program to protect its drinking water supplies in the Catskill Mountains and the Hudson Valley. The city is paying dairy farmers to improve their farming practices to keep field and barnyard runoff from entering lakes, rivers, and streams. The city also decided that purchasing conservation easements to keep farmland from being developed in the watershed was another good way to protect the water supply. This policy choice demonstrates that the harmful effects of some farm practices can be corrected and more sustainable practices established. But once land is developed and impervious surfaces created, the damage to the environment is hard to reverse and the cost to society is steep.

Protected farmland can help buffer parkland.

One of the potential dangers in protecting farmland and open space is that protection may create islands of protected land. Housing and commercial development then encroach upon these islands and render them almost useless for farming, forestry, or wildlife habitat.

It makes good sense to coordinate protection efforts with existing and planned park systems. Montgomery County, Maryland, just north of the nation's capital, has preserved over 40,000 acres of farmland and zoned another 38,000 for farm use. Both the preserved and farm-zoned acreage are nicely connected with the county's 13,000 acres of public parks. The farmland buffers the parkland and vice versa, keeping development at a manageable distance and providing contiguous wildlife habitat and watershed protection.

Keep your goals and expectations realistic.

Just as there are many good reasons to protect farmland and open space, there are sensible ways to promote farmland protection in a community. Below are some "don'ts" that will help sort out unrealistic expectations and unattainable goals.

- Don't attempt to protect farmland without farmers behind you. The key ingredient in any protection effort is the landowners themselves. It will be difficult, if not impossible, to protect farmland if your local farming community is overwhelmingly opposed to protection. Farmers are very independent people who are generally suspicious of government.

> **BOX 2.1**
> **What You Can Do to Help Protect Farmland**
>
> 1. *Buy* locally grown produce in stores, at farmers' markets, at roadside farm stands, and at U-pick operations.
> 2. *Join* a community-supported agriculture project.
> 3. *Encourage* your friends and neighbors to buy locally grown produce and join a community-supported agriculture project. Word of mouth is powerful advertising.
> 4. *Educate* yourself about farming and farmland protection. Talk to farmers, the extension service, the conservation district, and the local planning department.
> 5. *Attend* meetings of the planning commission and elected officials in your community to express your views about the local comprehensive plan, zoning, and development proposals that would impact farmland.
> 6. *Support* farmland protection programs, even if your taxes will increase to pay for them. Encourage your friends and neighbors to support these programs, too.
> 7. *Organize* a land-use watchdog group to keep track of and comment on development projects that would impact farmland.
> 8. *Work* with the development community to encourage more compact development and development in appropriate places, away from farming areas.
> 9. *Lobby* your state and federal representatives to support farmland protection legislation. Letters, faxes, e-mail, and phone calls do not go unnoticed.
> 10. *Keep* at it! Farmland protection doesn't happen overnight, in six months, or six years. It is an ongoing effort among farmers, public officials, and concerned citizens.

Although you should not expect total support from the farm community, you should have at least a few farmers who are willing to speak in favor of protection.

Farmland protection efforts have been most successful where farmers want to continue to farm and have recognized the value of incentive programs and land-use controls in limiting development and its potential conflicts. You should be prepared to discuss farmland protection with the farming community and attempt to arrive at a consensus for protection.

- Don't expect farmland protection to produce big changes overnight. All too often, a community will begin to talk about farmland protection after a large development is proposed or has already been built. Farmland and open-space protection must anticipate development needs and channel growth to places where it can be accommodated with a minimal loss of farmland and open space.
- Don't call for protection programs that lead to unrealistic expectations. In some communities, the land base has been seriously fragmented into parcels that are suited only for intensive vegetable and nursery crops. Advocating large-minimum-lot-size zoning of one dwelling per 50 acres,

for example, is not realistic and could result in the rejection of more workable techniques.
- Don't attempt to create programs that may make farming difficult. Some suburban communities seem to want farmland without farmers! There are many examples of farm operations that have been opposed by local residents with an attitude of "not in my back yard" (NIMBY). Farming is not just pleasant scenery or a way of life; it is a business. Just as in any other business, if it becomes difficult to operate in one community, the owner will look to move or liquidate the business.

  You must work with farmers and learn to understand farming from their point of view if you are to be successful in advocating farmland protection.

  Your community must decide whether it wants to protect working farms or just open space. This decision will influence the protection tools your community should use (see chapter 13).
- Don't recommend protection programs just because they are being used or have succeeded elsewhere. Protection programs must be tailored to the local farming community and local political realities. Although we present many case studies and anecdotes in this book, each community has different circumstances and certain people that make them special. While one technique, such as purchase of development rights, might succeed in one place, it might not in another.
- Don't support the creation of farmland or open-space protection programs that will last only a few years. Protection programs need to be part of a community's ongoing planning and growth-management efforts. A community should take a long-term view and be prepared for a long-term commitment of money and personnel.

  Take the example of King County, Washington, around Seattle. In the early 1980s, the voters of King County agreed to spend $50 million to purchase development rights to farmland and open space. By 1985, the county had preserved 12,500 acres, and the program ended. In 1995, the county realized that more protection was needed. But ten years of booming development had taken its toll: tens of thousands of residents have added even more pressure on the remaining farmland.
- Don't make the claim that America's food supply will be threatened if farmland isn't protected. America's farmers continue to produce surplus amounts of several products, including milk, corn, soybeans, and wheat. Approximately 1 out of every 5 acres in crops goes to the export market. And as recently as 1995, 60 million acres of potential cropland were idled under a variety of government programs. These figures suggest an abundance of farmland.

  But on the local, state, and regional levels, farmland loss in urbanizing areas is a very real problem. Most American cities began in places with rich farmland and ample water supplies; and even today about one-third of the nation's most fertile land is within or close to metropolitan areas. The continued loss of farmland could threaten farming as a local or regional industry.

A 1993 study by the American Farmland Trust entitled *Farming on the Edge* found that *more than half* of America's food production takes place in metropolitan counties, in counties adjacent to major cities, and in counties with both high population growth and higher than state average agricultural production. The study reported that 90 percent of the U.S. population lives in these 1,549 counties and the farmers there account for a large percentage of the nation's food output, including:

- 79 percent of milk
- 45 percent of meat
- 47 percent of breads
- 86 percent of fruits
- 87 percent of vegetables

According to the U.S. Department of Agriculture, Americans need to eat more of these last three food groups, making farming in metropolitan regions even more vital.

The AFT study then cited twelve important farming areas, a total of 157 counties, that are threatened because of heavy population pressure:

1. The Central Valley of California, where much of the nation's vegetables are grown. According to AFT president, Ralph Grossi, population in the Central Valley is expected to triple by the year 2040, and if the development occurs at current densities, half a million acres of unique farmland would be lost from production. Fresno County, the nation's leading county, with $3.2 billion in farm products, is expected to have almost twice as many people by the year 2025.
2. South Florida, a leading citrus region.
3. The "collar counties" around Chicago and up into Wisconsin and Michigan, where farmers produce milk and a variety of grains and small fruits.
4. Southeastern Pennsylvania, which contains most of the state's prime farmland and accounts for over $2 billion in farm output.
5. The Champlain Valley of Vermont and the Hudson Valley of New York, a major dairy region.
6. The Piedmont of North Carolina, a milk and tobacco area.
7. Counties around Seattle, Washington, which grow specialty crops.
8. The California Coast, south of San Francisco.
9. Oregon's Willamette Valley.
10. The Shenandoah–Cumberland Valley of Virginia.
11. The Minneapolis–St. Paul Metro Area.
12. Western Michigan, a major dairy and small fruit-producing region.

- Worldwide, evidence suggests we should take a cautious approach in favor of farmland protection. In 1992, the United Nations reported significant losses in the world's arable land due to urbanization, water and wind erosion, and poor farming practices. Recent estimates by the United Nations suggest that by the year 2020 world population will exceed 8 billion, up 45 percent from today's 5.7 billion. The demand for food is expected to rise even faster because as more people escape poverty, they eat more, and they have more meat in their diets. For instance, China, with one-quarter of the world's population, has become an importer of grain, in part to feed livestock. The combination of growing populations, higher living standards, and loss of productive farmland overseas could make American farmers the source of greater world food supplies.

    But some experts believe there will be no worldwide food shortages. Per Pinstrup-Anderson of the International Food Policy Research Institute claims that the world is capable of feeding twelve billion people. Others look to advances in biotechnology to greatly increase food supplies.
- Don't call for farmland protection programs because you have a romantic notion of what farming is or used to be. Green cornfields, red barns, and white farmhouses are certainly scenic; but the business of farming involves hard, dirty work, long hours, a rather low return on investment, and plenty of weather risk. No matter what protection techniques you advocate, you cannot compel a landowner to keep land in active farm use. Also, local restrictions on normal farming practices will tend to drive farmers out of your community. Farming typically produces noise, odors, and dust. Your community must recognize and accept this fact.
- Don't attempt to use the protection of farmland and open space to exclude certain ethnic groups or income groups. For example, in a suburban community, large-lot agricultural zoning of one dwelling per 10 acres could be overturned by the courts as discriminatory if that zoning is not aimed at protecting a viable farming industry.
- Don't use the protection of farmland and open space as a "no growth" technique. Many newcomers to the urban fringe adopt a "pull up the ladder" or "last one in" mentality. As we discuss in the next chapter, it is illegal to impose a population cap for a county, state, or municipality. The challenge under American law is how to accommodate growth without losing the qualities that make your community a pleasant place to live.
- Don't rush into farmland protection programs. Orange County, North Carolina, officials learned the hard way that public education and acceptance take time. In 1994, three months before the election, a $2 million bond referendum was put on the ballot to start a purchase of development rights program. There wasn't sufficient time to educate the voters and the farmland owners, and the bond measure barely lost, setting back protection efforts for at least two years.

It takes time to put together a farmland protection program. In simple terms, you need to learn, listen, organize, advocate, and lobby. Read up on what techniques have been used and how they work. Talk to experts from around the country. Set up meetings with farmers and non-farmers. Make farmland protection an issue for discussion in the community. Get the attention of the local newspapers, radio and television stations. Get local officials to appoint a farmland protection task force. Gather information, collect opinions, and weigh the evidence. Write up a report with practical recommendations. Then lobby elected officials to implement the recommendations.

- Don't retire from public life once a farmland protection program is in place. Any program that is established will need continued public support and understanding. It is a good idea to create an organization that supports the program, becomes the program's "booster," and makes sure that public officials don't cut funding or reverse zoning and development standards. An organization can also help to "sell" to the voters bond referendums that include funding for farmland protection. In short, even a small group of diligent citizens can bring about big changes.
- Don't be afraid to ask for help or advice. Many small towns and counties or rapidly growing places do not have the money or people available to set protection efforts in motion or to review and improve their protection programs. In the back of the book, we have listed several model documents and organizations to contact for help. Throughout the book, we discuss protection programs in several communities and states. Don't be shy about obtaining the information you need. A letter, phone call, e-mail, or fax can produce valuable results well worth the time and cost.

The Politics of Land Protection

It is important to understand the major political obstacles to effective land-use protection. The first obstacle is the large number of political jurisdictions that are reluctant to share power or cooperate for region-wide land protection. In the Northeast, for example, there are anywhere from dozens (in Rhode Island) to more than a thousand (in New York) townships in each state. Each township covers about 15,000 to 20,000 acres and has its own power to conduct planning and zoning. It's not difficult to imagine that the quality of land protection efforts can vary greatly. Throughout the United States, cities, suburbs, and rural areas compete for economic growth: jobs, businesses, and property tax base. Land protection is often not high on the political agenda. But the more people try to protect land resources, the more they will see the need for several townships or even several counties to work together.

A second obstacle to land protection is the lack of political will on the part of elected officials to establish more restrictive zoning that would better protect farmland from development. As we will see, the planning of land uses in a community is a very political process. Limitations on land use run the risk of a backlash from landowners and the threat of lawsuits claiming loss of equity. In fact, local government attorneys often advise against tougher land-use controls for fear of legal challenges, even though case law may be on the side of local government. But most politicians fear the costs, both monetary and political, that court cases can generate. It's important to find politicians or candidates who will stand up for land protection and support them.

Another political challenge is forming and maintaining an effective citizen group to participate in the land-use process. Most citizen groups are formed in response to a perceived development threat: a mega-mall, the conversion of a farm into a housing development, or the construction of a highway through a natural area. Once the development is either allowed or denied, the citizen group disbands. Then another threat arises, and a new organization must be hastily patched together.

A formal organization, especially a nonprofit corporation to avoid personal liability, with a recognizable name and officers is more likely to be taken seriously by elected officials, developers, landowners, and the general public. An organization can recruit members, monitor development proposals and land regulations, and raise funds for studies, consultants, and legal help.

To succeed in the political aspects of farmland and open-space protection, it's important to form partnerships between local government and private landowners and citizen organizations. Strong links must be forged among those with political power, the public, and citizen groups. Strong public and private leadership can produce solid, sensible land-use planning of which the protection of land resources and managed growth are the cornerstones.

Finally, it is far too easy to be cynical, to let change in a community or region take its course. It is a much greater challenge—and, ultimately, more rewarding—to take part in shaping change.

Making Choices

Citizens and elected officials must decide on how to manage the community's land resources just as private landowners decide how to manage their property. Making land-use policy means making choices. The four choices discussed here recognize that something needs to be done and soon. These four choices may not be mutually exclusive; some elements of one choice may be combined with parts of another. But before a choice can be made, the community must be aware of the options.

Private responsibility for land conservation. Land conservation is not a no-growth ethic; rather, it advocates the wise use of resources so that development is accommodated and valuable farmland and natural areas are retained as much as possible. The emphasis is on providing landowners with voluntary financial incentives, such as use-value property taxation, the purchase of development rights, and the donation of conservation easements. These may provide landowners with an alternative to selling land at high prices for development.

This strategy does not infringe on private property rights but does require public funds. The danger with this approach is that voluntary actions cannot guarantee the protection of critical farms and natural lands. Each landowner may respond differently to incentives and to the temptation to sell for development. Thus, protection could result in a patchwork of open space mixed with development that would not be conducive to farming in the long run.

Long-term protection for "sacred lands." This choice entails strong government action for long-term protection of farmland and natural areas. Agricultural zoning, purchase of development rights, and urban growth boundaries have been used to protect farmland and natural areas. The flip side is a need for more and better development in cities and villages to accommodate growth.

Attractive urban development to take the pressure off of rural lands. Many American cities have become less livable over the past few decades. An exodus to the suburbs has robbed cities of vitality, tax base, and economic activity. Meanwhile, the suburbs continue to sprawl, often in haphazard patterns. Urban communities are in need of revitalization, and the space for growth and in-fill development on empty parcels already exists.

The state of Oregon has required its cities and counties to cooperate in creating urban growth boundaries. Within the boundaries there is supposed to be enough buildable land to accommodate development for the next twenty years. The more compact style of development is cheaper to service, reduces sprawl, and is a more efficient use of land.

Regional land-use control. Many land-use and public services problems are beyond the control of one local government. Local governments often compete for development to broaden the local property tax base. The actions of one jurisdiction can easily spill over to another. For example, one community attracts a factory that generates handsome property taxes for that community. But the community next door gets the houses of the people who will work in the factory and must pay for the education of their children.

No community stands alone. Every town, city, and rural area is influenced by other places in the region. Two common threats to farmland and open space are sprawl and developments of regional impact. Sprawl typically begins as a steady increase in the number of houses and businesses placed along or just off major roads. Developments of regional impact include major shopping malls, manufacturing complexes, office buildings (usually with corporate headquarters), and large housing subdivisions. A single community often does not have the planning expertise to judge the impact of such huge projects beyond its borders.

Regional planning has not enjoyed much success in America. Individual counties and townships jealously guard their authority over planning and zoning and property taxation. In only a few places, notably in the Minneapolis–St. Paul region, is there tax-base sharing between municipalities as a means of overcoming development imbalance. Tax-base sharing to reduce local competition for economic development and regional land-use planning to limit farmland conversion and sprawl may yet enter the mainstream of American local government administration.

Summary

There are many reasons for farmland protection. American communities are losing a vital resource while historic landscapes, traditional communities, and local economies are dramatically changed. Unless farmland protection is taken seriously, population growth and development pose true threats to some of the nation's most valuable farmland.

Land protection is a political process. This means that ordinary citizens, as well as major landowners, developers, and elected officials need to be involved in the decision making. In the next chapter, we look at the legal issues of land protection and how a community can plan for and regulate land use.

CHAPTER 3

Managing Community Growth

WHAT YOU CAN AND CANNOT DO

Where there is no vision, the people perish.
—Proverbs 29:18

Farmers benefit from land use planning for one simple reason: stability.[1]
—Ken Buelt, farmer
Washington County, Oregon

The Legal Foundations of Managing Growth

Whether you are updating your community's farmland protection program or considering options for managing the growth of your community, you should be aware of what you can and cannot do under the law. In discussing matters of law, we do not presume to give legal advice. Our aim is to provide a basic understanding of the land-use planning process. We highly recommend that you obtain a copy of your state's planning and zoning enabling act. We also strongly encourage you to consult a private attorney who specializes in land-use law; the attorney for your town, city, or county; or your state attorney general's office for specific advice. Land use laws vary from state to state, as do some court rulings on land-use issues. Laws and rulings do change over time.

Private Property Rights and the Public Interest

Americans have long debated how much freedom landowners should have to use their property as they desire. There is intense disagreement over how to weigh the public interest against private property rights. Yet it is important to realize that state, local, and federal governments have long regulated, at least to some extent, the use of private property. While private property and the right to use it in productive ways are essential parts of our capitalist economic system, landowners must recognize that the attitude of it's-my-land-I-can-do-what-I-want-with-it is not realistic. A landowner does not have a right to use land in ways that harm others, such as building a hog barn in the middle of a residential area. Community land-use planning and regulations are meant to protect the public health, safety, and welfare, mainly by separating land uses—hogs and houses, for example—that don't belong together. But no level of government has the right to restrict private property so rigidly that no reasonable economic use remains.

Private property rights are a creation of government.[2] When you own land, what you really own is a bundle of property rights. These include water rights, mineral rights, air rights, the right to use the land in many different ways, the right to develop the land, the right to sell the land, and the right to pass the land on to heirs. This bundle of rights is known as the *fee simple*. Each right in the bundle may be separated and sold or given away. For example, it has long been common for landowners to sell mineral rights to mining companies. The landowner retains all other rights to the land, but the mining company owns the right to dig up any mineral or energy deposits under the land.

The sale or donation of the development right in the bundle is a popular farmland protection tool, discussed in chapter 9. The landowner sells or donates the right to develop the land through a deed of easement, which then "runs with the land"; that is, it applies to every future owner of the land. The landowner retains all other rights to the land, including, of course, the right to farm the land and to sell it.

The protection of private property rights comes from the Fifth and Fourteenth amendments to the United States Constitution. The Fifth Amendment requires a government to pay "just compensation" to a landowner if it "takes" private property. Clearly, when a government condemns farmland for a highway, it is physically taking the farm to put a road through it; and the government must pay the landowner a fair price for taking the land. Controversy may arise when a government regulation limits the use of private property. Property rights advocates say a regulation that reduces land value is a partial taking and the landowners should receive compensation, even if the land can still be put to a beneficial use. When a farmer buys a wetland expecting to be able to drain it and farm it and then finds out government regulations won't let him, should the farmer be compensated?

The reality is that some land investors are speculators, hoping for a quick gain from buying and selling land. Even some long-term landowners count on being able to sell their land for development. Governments are under no obligation to guarantee a capital gain to a landowner. Government regulations are supposed to achieve a public purpose. Ultimately, state courts or the United States Supreme Court will decide if a government regulation has gone too far and resulted in a taking of private property rights.

The Fourteenth Amendment contains two provisions that are very important for landowners and governments. First, a government must treat people equally and fairly according to the "due process" of law. Second, all Americans have the right of "free travel," which means that a city, town, county, or state may not impose population caps. This has created a bias in favor of growth and development, which made sense in the early days of the nation when conquering the wilderness was a priority. Today, relentless population growth and development run contrary to the concept of carrying capacity: the point at which the land and environment become polluted and quality of life declines.

Defining the Public Interest in Land

Local and state governments may limit a landowner's right to use or develop private property according to the Tenth Amendment to the United States Constitution. Government may impose land-use controls to protect public health, safety, morals, and welfare. And the government need not pay compensation to landowners for limitations on the use of their land as long as an economic use of the property remains.

Public health and safety situations may be fairly simple to identify; for example, it makes little sense to locate a toxic waste dump next to a public water supply. Morals are more difficult to define but generally refer to accepted norms of behavior. Public welfare is also hard to pin down. What is the public welfare, and who decides what is (or isn't) in the public welfare? Usually, voters or elected officials determine what is desirable and what is not. For example, the fact that all fifty states have enacted programs to protect farmland suggests that protecting farming promotes the public welfare.

In devising land-use controls aimed at furthering the public welfare, governments must be careful not to be overly strict. If a government regulation removes all economic use of the land, the regulation could be considered a taking of private property. The regulation could be ruled invalid, or the government could be required to buy the land from the landowner.

By the end of 1995, eighteen states had passed laws requiring governments to compensate landowners for any regulation that reduces the value of property by more than a certain percentage; and the landowner would continue to own the property! For example, a farmland owner in Louisiana could sue the government for compensation if a regulation resulted in a loss of more than 20 percent of the property value. Most of the laws require the state government to determine the impact of proposed state regulations on private property values. The cost of paying compensation and the administrative burden could be overwhelming.

Much of the drive for property rights laws has come from several wetlands cases in which regulators have made farming difficult, if not impossible. The size of a wetland, whether it exists throughout the year, and whether it can be drained for farming purposes are issues that need to be more clearly defined for both regulators and landowners.

How local land-use ordinances, such as zoning, fit in with these property rights laws is still uncertain. Yet local governments in states with compensation laws may shy away from stricter zoning of farmland for fear of compensation claims. But as any realtor can tell you, one of the first things you do when looking to buy land is find out what the zoning allows you to do. The government is not in the business of guaranteeing a landowner a return on a land investment. There is no obligation for a local government to grant a landowner a rezoning from agriculture to residential use just to make the landowner rich.

Government land-use regulations must be reasonable and designed to achieve a public purpose. They must guide developers to create developments that fit with the public interest and help avoid or mitigate negative impacts. For example, a county cannot require a developer who wants to build fifty houses on farmland to put in a roadside vegetable stand. Even though a vegetable stand might remind people that the housing development was once a farm, whether not there is a vegetable stand should not influence the county's decision to allow the developer's housing project—that would be unreasonable. On the other hand, the county could have an ordinance that says that any housing development of more than ten units must be connected to a public sewer. This is a public health issue, designed to protect ground water drinking supplies.

Growth Management and Citizen Participation

A community with a viable farming industry needs to discuss how to include farmland protection in its efforts to manage growth and development. We have seen too many examples of communities that do not plan for farmland protection until there is literally one farm left. The key to good land-use planning is anticipating and accommodating change to fit in with the goals and objectives of the people who live in the community. The lack of certainty over the outcome of land-use decisions is frustrating and often expensive for both developers and neighborhood landowners, and sometimes for the entire community. Good land-use planning provides greater certainty about where development is and isn't desired and the infrastructure costs the community is willing to bear.

The protection of farmland and open space involves many players—landowners, developers, taxpayers, voters, elected officials—and the land-use planning process. It is essential to have a good understanding of who is important in the process and how it works.

Land-use planning is economic planning. It influences the value of land by designating, for example, the type and number of homes, or the size of commercial units such as stores and businesses, that are allowed throughout the community. Land-use regulations have a major impact on the appearance of a community and on the delivery and cost of public services.

Land-use planning is also a political process. The elected municipal or county officials can enact or scrap a variety of land-use controls. A shift in the political make-up of local government can result in changing regulations and decisions about proposed development. This fluid situation means that people who feel strongly about their community need to be active in politics and land-use planning.

It may sound naive to speak of putting your trust in the planning process if you believe as Chicago journalist Robert Heuer does that "historically, plan-

PHOTO 3.1
Farmland slated for development.

ners' bread and butter has been planting subdivisions on farmland."[3] But in the American legal and land-use systems, the planning process is the best opportunity you have to influence the future of your community.

The Planning Commission and the Comprehensive Plan[4]

The planning process begins with local elected officials appointing a planning commission to fulfill three main functions.[5] First, the planning commission must draft a *comprehensive plan* (also known as a master plan or general plan) for the community. The purpose of the comprehensive plan is to guide growth over the next ten to twenty years. Second, the planning commission drafts the zoning ordinance, which puts the comprehensive plan into action. Third, the planning commission makes recommendations on development proposals to the elected county or township officials, based largely on the comprehensive plan and the zoning ordinance.

These tasks may already have been in the process for many years, or they may not have been done at all. Localities differ from state to state and within each state in the amount of planning conducted.

Recommendations made by planners, whether the planners are citizen appointees or professionals, and the decisions that elected officials make based on those recommendations have a long-term impact on the community and will affect the daily lives of residents in many ways. So it is a good idea for concerned citizens to be involved in this process as early as possible.

Find out how the planning process occurs (or does not occur) in your county or township. If there is a land-use plan, read it. Attend some meetings of the planning commission or advisory board that deals with land use. *These meetings are rarely exciting, but they are where the day-to-day changes to the land and community originate.*

If you are a professional planner, determine whether your planning department is user friendly for citizens other than developers and builders. Are citizens involved in the planning process enough to make a difference? If not, the process needs to be opened up in a way that makes it easy for citizens to learn and to participate.

The Comprehensive Plan

The comprehensive plan has four important purposes:

1. The plan expresses a realistic vision of how and where the community should grow.
2. The plan presents facts that help to educate residents about the community's land uses, public services, economy, and population trends. In short, the plan shows the community's strengths and weaknesses as well as opportunities and threats.
3. The plan guides public decision makers and private landowners and investors as to where different types of development are appropriate.
4. The plan provides legal support for land-use regulations and direction for capital budgeting, both of which influence where the community will accommodate growth.

It is in these last two functions that the comprehensive plan can influence the future and provide an opportunity for citizens to examine a community's land-use policies. Any land-use decision officials make is open to public scrutiny to determine whether it is consistent with the comprehensive plan.

The comprehensive planning process begins with an inventory of the physical features and economic assets of the community, with an emphasis on the pattern of development and open lands. Next come studies of trends and land needs that will affect the use of land in the community over the next ten to twenty years. Then general goals and specific objectives should be determined for each section of the plan, including:

- a projection of expected future population
- housing
- transportation
- the local economy
- land-use patterns
- public facilities, such as sewerage, water, and schools
- agriculture, open space, and natural areas

The plan contains community goals and objectives that recommend certain land-use changes and can be used to justify zoning changes.

You may find that the comprehensive plan in your community is more than ten years old and has not been updated since it was first adopted. The first thing to do is let your officials know it is time for an update. For the purpose of land protection, however, look at how the plan treats agriculture and open space. Do not be surprised if there is little mention of agriculture and farmland. Most of the professional planners who write comprehensive plans have limited knowledge of farming. It is not uncommon to see farming areas labeled "vacant land" on a map of current land uses in comprehensive plans!

A good time to begin or to improve farmland protection measures is during revisions to the comprehensive plan or when a new comprehensive plan is being drafted.

Agriculture and the Comprehensive Plan

The agricultural section of the comprehensive plan might follow the example below:

I. Agriculture in _____ County.
   A. Overview of land in farm use, soil quality, number of farms, value of farm production, and type of crops and livestock.
      1. County soils maps from the Natural Resources Conservation Service.
      2. Data from the U. S. Census of Agriculture and the state Department of Agriculture.
   B. Contribution of agriculture to the local economy: jobs, value and type of products produced, manufacture of food and fiber products, farm support businesses, tourism.
   C. Threats to and opportunities for Agriculture in _____ County: loss of farmland since 1982, population growth since 1980, location of farming areas and growth areas, problems of incompatible nonfarm land uses in farming areas.
II. Goals and objectives for farmland protection.
   A. Goal 1: To encourage farming as an important part of the local economy.
      1. Objective 1: the county helps fund the creation of three farmers' markets.
      2. Objective 2: the county planning commission reviews local zoning ordinances to ensure that they do not discourage normal farming practices and do allow for some farm-based businesses.
   B. Goal 2: To protect farmland from conflicting nonfarm development

by keeping large-scale residential subdivisions and commercial developments out of the countryside.
1. Objective 1: The planning commission monitors plans to locate or extend public sewer and water lines to make sure that, to the extent possible, they do not enter the main farming areas of the county.
2. Objective 2: The planning commission permits some low-density rural residential zoning to accommodate the growth of the rural population.
3. Objective 3: The county works with the cities and villages to limit commercial strip development and to form growth boundaries so that most of the projected population growth and mixed-use development can be accommodated in and next to built-up places.
4. Objective 4: The planning commission works with farm groups, individual farmers, and nonfarmers to review or adopt agricultural zones.
5. Objective 5: The county government supports financial incentives, such as preferential property taxation, the purchase of development rights, and the donation of conservation easements, to protect farmland.

Information on past trends and the current agricultural economy, as well as on the location and quality of farmland, provides a factual base for the general goals and specific objectives. The agricultural section is especially important if the community is interested in implementing agricultural zoning. Zoning is greatly strengthened when there is a direct connection between the zoning regulations and the goals and objectives of the comprehensive plan.

It is a good idea to point out the positive contributions that agriculture makes to the community. The following policy statement from the comprehensive plan of Shrewsbury Township in York County, Pennsylvania (1980), lets the reader know that agriculture is an important part of the local comprehensive plan and that agricultural zoning is one way to put the comprehensive plan into action:

> In planning for agricultural land, it is the Township's policy not to consider agricultural land as "undeveloped farmland" awaiting another use. Farmland must be considered as "developed land." It is being used to produce a product. Farming is a land-intensive manufacturing process which converts raw materials into a product, comparable to other industrial operations, with occasional accompanying noise, odor, and dust. The agricultural zone should not be considered as a holding zone, but as a zone having a positive purpose of utilizing the Township's natural resources for the benefit of the entire community. The Township should protect the agricultural zone from interference by incompatible uses which break down the integrity of the zone and also interfere with normal and customary [farm] operations within the zone.[6]

There are published maps to help identify where the good farmland is in a community. The Natural Resources Conservation Service (formerly known as the U.S. Soil Conservation Service) has published a soil survey for every county. The soil survey is a series of aerial photos on which soils have been identified. The soil survey also explains the quality of each soil and any limitations for farming or for development. The survey helps to indicate those farming soils that are worth protecting.

County-wide maps of tax parcels are often available at county government offices. And in some counties, private companies have published "plat books" showing the location, parcel size, and even the names of landowners. These two sources are useful for determining where farmland and open space are held in large blocks that would be good candidates for long-term protection.

An excellent way to store, update, and evaluate map information is with a computer. Most county governments and hundreds of municipalities have developed a geographic information system (GIS) to make an inventory of land resources, to track changes in land-use patterns, and to portray impacts of future growth. GIS maps can show layers of information far beyond ordinary road maps (see box 3.1 and figure 3.1). GIS maps can show features of a community not seen before by citizens, such as the proximity of public sewer and

FIGURE 3.1
GIS Map of Farms under Easements and Easement Sale Applications in Northwestern Lancaster County, Pennsylvania

 Farms under easement

 Easement sale applications

> **BOX 3.1**
> A Sample of GIS Information Layers
>
> - Soil quality
> - Tax parcels (size of parcels)
> - Land use
> - Public sewer and water lines
> - Roads and streets
> - Zoning
> - Rivers, streams, lakes, and ponds

water lines to farmland. GIS is perhaps the best tool to illustrate potential problems, such as a map of proposed sewer lines extending into prime farmland miles beyond city boundaries.

Citizen Participation and Creating a Community Vision

Public participation in the comprehensive planning process is especially important to form a *vision* for the community. A vision provides direction for action. A vision is a set of ideas about how the community should look, change, and function over the next ten to twenty years. A vision helps to identify which physical, economic, and even social features of the community are worth retaining and which need to be changed. Often people are concerned about rapid growth that is producing (or threatens to produce) a crowded landscape, traffic congestion, social tensions as new residents arrive, and the loss of rural character.

When the general public participates in the visioning and comprehensive planning process, the results are likely to have popular support. This is planning from the bottom up rather than from the top down. In creating a vision, the community must decide which parts of the countryside should be mostly for people to live in and which parts should be maintained as a working landscape of farms, forestry operations, and other small businesses.

Some communities have conducted surveys to find out what residents prefer in new housing, street design, and the rural landscape. As a follow-up, the community can hold a design workshop at which teams of planners, landscape architects, and historic preservationists work with local officials, landowners, and other citizens to create pictures and designs of how the countryside could appear. These pictures can display different design and planning scenarios that in turn suggest options for the community. For example, what would the countryside look like with one house every 2 acres? Every 10 acres? Every 30 acres?

But one thing to keep in mind is that a rural "look" is not necessarily the same thing as a working rural landscape of farms and forestry operations.

An increasingly popular technique for planning communities is a *planning charette*. The charette is a community meeting that lasts for two to four days (make sure you have lots of food!) and involves a variety of people working together in small and large groups to develop a common vision for the community, set goals for future development, and identify actions to achieve those goals.

The charette can be an excellent way to establish a new comprehensive plan and zoning ordinance for a community. The charette process is best helped along by outside consultants or faculty and students from a nearby graduate planning program.

Another good way for a community to involve citizens in the planning process is for the planning commission to appoint a planning advisory committee. The advisory committee should have wide representation from the community but probably no more than twelve to fifteen people. The committee's purpose is to review proposed changes to the comprehensive plan and report to the planning commission.

Citizens and elected officials must decide on how to manage the community's land resources to make the community vision a reality. An important part of the comprehensive plan is to make a map of the current land uses in the community and then display the community's vision for growth on a map of future desired land use.

The debate over land use is difficult because so many interests are at stake. Often a single, loud, opposing voice will scare off those who are interested in farmland protection. A professional facilitator can assure that all voices are heard. Good debate is healthy. It brings out all sides of an issue. But good debate should lead to action. Some communities try to put off making difficult land-use decisions, but "not to decide is to decide," and a rural community can become a congested suburb by default.

Zoning: Encouraging Growth and Limiting Growth

Once a community has drafted a comprehensive plan, it needs to adopt land-use regulations to put the plan into action. These regulations should encourage growth in some places and limit growth in other parts of the community, according to the future land-use map of the comprehensive plan.

Zoning is the most common technique local governments use to influence the location and density of development. A zoning ordinance is a set of local legal regulations that spell out what a landowner can do with land and buildings. There is no national standard for zoning. Zoning ordinances vary from city to city, county to county, and state to state.

A zoning ordinance has the force of law, while a comprehensive plan generally does not. However, the comprehensive plan provides a legal foundation for the zoning ordinance; and the zoning ordinance should carry out the goals, objectives, and vision expressed in the comprehensive plan as well as in the future land-use map.

Too often the comprehensive plan sits on a shelf collecting dust because the planning commission and elected officials do not find it useful or do not know how to implement it. Many people mistakenly believe that planning is simply zoning. Others think all they need to do is draft a comprehensive plan to solve their problems. In fact, both a plan and zoning are needed.

Some communities have drafted or amended their zoning ordinances without referring to the comprehensive plan. Ignoring the comprehensive plan can render the zoning changes invalid if they are challenged in court. The comprehensive plan, the zoning ordinance, and other land-use regulations must work together to achieve orderly growth and farmland protection.

What Is Zoning?

Zoning designates how a parcel of land in a county, city, or township can be used. Zoning regulates how many buildings or "dwelling units" are allowed per acre (also referred to as a density allowance) and the siting, height, and area covered by buildings.

The purpose of zoning is to protect the public health, safety, and welfare. Zoning should strike a balance between the right of property owners to use their land and the right of the public to a healthy, safe, and orderly living environment. Zoning became popular in cities in the 1920s, but most suburban communities did not adopt zoning ordinances until the 1960s, and zoning in most rural areas did not occur until the 1970s. Many rural places even today do not have zoning ordinances.

Much of the opposition to zoning in rural areas comes from the fact that open rural land is a major store of wealth and source of income. An urban dweller does not have the option of selling off his front lawn to earn income, but farmers have traditionally sold off some land when they needed money. Zoning does not preclude farmers from selling land, but it does place limits.

The power for local governments to implement zoning comes through the Tenth Amendment and through state laws that empower local governments to use planning and zoning. States differ in how much power they grant to their localities.

The U.S. Supreme Court has consistently upheld the legality of zoning, beginning with *Village of Euclid, Ohio v. Ambler Realty Co.* (272 U.S. 365, 1926). But zoning must be reasonable and be designed to achieve a clear public purpose. A community cannot use zoning arbitrarily within land uses, or exclude any group or class of people from a certain area, and zoning cannot result in a taking of private property.

Must be reasonable. An agricultural zone with a 100-acre minimum lot size makes little sense in an area where the average farm size is 50 acres. Hardly anyone would be able to purchase a 100-acre tract, and few landowners would own the 200 acres needed to create two 100-acre tracts. But a 25-acre minimum lot size might be considered a reasonable zoning standard in this area because no tracts smaller than the average farm size (50 acres) could be divided. In short, there must be a clear connection between the purpose of the zone and the way the zone achieves that purpose.

Must achieve a public purpose. The public purpose of zoning is to protect the public health, safety, and welfare and to reduce conflicts between neighbors. For example, one purpose of agricultural zoning is to protect owners of farmland from conflicts with nonfarming neighbors. Local governments have adopted zoning standards to separate feed lots and housing developments.

Must be nonarbitrary. Zoning must be consistent where like circumstances exist, and standards must be based on sound reasons. In an agricultural zone, for instance, all new farm buildings might be required to be set back at least one hundred feet from the road and all property lines to protect adjacent properties and the traveling public from possible annoyances or hazards. Allowing one landowner to put a barn five feet from the road would be inconsistent, arbitrary, and also dangerous.

Must be nonexclusionary. Zoning may not be used to keep people out of a community. Some places have tried to require a landowner to own at least 5 acres to build a house. This can be prohibitively expensive for most people, and the courts may judge such zoning to be exclusionary. An agricultural zone, which tightly limits nonfarm development, is not exclusionary as long as the community provides a zone for small, affordable residential lots and multifamily housing. Requiring a 5- to 10-acre minimum lot size under the name of agricultural zoning cannot really protect farmland and may be ruled exclusionary by the courts.

Must not result in a taking of private property. A zoning ordinance cannot regulate land so that it does not have any economic value. The courts have ruled that such zoning confiscates private property without paying compensation and is unconstitutional. For example, zoning an area of mostly 5-acre parcels for exclusive agricultural use would probably be ruled a taking of private property. It would be difficult to farm such properties in any way other than greenhouses or intensive gardens. Animal agriculture or extensive grain operations would be impossible.

There are several different types of zones and zoning districts, which serve to separate different land uses. The most common zones are agricultural, residential, commercial, and industrial. Zoning districts refer to different densities of a similar land use, such as R-1 for a single-family residential district and R-2 for a district where duplexes and triplexes are allowed. Many communities have an A-1 zoning district that allows farming, ranching, and agriculture related uses. Some places also have an A-2 zoning district for agricultural manufacturing and processing and farm support businesses.

> **BOX 3.2**
> What Zoning Can and Cannot Do
>
> Zoning Can:
> - Help prevent residential developments from moving into agricultural areas in an unplanned fashion
> - Maintain the attractiveness of a community by protecting open space and natural terrain features
> - Protect individual property owners from harmful or undesirable uses of adjacent property
> - Help put into effect plans for future development in the "right" places; coordinate with the location of public services to "phase in" development over time
> - Allow for important community decisions on growth and development to be made within the community
>
> Zoning Cannot:
> - Interfere in farming decisions, such as crop or livestock selection
> - Change or correct land uses already in existence
> - Assure that land zoned a certain way (e.g., for agriculture) will permanently stay in that use. Rezoning of land for another use (from agricultural to residential, for example) is possible in response to changing conditions and unanticipated opportunities
> - Assure the proper administration of the zoning ordinance. The effectiveness of any zoning ordinance depends on the people who administer it

The Zoning Ordinance

A zoning ordinance is drafted by a community's planning commission or planning department or by a hired attorney or planning consultant. Once the draft is ready, the zoning ordinance is recommended to the elected officials, who may reject, amend, or approve the ordinance after holding a public hearing.[7]

A zoning ordinance has two parts: a text and a map. (For an example of a zoning text, see appendix A.) The text includes:

- The different land-use zones (e.g., agricultural, residential, industrial, and commercial), the districts within each zone (e.g., R-1 single-family residential, R-2 multifamily residential), and the purpose of each district.
- For each zoning district, density standards (such as one dwelling per 20 acres in the agricultural zone), minimum lot sizes (such as 1 acre on which to build a house), permitted land uses, and land uses that require the landowner to obtain a special exception from the zoning board of adjustment or a conditional-use permit from the elected officials. Each district will also have uses that are not allowed (such as a feedlot in a residential R-1 district), though these prohibited uses are often assumed to be those not expressly permitted in the text. And there are preexisting uses that are not normally permitted but are allowed to continue as

"nonconforming uses" (such as a dentist's office in an agricultural zone) as long as they don't expand.
- Some general development standards, such as how far from a property line a house must be built; the maximum coverage of a lot by buildings; height limitations; parking; signs; and road frontage.
- How the zoning process is to be administered. An important process is the rezoning of land from one use to another, such as the rezoning of land in the agricultural zone to the residential zone. Another important change might concern amending the minimum lot size in the agricultural zone, say from one dwelling per 2 acres to one dwelling per 25 acres.

The zoning map shows the location of the different zoning districts (see figure 3.2). It should be based on the future land-use map from the comprehensive plan. Changes to zoning may refer to the map (if the size of a zoning district is altered), to the text (if the provisions of a zoning district are amended), or both the text and the map (if a new zone is adopted, such as an

FIGURE 3.2
The Zoning Map

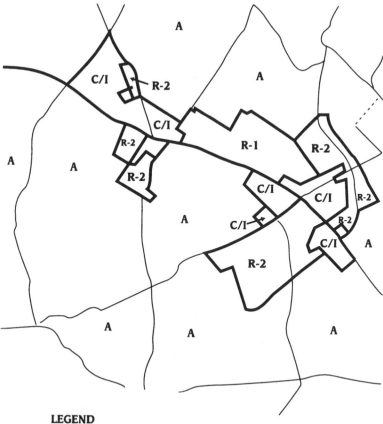

LEGEND
A = Agriculture
C/I = Commercial/Industrial
R-1 = Single Family
R-2 = Multi-Family

agricultural zone replacing a rural residential zone). A change to the zoning map should occur either after or together with a change in the future land-use map.

Zoning Administration

There are several professionals, appointees, and elected officials in local government who make recommendations or decisions on zoning matters. A zoning administrator may be hired by local government to administer, enforce, and answer questions about the zoning ordinance. When a landowner or developer proposes a development, the zoning administrator checks to see if the proposal meets the local zoning requirements. When the zoning administrator finds a violation of the zoning ordinance (for example, a landowner operating a feedlot in a residential zone), the administrator will notify the violator in writing and can require the violator to cease operation immediately.

In most states, a local zoning board of adjustment or appeals is established to make decisions on special zoning matters. The board hears appeals of decisions made by the zoning administrator and reviews applications for special exceptions to the zoning ordinance. A special-exception use is considered to have some impact on the local neighborhood but not on the entire community. For example, a home occupation involving a beauty salon might be a special exception in an agricultural zone. The beauty salon is incidental to the farm operation but helps to bring in extra income. The ordinance might require no additional structures to be built to house the home occupation and no more than two full-time employees who live off the farm.

Zoning boards may also grant variances from the zoning text or map in unique, hardship situations. An area variance allows minor changes to building setbacks, minimum lot sizes, building height, and lot coverage requirements so long as there is no conflict with neighboring properties. A use variance permits a land use not normally allowed in a zone, such as a farm implement dealership (a commercial use) in an agricultural zone. We recommend that use variances be allowed rarely if at all. A use variance is really a rezoning, and a rezoning is a decision for the planning commission and the local governing body to make.

The zoning board is not found in all states. Some states allow the local governing body to assume the board's duties; others let the planning commission make those decisions.

The planning commission, usually with the help of planning staff, is responsible for drafting and recommending the comprehensive plan, the zoning ordinance, and any amendments to the elected governing body, which officially adopts the plan, the zoning ordinance, and any amendments. The planning commission reviews development proposals to ensure that they comply with the zoning ordinance. Another task of the planning commission is to evaluate proposals for conditional uses. A conditional use is a land use allowed in a zone only after review by the planning commission and approval from the governing body. A conditional use can have community-wide impacts and must meet very specific standards. For example, a radio antenna might be allowed as a conditional use in an agricultural zone if it is designed, sized, and located as specified. Conditional uses are more controversial than special-exception uses.

The planning commission also may initiate changes to the comprehensive plan or the zoning text or maps, which must then be approved by the governing body. A zoning amendment can also begin when a group of landowners or a developer files a petition to change the zoning ordinance. The zoning administrator receives the petition and passes it on to the planning commission. The planning commission holds a public hearing on the proposed amendment and makes a recommendation to the governing body. The governing body then holds a public hearing and makes a final decision.

A zoning amendment that changes a zone boundary or the uses allowed on a specific property is called a *rezoning*. For example, a farm in an agricultural zone could be rezoned to the R-1 single-family residential zone and developed into houses. A density change in a zone from one dwelling per 2 acres to one dwelling per 25 acres could be called a rezoning if the zone's name is changed (such as from rural residential to agricultural), or it could be called a *downzoning*.

Rezoning and downzonings can be controversial and can produce heated debate, especially where there is a perception of a potential loss in property value. On the other hand, a rezoning from agriculture to residential use or an *upzoning* from one dwelling per 20 acres to one dwelling per 2 acres can give the impression of favoritism. Zoning is a legal process with political overtones. Even after a governing body rules on a rezoning, the decision can be taken to court.

How well a zoning ordinance works depends on the decisions of the planning commission, the zoning administrator, the zoning board, and elected officials in responding to development proposals and proposed changes to the zoning text and map. Administration should be fair and thoughtful. Too many changes to the zoning ordinance or map can cause instability and reduce the public's confidence in the reliability of zoning. For example, the frequent granting of rezonings from agricultural to residential use can undermine agricultural zoning as well as farmers' confidence in the long-term survival of farm operations.

See chapter 7 for an in-depth discussion of agricultural zoning and rezoning decisions.

Subdivision Regulations

A subdivision is the dividing of land into lots (see figure 3.3). Local governments adopt subdivision regulations, which set standards for new lots and the services (roads, sewer, and water) the subdivider must provide before any lots can be sold or building can begin. Some rural areas do not have subdivision regulations, and private landowners create and sell lots without local government review—and with no forethought about the loss of farmland. The lack

FIGURE 3.3
Subdivision: The
Creation of New Lots

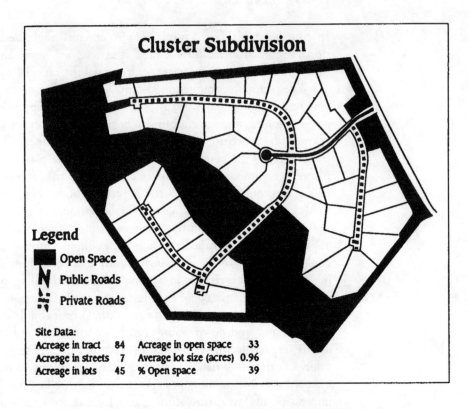

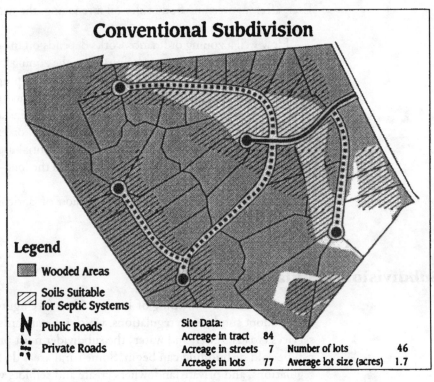

> **BOX 3.3**
> Stormwater Management Ordinance
>
> More and more often, a stormwater management ordinance is linked to the subdivision ordinance. Managing stormwater is especially important in rural areas where rain runoff from developed properties can damage neighboring farmland and overall water quality. The ordinance can include suggested stormwater management facilities, such as temporary basins, year-round ponds, meadows, and grass strips designed to slow runoff and increase the recharge of groundwater.

of subdivision regulations can cause legal problems in that unbuildable lots could be created and sold and accurate records of who owns what land might not be maintained.

The subdivision ordinance should help to achieve the goals and objectives of the comprehensive plan and should work in tandem with the zoning ordinance to ensure orderly development. Subdividers must meet all relevant requirements of the zoning ordinance in proposing developments. For instance, the size of any new lot must meet the minimum lot size in that zoning district. If an agricultural zone requires that any new farming lots be at least 25 acres in size and a subdivider proposed to create a 5-acre farm lot, the request would be denied. The agricultural zone might allow a nonfarm dwelling on a 1-acre to 2-acre lot; thus, if the subdivider proposed to create a 1.5-acre lot on which to build a nonfarm dwelling, it would be approved, assuming all other regulations were met.

The Subdivision Process

A subdivider is required to make three trips before the planning commission to obtain approval of a subdivision. The subdivider first submits a sketch plan of the proposed subdivision to the planning commission. The subdivider usually sits down with a staff planner to review the sketch plan or else the planning commission reviews the sketch plan at its regularly scheduled meeting. The staff planner or planning commission will explain any changes to the proposal that need to be made to comply with the subdivision regulations, the zoning ordinance, or the community comprehensive plan.

It is a good idea for concerned citizens to review the sketch plan at a public hearing held by the planning commission. The subdivider needs to hear feedback from the planning commission and the public on the subdivision proposal. The sketch plan is essentially a concept plan, and early input can help the developer avoid problems later on.

Farming neighbors may want to ask that home sites be placed as far away as possible from their boundaries to help avoid complaints about noise, dust, and odors. Or they may suggest a forested buffer area between the development and their farms. A group of community residents may request that the developer make some road improvements to prevent traffic congestion or make landscaping changes to make the site more attractive.

PHOTO 3.2
House lots subdivided off a farm.

Within a certain period, usually six months to a year, the subdivider must file a *preliminary plat* or the sketch plan will be declared invalid.

The preliminary plat application features a plat or survey of the proposed lots prepared by a professionally licensed surveyor. The survey will show any existing and proposed roads, neighboring landowners, contours, and right-of-way easements for utility lines or pipes. The plat must also list any conservation easements that restrict the use of the land.

The planning commission will review the preliminary plat to ensure that it is technically correct and that the subdivider has made any changes required at the sketch plan stage. The planning commission then holds a public hearing on the preliminary plat. This is the last opportunity for public comment and for the planning commission to make changes to the proposed subdivision.

If the preliminary plat is approved, the subdivider will submit a final plat, usually within one year. The planning commission checks the final plat to make sure that everything is in order. The chair of the planning commission and the governing body sign the final plat, which is then recorded at the recorder of deeds office. The planning commission or governing body may require a subdivider to put in utilities or post a bond in lieu of facilities before any lots are sold.

Capital Improvements Program

It is a common saying that "Civilization follows the sewer line." Public services such as sewerage, water, roads, and schools play an important role in determining the future growth and location of development in a community. A community's *capital improvements program* spells out:

- which services a community will build, repair, or replace
- where those services are located or will be located
- a schedule of when those services will be built, repaired, or replaced
- the anticipated cost of each project
- how the community will pay for each project

A capital improvements program is part of a community's overall budgeting process. The planning commission may draft a capital improvements program and submit it to the governing body, or the governing body may develop its own program.

A community drafts a capital improvements program in order to anticipate public service needs and to provide quality services at a reasonable cost. Most capital improvements programs look five to ten years into the future and include estimates of current capacity and future needs. The location of current and future needed services is especially important and should mesh with the community's comprehensive plan and zoning ordinance.

A good rule of thumb is to require that public services be in place before development can occur. This concept—known as *concurrency*—has been adopted by Florida and Oregon to promote orderly development that will not overburden a community. In addition, communities in several states have adopted ordinances requiring that specific "adequate public facilities" be in place before new development begins. For example, are local schools already overcrowded, or will they be if new homes are built?

Finally, a community must join its land-use goals and objectives with the capital improvements program so that public facilities that are known to induce growth are available in areas where the community wants growth and kept out of those areas, such as farm and forest lands, that the community wants to protect.

Monitoring Changes in the Land Base

If your community has been trying to protect farm, forest, and natural lands for several years, it is not too early to examine the progress of those efforts. What is actually happening on the ground? How much farmland and forest land is being lost each year? How much can be maintained, and will it be enough to sustain an industry?

These are important questions. To answer them, a community must have a way to keep track of land changes in the past and present and to anticipate land changes in the future. Some of this monitoring of land changes can be done by local or state government, but citizen groups and nonprofit land trusts can also play a part. In New England, several townships have appointed citizen environmental commissions to review the potential impacts of proposed developments.

In the planning process, a community makes an inventory of land resources that can be shown on maps. A geographic information system can display a

> **BOX 3.4**
> A Rezoning Reversal
>
> Sietsema Farms is a multimillion-dollar turkey producer in Kent County, Michigan. In the early 1990s, a proposed development placed a cloud over the future of the turkey operation.
>
> According to Harley Sietsema, "The township rezoned adjacent property from agricultural to residential and approved a seventy-unit housing development within 450 feet of my grain-drying system and buildings containing 30,000 turkeys. I am very concerned about the impact it would have on my long-term ability to operate my business. These individuals that move into the country don't have an appreciation for some of the activities that we do in the agricultural business. We have a major investment here. We would like to see growth in our township, but they have got to be more concerned that it does not occur in a manner that interferes with the ability for us to operate our farms" (*Planning and Zoning News*, April 1996, p. 7).
>
> Mr. Sietsema got together with concerned citizens and petitioned to have the rezoning put to a public referendum. By a three-to-one margin, the voters turned down the rezoning, thus supporting Mr. Sietsema's petition.

variety of what-if scenarios. For example, suppose the zoning is changed and a certain farm is developed. What pressure does that put on remaining farms? Or if a particular group of farms were zoned for agriculture, could they act as a buffer against the extension of development into a prime farming area?

GIS maps are often available from government agencies or private firms. These maps can show a remarkable amount of information—from public facilities to zoning, land ownership, and soil quality. The cost of these maps has been coming down along with the cost of computers and GIS software. And public interest groups and land trusts are beginning to set up their own GIS databases. Since 1982, the state of California has operated a GIS-based Farmland Mapping and Monitoring Program to monitor both the amount and the quality of land converted to and from agricultural use.

Tracking the Real Estate Market and Development Proposals

To keep track of changes in the land base, it is necessary to follow the local real estate market. Land sales indicate where properties are being sold, the type of property, and the price. This information can be matched with the zoning designation to find out if specific farmland or open-space protection techniques (such as a conservation easement) apply to the property. For example, Lancaster County, Pennsylvania, publishes an annual sales analysis of all farmland parcels of 20 acres or more that sold in the calendar year. The analysis shows where in the county land sold, the size of the farm, the total price and price per acre, and any farmland protection techniques that apply to the land. The analysis helps to pinpoint where land is being sold for farming or development.

Land sales information is a matter of public record and is available from the office of county recorder of deeds. In some communities, land transactions are published in the local newspapers.

Land development and rezoning proposals also must be followed once they are submitted for review to municipal and county planning commissions. Public comment can be made on any development or rezoning proposal (see box 3.4). But the amount of public input and the political pressure landowners or developers may feel to respond to community concerns differ from one locality to another.

As stated previously, the regular monthly or bi-monthly meetings of the planning commission and elected officials are rarely exciting, but these meetings are where the day-to-day changes to the land in community occur. To influence the future of your community, these meetings are where to start and maintain a presence.

Sometimes it may be necessary to go to court to challenge local government decisions on development. Lawsuits are expensive, and it is a good idea to have many people and even farm groups contributing to the cause. For example, on several occasions in the early 1990s, the California Farm Bureau joined with county Farm Bureau chapters to sue county governments that had approved developments on farmland. Farm Bureau attorney Carolyn Richardson explained, "We will fight vigorously to protect farmers and ranchers near urban communities against condemnation to supply land for urban growth."[8]

Common Problems and Misperceptions in Managing the Growth of a Community

There are some common development problems and community misperceptions that should be addressed before they get out of control. As a rule, we recommend that interested people focus on influencing the comprehensive plan, the zoning ordinance, the subdivision regulations, and the capital improvements program. It is important to comment on specific development proposals; however, opposing a certain project can be a) expensive, if lawyers and experts are involved; b) bitter, if drawn out over a long time; and c) polarizing, if members of the community feel compelled to take sides.

While it is usually easy to get people involved in opposing a particular, imminent threat, such as a large residential subdivision, it is difficult to get people mobilized to change the law that allowed that subdivision in the first place. Often, careful, clear-headed planning could have prevented such a subdivision.

Below are some common problems to watch for:

Building lots larger than necessary. The move away from cities and inner suburbs has been accompanied by the desire of many people to build a house on a lot of an acre or more, which is often larger (and more expensive) than necessary.

These lots eat up farmland and contribute to sprawl. For example, Michigan's subdivision control act allows lots of 10 or more acres to be subdivided by right. This has led to the removal of tens of thousands of acres from farming. A great deal of land could be saved by requiring new development to occur adjacent to existing communities with public sewer on smaller lots.

Unrealistic zoning in the countryside. In the comprehensive plan process, it is common to hear many people speak about the need to protect "rural character." By this, they mean open space, the farms, the unhurried pace of life, and the older buildings. Yet when decision makers suggest strong zoning measures, landowners and others often respond, "We want to keep our rural character without land-use controls or with as little control as possible."

Much of the nation's rural land carries either no zoning or zoning that allows a house for every 1 or 2 acres owned. Land-use consultant Joel Russell tells the all too common story of the out-of-town developer who proposes to put fifty houses on a 100-acre farm, completely in keeping with the local zoning ordinance. The local planning commission asks the developer for water, sewage, and traffic studies. The developer spends a large amount of money on the studies. What the planning commission really wants is for the developer to go away, and the local government denies the development proposals. But the developer persists and eventually takes the local government to court and wins. The development is built, and the community is changed.

Such a community had probably adopted 2-acre zoning because it wanted to allow local landowners the option of selling off a few lots to family members or to raise cash for education, retirement, or an emergency. The community did not envision that a large residential subdivision could occur, or that the entire community could be divided into 2-acre lots. As Joel Russell warns, "What you zone for is what you get, sooner or later."[9]

Be prepared to do a *build-out scenario* of the amount of development the zoning ordinance would allow. Rural areas of mainly farming, forestry, or conservation lands should be zoned for low-density development. Areas for residential growth should be separated from resource lands and zoned to encourage development.

Rigid, single-use zoning in cities and villages. One of the most frequently heard criticisms of zoning is that local governments have used it to create single-purpose zones and have forced reliance on the automobile to travel from a single-family area to work or shopping. Villages built in the eighteenth and nineteenth centuries featured a mix of shops, businesses, and homes. Under many current zoning ordinances, those villages could not be built today.

Cities and villages should adopt mixed-use zoning as a way to encourage redevelopment of downtowns and to discourage sprawl onto farmland.

Assuming that land conservation lowers land values. Several studies have shown that protecting land can increase land values in the community.[10] A community with an attractive appearance that grows carefully is a desirable place to live and work.

Blindly assuming that growth is good and expands the tax base. Studies conducted by the American Farmland Trust in several states show that for every dollar generated in property tax revenues, farmland requires only 21 to 75 cents in public services. By contrast, residential development requires $1.05 to $1.67 in services for each property tax dollar collected (see figure 3.4 and table 3.1).

These studies shed light on the local debate on growth management and tax base. First, farmland protection is fiscally responsible; farmland pays more in taxes than it demands in public services. Therefore, farmland property tax breaks are justifiable. Second, residential growth does not pay its own way. This finding is really not surprising because educating children makes up the largest segment of most local budgets. So the more residential development, the higher property taxes must be raised to pay for education costs.[11]

The purpose of a cost-of-community-services study is to give local officials a better understanding of the tax base and how decisions about the amount, type, and location of development will affect the balance between property taxes and the cost of providing services.

A 1990 fiscal analysis study in Vermont found that "the higher the population in a town, the higher the tax bill per household. A lower tax burden results from lower population and [less] development," according to coauthor Jim Northup.[12] The Vermont Natural Resources Council and the Vermont League of Cities and Towns initiated the study to counter the public perception that farmland was a drag on local finances.

As more development occurs in farming areas, the pressure on the remaining farm operations becomes intense. Farmland values are bid up to prices at which farmers cannot compete. As farmland is developed, the remaining farmers then face higher property tax bills to help pay for the increased public services. A breakdown of the capital improvements program will show this. Growth must be accommodated selectively, the right development in the right place.

FIGURE 3.4 Comparing the Fiscal Impacts of Farmland and Residential Development

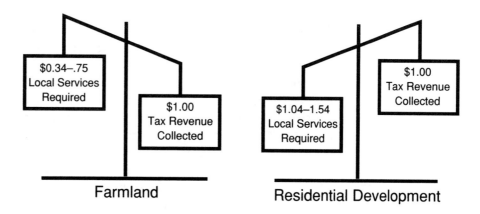

TABLE 3.1
The Cost-of-Community-Services Studies

	Ratio of Revenue Dollars to Expense Dollars by Land Use		
	Residential	Commercial/Industrial	Farm, Forest, and Open Land
Hebron, CT	1.00:1.06	1.00:0.42	1.00:0.36
Agawam, MA	1.00:1.05	1.00:0.41	1.00:0.30
Deerfield, MA	1.00:1.16	1.00:0.37	1.00:0.29
Gill, MA	1.00:1.15	1.00:0.34	1.00:0.29
Beekman, NY	1.00:1.12	1.00:0.18	1.00:0.48
Northeast, NY	1.00:1.36	1.00:0.29	1.00:0.21
Madison Village, OH	1.00:1.67	1.00:0.20	1.00:0.38
Madison Township, OH	1.00:1.40	1.00:0.25	1.00:0.30

Source: Data from studies conducted by American Farmland Trust.

Summary

The land-use planning process was originally created to promote orderly and safe development. The idea that land-use planning could be used to protect valuable farmland and natural lands from development is a fairly new concept. In the United States, land-use issues are mainly governed by the local county, city, or township. A few states—notably, Oregon, Florida, and Vermont—have state-level controls.

There are four main elements to local land-use planning. First is the comprehensive plan, which includes an inventory of land uses, an analysis of future needs, a general vision for the community over the next ten to twenty years, and specific goals and objectives to make the vision a reality and assure that needs are met.

The zoning ordinance, subdivision regulations, and the capital improvements program are flexible tools that work together to put the comprehensive plan into action. The zoning ordinance serves to separate conflicting land uses, promote more orderly development, and protect important resource lands.

The subdivision regulations control the division of land into lots, usually for eventual sale and development. These regulations ensure that new lots will be capable of supporting development and that necessary services will be provided before lots are sold.

The capital improvements program is a community's five- to -ten year estimate of public service needs and costs. The location and quality of public services play a major role in determining the location and density of development. It is a good idea for the community to adopt a policy of concurrency, requiring public services to be in place before private development can occur.

Land-use planning is a continuous process of reviewing development proposals, drafting and revising the comprehensive plan, the zoning ordinance,

> **BOX 3.5**
> How to Do a Cost-of-Community-Services Study
>
> 1. Meet with local sponsors who will help pay for the cost of the study and/or help gather and analyze the data.
>   Work with the county or township planning commission, tax assessor, and auditor or comptroller to identify land in the following categories: commercial; residential; industrial; or farm, forest, and open land.
> 2. Collect data
>   A. Obtain reports:
>   - year-end financial statements
>   - audit report
>   - property tax distribution amounts
>   - annual allocation of tax revenues to services
>   B. Talk with local government officials and departments:
>   - county commissioners (or township supervisors)
>   - auditor
>   - county (or township) planning commission
>   - county treasurer or township clerk
>   - sheriff
>   - county engineer
>   - health board
>   - fire department
>   - school districts
> 3. Group revenues and allocate them by land use:
>   - property taxes
>   - local user fees
>   - state aid
>   - sales tax (if any)
> 4. Group expenditures and allocate them by land use:
>   - general government
>   - public safety (fire and police)
>   - education
>   - public works
>   - recreation
>   - health and human services
> 5. Analyze the data and calculate the ratio of revenue to expenditures for the four land-use groups under number 1 above.

subdivision regulations, and the capital improvements program. Many people would just as soon draft a community land-use plan and declare the job done. But community needs and desires change. The comprehensive plan should be reviewed and updated every three to five years. Planning is a process that helps the residents of a community learn more about how to shape the future of where they live. It is also a process through which the public can exercise its interest in the actions of private landowners and private landowners can decide how best to use their land.

CHAPTER 4

The Business of Farming in America

Ultimately, what determines the future of agriculture is the collective decisions of individual landowners.[1]
—Bob Berner, director
Marin Agricultural Land Trust

Farming will always be a way of life for those who treat it as a business.
—Anonymous

The family farmer is a tough species, and will find ways to continue.
—Mas Masumoto
Epitaph for a Peach

All too often, local governments and citizen groups focus on the tools to protect farmland without understanding the business of farming. As stated earlier, it makes little sense to protect farmland if farmers cannot make a living. A general overview of farming will show the number and size of farms and where they are located. Next it is important to understand how federal farm programs affect farmers' business decisions and how land plays several roles as a productive resource, a business asset, a retirement account, and an insurance policy. All of these factors influence the future of farming.

Many people have a romantic view of farming. They think of farmers plowing wide-open fields, ranchers riding the range with their cattle, green cornfields, red barns, and a tight-knit family sharing in the chores. Indeed, farming thus seems more like a way of life than a business. But the reality is quite different.

A hundred years ago, about one out of every three Americans lived on a farm, and farming was very much a way of life. Today, there are hardly any small subsistence farms of the sort Thomas Jefferson championed. A farm is a business. The health of the business depends on the ability of the owners to make a profit. Nearly all farms are owned and operated by families, including extended families. Farming today means long hours, hard work, and, for a family or corporation, a large investment in land and equipment. Most farm families depend on off-farm jobs to supplement their farm earnings.

It is often said that there are two ways to get a farm: marry into one or inherit one. There is virtually no way that a young person fresh out of high school or college is going to buy a farm and make a go of it. Meanwhile, many of America's farmers are approaching retirement age. One of the challenges facing the nation's farms is who will do the farming in the next generation.

Farmers are business people. They are not the yokels and hayseeds of cartoons and traveling salesmen jokes. As journalist Robert West Howard points out, "The agriculturalist of today and tomorrow has become as sophisticated as any urbanite."[2] Farmers manage over $1 trillion in assets, an average of more than $500,000 per farm. They must be trained in agronomy, animal husbandry, finance, marketing, computers, and mechanics. Farmers use the latest technologies, and, because farm exports are a multibillion dollar market, they stay aware of what is happening in the world. One of the biggest concerns farmers have is how to comply with government regulations and paperwork—on everything from when to spread manure, to workers' compensation forms, to how much water can be used—and still earn a profit.

But farmers are far from self-sufficient. They have to buy machinery, fuel, fertilizer, feed, seed, livestock, and a variety of chemicals and often need to borrow money. They mainly provide land, labor, and management in the production of food and fiber. John F. Kennedy, when he was running for president, noted that farmers were the only business people who bought their inputs at a retail price and sold their commodities for a wholesale price.

Farmers are at the bottom of the food chain. They are "price takers." They have little power to set prices for their crops and livestock. Prices are determined in regional, national, and even international markets. In the middle and at the top are huge agribusiness corporations that process, package, and sell food and

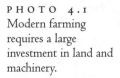
PHOTO 4.1 Modern farming requires a large investment in land and machinery.

fiber products. (See figure 4.1.) Farmers are in a high-volume business with low profit margins per bushel of wheat, per gallon of milk, and per chicken. A farmer receives only a small fraction of the consumer's price of a loaf of bread and less than half the retail price of a gallon of milk. Most of the chickens and many of the hogs in the United States are raised under contract between farmers and the big processors. The four leading beef packers make up 85 percent of the market, suggesting a power to control prices.

Farming is a business that usually generates a rather low return on investment, and income can vary widely from year to year. Weather is always a factor. There is always a risk of rising input prices and a fall in prices paid by the food processors. While federal farm programs have supported several farm commodities—such as wheat, corn, milk, soybeans, cotton, rice, sugar, and peanuts—the federal government is eliminating most crop subsidies and moving to "free market" pricing, which will favor the larger and more efficient farmers.

So why do people farm? Some make money at it. Some don't know what else to do. But nearly all of them love it.

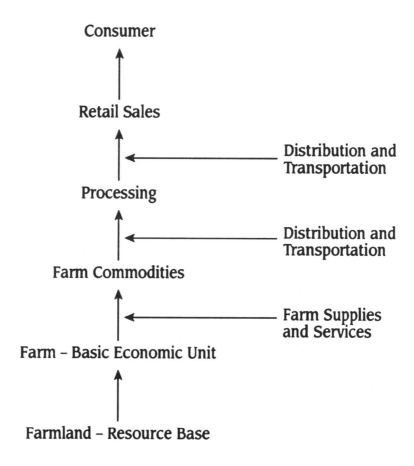

FIGURE 4.1
The Food and Fiber Industry

BOX 4.1
Challenges to a Farming Business

Weather: It's unpredictable for more than a few days. Wet weather can delay the planting or harvesting of crops. Hot, dry weather can shrivel crops and livestock. Snow and ice storms can bring down barn roofs, as can tornadoes. Lightning is a threat to barns in the summer.

Costs of production: Farmers try to manage production costs, but some costs, such as those for machinery, property taxes, feed, seed, and health insurance, are beyond farmers' control. Farmers typically borrow money to buy machinery, livestock, and land. Interest rates, set by the Federal Reserve in Washington and by individual banks, determine the cost of debt.

Prices received for crops and livestock: The farmer has little control over these prices.

Size: Farms need to be a certain size to function efficiently. It makes little sense to grow wheat on only a few acres if the farmer must supply the tractor and combine needed to produce the wheat. On the other hand, intensive production of fruits and vegetables on small farms can be profitable.

Water supply and water quality: Farms use most of the water consumed in America. They need a reliable supply of clean water to produce healthy crops and livestock.

Air pollution: Dirty air can reduce crop yields.

Fragmentation of the land base: Many farmers rent land, and others look to expand their farms by purchasing additional land. As land is broken into smaller parcels for building lots and "ranchettes," the price of land rises and land is taken out of farming. Farmers must travel farther to find rental ground, and they must compete against nonfarm buyers to purchase land.

Getting along with nonfarm neighbors: Farms may use chemical sprays or produce noise, odor, and dust, which impact nearby nonfarm properties. Complaints from neighbors can be a headache. Educating nonfarm neighbors takes time and patience.

Following government regulations: There may be good health and safety reasons for regulations, such as for handling pesticides, but complying with them takes time and costs money.

Maintaining soil quality: Good stewardship of the land requires a long-term investment in soil conservation practices, such as strip cropping, terraces, shelter belts, and grass waterways.

Biosecurity: Animal diseases can be devastating, particularly for confinement hog and chicken operations. Controlling disease is a must.

Investing for the long-term: A farm is expensive to buy, and buyers normally borrow money to pay for land and equipment. These expenses are paid off over several years from farm income.

The Structure of American Farming

The U.S. Department of Agriculture defines a farm as any agricultural operation with at least $1,000 a year in gross sales. By this definition, there are slightly less than 2 million farms in the United States, down from nearly 7 million in the 1930s. The Department of Agriculture predicts that farm numbers will continue to decline to only 1.7 million by the year 2002.

Today, fewer than one-third of all farms can be called true commercial operations, with annual sales of $40,000 or more. These farms produce over four-fifths of the nation's farm products (see tables 4.1 and 4.2). A more elite group of the top 200,000 farms produces about two-thirds of all farm commodities and earns 85 percent of total farm income. At the other end of the production spectrum are hobby farms, which make up about half of all farms but produce less than $10,000 a year in sales and account for only 3 percent of total farm

TABLE 4.1
U.S. Farm Size and Farm Product Sales, 1992

Farms by Size	Value of Farm Products Sold
Small	
1–9 acres	$4.9 billion
10–49 acres	$11.0 billion
Medium	
50–179 acres	$15.7 billion
180–499 acres	35.7 billion
Large	
500–999 acres	$32.2 billion
1,000–1,999 acres	$26.5 billion
2,000 or more acres	$31.4 billion
Total	$157.4 billion

Source: *1992 Census of Agriculture* (Washington, D.C.: U.S. Department of Agriculture), 89–90.

TABLE 4.2
Farms by Value of Products Sold, 1982–92

	Number of Farms		Percentage Change
Annual Sales	1992	1982	1982–92
Large			
$500,000 or more	46,914	27,800	+68.8%
$250,000–499,999	78,546	58,668	+33.9%
$100,000–249,999	208,405	215,912	–3.5%
Medium			
$40,000–99,999	248,532	332,751	–25.3%
$20,000–39,999	204,319	248,825	–17.9%
Small			
$10,000–19,999	232,067	259,007	–10.4%
$5,000–9,999	251,883	281,802	–10.6%
Less than $5,000	654,634	814,535	–19.6%
Total	1,925,300	2,239,300	–14.0%

Source: *1992 Census of Agriculture*, 89–90; *1987 Census of Agriculture*, 9.

output. On average, hobby farms show a negative farming income. Their owners earn their primary incomes elsewhere.

Since World War II there has been a trend toward bigger commercial farms and many hobby farms. The small to medium-size family farm has been caught in the middle, and many of these farms have kept going because of part-time or full-time off-farm jobs. Table 4.3 shows the change in the number of farms of different sizes between 1982 and 1992. The number of small and medium-size farms fell by over 300,000 in just ten years. Meanwhile, large farms slightly increased in number.

Several factors have caused this trend. Advances in machinery, plant and animal genetics, fertilizers, pesticides, and herbicides have greatly increased the output per acre of cropland and per animal. Small family farms do not have the acreage to use these technologies as efficiently as large farms. Also, large farms can pay for long-term farm labor, and small farms often cannot. Finally, federal farm subsidy programs have favored larger farms. Because the subsidies are

TABLE 4.3
U.S. Farm Size, 1982–92

Farms by Size	Number of Farms 1992	Number of Farms 1982	Percentage Change 1982–92
Small			
1–9 acres	166,496	187,665	–11.3%
10–49 acres	387,711	449,252	–13.7%
Medium			
50–179 acres	584,146	711,652	–17.9%
180–499 acres	427,648	526,510	–18.8%
Large			
500–999 acres	186,387	203,925	–8.6%
1,000–1,999 acres	101,923	97,395	+4.6%
2,000 or more acres	70,989	64,577	+9.9%
Total	1,925,300	2,240,976	–14.1%
Average Farm Size	490 acres	440 acres	+11.4%
Total Farm acres	943,397,000	986,029,440	–4.3%

Source: *1992 Census of Agriculture*, 89–90.

TABLE 4.4
Farms by Size and Share of Farm Ouput and Government Payments

Farm Size by Annual Sales	Number of Farms in Size Class	Percentage of U.S. Farm Production	Percentage of Gov't Payments
over $500,000	46,914	40%	14%
$250,000–500,000	78,546	16%	19%
$100,000–249,999	208,405	21%	33%
$40,000–99,999	248,532	12%	19%
under $40,000	1,342,903	11%	15%
Total	1,925,300	100%	100%

Source: *1992 Census of Agriculture*

based on the volume of crops, land left fallow, or milk produced, large farms garner the majority of federal farm payments (see table 4.4). But roughly two-thirds of the $160 billion in annual farm commodity sales comes from crops and livestock not covered by federal programs.

Most of the medium-size family farms are located in the Midwest, the East, and the South. Most of the large farms and ranches are found in the Great Plains and west of the Rockies. Cash grain farms, found mainly in the Midwest and Great Plains, make up about one-third of all U.S. farms and account for 45 percent of the farmland. Livestock operations are about one-quarter of all farms and vary greatly in size from cattle ranches of 10,000 acres to hobby farms of 10 acres. Dairy farms exist in every state and make up about 12 percent of all farms. The remaining 30 percent of farms range from vegetable and fruit operations in California and Florida to rice, cotton, and peanut enterprises in the South and Southwest to the unspecialized general farm.

Federal Farm Programs

Since the 1930s, federal farm policies have directly influenced the business of farming. The federal farm programs have had two goals: 1) to ensure an adequate supply of food and fiber; and 2) to provide farmers with a stable income.

Today, American consumers pay an average of 12 percent of their disposable (after-tax) income on food, the lowest percentage of any nation. Farmers often complain that there is a "cheap-food policy" that holds down the prices farmers receive for their products. The reality is that federal farm subsidies have enabled many farmers to stay in business and some to become downright rich. About one-third of farmers receive government farm program payments.

The main method of supporting farm incomes has been through subsidies for the major crops: corn, soybeans, wheat, cotton, and rice. The Department of Agriculture sets a target price for each commodity based on the concept of parity, which means a price that would cover production costs and give the farmer a profit that would enable him or her to live as well as a city dweller. If the target price per bushel or bale is above the actual market price, the USDA will pay the farmer the difference, called a deficiency payment. To participate in these programs, farmers agree not to plant crops on some acres (known as a set-aside). This practice is intended to reduce surplus crops. Critics of farm programs have long charged that farmers get paid *not* to grow food.

Subsidy payments do not apply to livestock, though income tax advantages for livestock do exist. There is a complicated milk-price support program that has the federal government buying up surplus milk production to support the price.

In fiscal 1995, farm programs cost the federal government about $14.5 billion, more than $45 million a day, but less than 1 percent of the $1.68 *trillion* federal budget. In 1986, during the farm crisis, the federal government spent $25 billion on farm programs. These figures do not include the cost of food stamps and school breakfasts and lunches, which also increase the demand for farm products.

In administering the farm subsidy programs, the Department of Agriculture has made no connection between the farms receiving payments and local land-use policies, which in part determine the likelihood of the farms being converted to nonfarm uses in the near future.

Federal farm programs received a major overhaul in 1996 when Congress decided to end most crop deficiency payments and acreage planting restrictions gradually over the next seven years. Nonrecourse loans, which farmers can take out by pledging crops as collateral and need not repay, will continue to provide a safety net. Subsidies to peanut and sugar farmers will also remain untouched.

Farm Loan Programs and Helping New Farmers

Another government benefit to farmers is the federal farm lending programs created in the 1920s and 1930s to supplement financing from commercial banks. The Farm Credit System of farmer-owned lending cooperatives makes loans for land and equipment, usually requiring at least 30 percent down. The Farmers Home Administration (now the Farm Service Agency) has acted as a lender of last resort for farmers who do not qualify for financing from the Farm Credit System or commercial banks. The Farm Service Agency has three main lending programs:

1. direct loans to farmers for up to $200,000
2. loan guarantees for banks making loans to farmers with at least a 25 percent downpayment (if the loan is not repaid, the government will reimburse the bank for 90 percent of the loan)
3. beginning farmer loans of up to 30 percent or $200,000 of the cost of buying a farm (but the farmer must pay back the loan in ten years, put at least 10 percent down, and find financing for the remaining 60 percent)

One way to slow the decline of farming in a community is to bring new blood into farming. In addition to the federal funding sources, several states have created beginning farmer loan programs through the sale of so-called "aggie bonds." These bond-backed programs offer loans of up to $250,000 and typically require a downpayment of 25 percent. Aggie bond programs have been set up in Alabama, Arkansas, Colorado, Illinois, Iowa, Kansas, Minnesota, Missouri, Nebraska, Ohio, Oklahoma, South Dakota, Texas, and Wisconsin.

In 1994, the Iowa legislature formed a beginning farmer center at Iowa State University, and the extension service and the Iowa Department of Agriculture started a Farm-On program to match older farmers looking to retire with young farmers searching for a farm. The program has the potential to bring young people into farming at a lower cost and to increase retirement income for older farmers. Similar Land Link programs now exist in California, Illinois, Iowa, Kansas, Massachusetts, Michigan, Minnesota, Nebraska, New

York, North Dakota, Ohio, Oklahoma, Pennsylvania, South Dakota, Utah, and Wisconsin.

The Changing Role of Farmland

A farmer's land plays several roles (see box 4.2). It is a renewable resource for the production of crops and livestock. It is the farmer's loan collateral when borrowing from a bank, the farm credit system, or the Farm Service Agency. Over time, farmland acts as a store of wealth. It is a bank account and an insurance policy. If a medical emergency arises or children are to be sent to college, land can be sold. And finally, farmland is often the farmer's retirement nest egg; a farmer will look to sell the farm to raise enough money to live on through the golden years. Some farmers will sell the land to a family member or another farmer. Others will sell the land for development and get a higher price.

The Equity Issue

How much is my land worth? This is a question that every landowner has asked at least once. And the answer changes over time. The value of land depends on several factors, but the most important are:

- location
- size and quality of the land parcel
- feasibility of uses to which the land can legally be put
- interest rates
- property taxes
- timing

The closer land is to public services—roads, schools, sewer and water lines—and to private shopping and employment services, the higher the land value. The better the land quality—level, deep, with well-drained soils—the easier it is not only to farm but to develop for houses, stores, offices, and factories. Generally, the smaller the parcel, the higher the value per acre of land. For example, a 100-acre farm might sell to a developer for $15,000 an acre, but the developer could get $40,000 for each 1-acre building lot.

BOX 4.2
Farmland Roles

- Productive renewable resource
- Loan collateral
- Bank account
- Insurance policy
- Retirement nest egg
- Legacy to heirs

> **BOX 4.3**
> Equity vs. Expected Development Land Value: An Example
>
> Farmer Brown's equity in a 200-acre farm:
>
> | | |
> |---|---|
> | Downpayment: | $40,000 |
> | Mortgage Principal Payments: | $220,000 |
> | Building Improvements: | $100,000 |
> | Land Improvements: | $20,000 |
> | **TOTAL EQUITY:** | **$380,000** |
>
> Expected land value if farm is zoned rural residential at 2 acres per dwelling:
> 100 building lots × $30,000 per lot = $3,000,000

The uses to which land can legally be put is mostly governed by local zoning ordinances. Zoning is not set in stone and can be changed. But what uses are allowed in a zone has a major impact on the value of land (see box 4.3). For instance, 50 acres zoned industrial could be worth $30,000 an acre, or the same 50 acres zoned for agriculture could be worth $2,500 an acre. The difference in value makes sense only if there are adequate sewer, water, and highway facilities to support a factory on the 50 acres. Many rural communities have learned that zoning land for industrial use does not guarantee that the value of the land will increase or that factories will be built.

Zoning, as we discussed in chapter 3, is part of a community's overall growth-management program. Zoning must be reasonable, must be designed to achieve a clear public purpose, and must not result in a taking of private property. Often zoning is not designed to protect farmland in large blocks. Even in the countryside far away from public sewer and water, it is not uncommon to find farmland zoned for 1-acre and 2-acre minimum lot sizes. This type of zoning is really large-lot residential zoning, which is aimed at enabling the farmer to sell the farm piece by piece until it is all gone from farming. It is this sort of zoning that invites sprawl, sewage disposal problems, and water pollution. One of the biggest problems with rural residences is managing on-site septic systems and keeping sewage away from wells. Once groundwater is contaminated, it is very hard to clean up.

The challenge in zoning land for agriculture comes when farmland has been zoned for many houses and then the community proposes to zone it for a low density of, say, one house per 20 acres or one house per 40 acres. Farmers complain that they are being downzoned and that their equity in the land is being taken away. The first complaint is correct: agricultural zoning does not allow as many houses as rural residential zoning. The land is worth more as building lots than as farmland. Some farmers feel that zoning pressures them *not* to develop their land in order to slow the pace of development in the community and to maintain someone else's open space.

The complaint about equity is misleading. The equity you have in a property consists of the downpayment you made when you bought the property, plus the mortgage principal payments you have made, plus any improvements you have made in the land or buildings.

As Pennsylvania master farmer John Barley notes, "Having debt structured properly means you must have sufficient cash flow to cover interest and principal payments. Net worth or equity is less important than your ability to amortize debt."[3] In other words, farmers should determine that they can operate their farms profitably and service debts from the farm income.

Farmers often have *expectations* about what their land is worth, whether based on rural residential zoning at a house for every 2 acres, or industrial zoning, or agricultural zoning at one house per 40 acres. A change from rural residential to agricultural zoning does not impair the farmer's ability to farm. In fact, that ability may be enhanced because fewer nonfarm houses can be built in the area. Rural residential zoning tends to raise unrealistic expectations from farmland owners. If a large number of houses are allowed to be built, then providing adequate sewerage and water supplies can be a problem. Roads and streets might be needed. A developer would have to consider all of these costs before making an offer for the farmland.

In urban fringe areas over the last thirty years, the market value of farmland has increased more because of public investments in roads, schools, and sewer and water lines than because of farmers' investments in farm buildings and land improvements. As a result, some farmers hope to sell for development to capture an "unearned increment" in the value of their property caused by nearby public services.

The farmer's attempt to cash in on rising land values is neatly explained by Professor Dixon Esseks as follows:

> With the chance to sell or develop their lands for a return of several thousand dollars per acre, farmland owners often challenge restrictive zoning ordinances before the relevant legislative body and then, if need be, in the courts, using the argument that denial of the right to convert their land out of agriculture is an unconstitutional "taking" of property.[4]

The legal response to this claim comes from the U.S. Supreme Court in *Agins v. City of Tiburon:*

> The application of a general zoning law to a particular property effects a taking if the ordinance does not substantially advance legitimate state interests or denies the owner economically viable use of his land.[5]

In short, a carefully crafted county or municipal zoning ordinance to protect farmland is legally defensible. Agricultural zoning need not allow the landowner to put land to its "highest and best" (i.e., most profitable) use. Agricultural zoning may reduce the value of land. And agricultural zoning is legitimate if it allows farming to continue and furthers a clearly expressed public purpose. But farmers often see zoning as one more regulatory burden among the many costly and time-consuming government rules that go with operating their business.

Land as Loan Collateral

Related to the equity argument, some farmers claim that if their land becomes zoned for agriculture, they won't be able to borrow enough money against the value of the farm. But banks and other lending institutions generally base their farm loans on the agricultural value of the property and the ability of the farm business to generate a profit. In appraising a farm for loan collateral, a bank will use three approaches: the replacement cost of the property, the value of the property based on income generated, and the value based on the market for similar properties. Bankers will also want to see the borrower's balance sheet and income statements for the previous few years, as well as cash-flow projections showing how the loan will be used in the farm business.

For a bank to lend money based solely on what a farm might be worth if sold for a housing development sounds like speculation and could easily burden the farmer with large, risky loans that a farmer couldn't pay off without selling land for development. Zoning can always change, as can interest rates and the economic climate. A banker tries to manage the credit risk in making a loan, not lend as much as possible.

According to Darvin Boyd, a member of the executive committee of the American Banker's Association's Agricultural Bankers Division, farm organizations argue against land-use regulations because many farmers are more interested in selling their land for development than in farming. "You've got to get beyond that," says Boyd, "and look at the quality of life and the role agriculture can play in the economic viability of the community. Land values continue to have a significant value for agricultural purposes, and although such land used for building purposes would have a higher value, there comes a point where you have to look at what value is reasonable to lend to an agricultural operation."[6]

At the heart of the farmers' equity argument is the fact that many farmers have nearly all of their wealth tied up in farmland and farm buildings, and they have little cash on hand. Agricultural finance experts agree that farmers need to diversify their investments. Still, the tendency is to invest available funds in the farm operation. One of the challenges in farmland protection efforts is how to get some of the equity out of the farm property and into cash. As we shall see, purchase-of-development-rights programs and transfer-of-development-rights programs are two ways to achieve this goal.

Land Tenure: Who Owns the Farmland

Who owns the nation's farmland and what those owners intend to do with it have important, long-term consequences for individual farms, local farming industries, and community growth. The term *land tenure* describes land ownership rights, who owns or controls the land, and the pattern of land ownership in a community.

There are three kinds of farmers: tenants who only rent farmland, owner-operators who own all the land they farm, and farmers who both own and rent land. Nearly all farms are family run, though they may vary from a few acres to several thousand acres. There are some corporate farms, managed much like the large industrial corporations, mostly in California and the western states. The type of farmer is important because the owner-operator tends to have the biggest personal financial and emotional stake in the farm and is the most interested in passing the farm on to heirs. The tenant farmer does not have deep roots and is relatively free to move, and the corporate farm owners will do what is in the best interests of the shareholders. In short, the more farm owner-operators in a community, the greater the likelihood that farming will continue there.

A good example of the importance of land tenure for a local farming industry is found in the Oregon's Hood River Valley, about sixty miles east of Portland. Valley growers produce Anjou pears, which are pooled and sold through co-op packing houses. The profit each grower makes depends on the total volume of fruit grown, as well as the price for pears, because the average cost of marketing a box of pears falls as the number of boxes increases. To maintain the volume of pears over the years, a grower needs to have about one-fifth of his or her tree stock in young trees that have yet to produce fruit. If even a handful of Hood River Valley pear orchards were to go out of business, all the pear growers would suffer.

There are three main kinds of renters: retired farmers, absentee landlords, and financial institutions. The more owners of rented land support farming, the more stable farming will be in a community. Retired farmers often still live in the neighborhood and are probably the most interested in seeing the land continue to be farmed. Absentee landlords run the gamut. Some are children who grew up on the family farm, then moved away, and have since inherited the farm. Others are real estate speculators who are renting out the land for farming while waiting for development to come closer and land prices to rise before selling for building sites. Financial institutions, like corporate farms, will do with the land what is in the best interests of the shareholders. Farmland for farming can be an attractive investment, especially as a hedge against inflation as in the late 1970s. Or farmland can have development potential, such as in Connecticut, where most of the farmland is owned by insurance companies.

Rented land makes up about 40 percent of America's farmland. For farmers, there are advantages and disadvantages to renting farmland. On the plus side, renting can be cheaper than buying land. Rented land can help new farmers get started and can enable existing farmers to maintain or expand their operations. Rented land can mean the difference between turning a profit and not turning a profit. On the negative side, most rental agreements cover a few years; some are even made by handshake on a yearly basis. This provides the renter with little long-term security. Within a few years, the land may be rented to another farmer who is willing to pay more, sold to another farmer, or even sold for development. For these reasons, it is not uncommon for renters to treat the land they rent differently from the land they own. Renters are less likely to make the kind of long-range soil conservation and fertilizer investments they would make to maintain the productivity of land they own.

It is often difficult and time consuming to identify who owns what land and which lands are rented rather than owned. Sometimes the only sources of information are property deeds and recorded leases filed in the county courthouse.

The pattern of land ownership is often clearly displayed on county or township tax maps. This includes parcel size and how the pieces of farmland fit together, and where farmland is isolated from other farms. This pattern of land ownership is essential to review and understand as part of drafting the community comprehensive plan, zoning ordinances, and capital improvements plan.

The Impermanence Syndrome: Deciding Whether to Stay or Sell

George Mielke and his extended family operate Trenton Mill Farms in Baltimore County and Carroll County, Maryland. Most of the 3,000 acres they farm is rented, and each year they lose some land to development. This often means that they must transport farm machinery to land located farther away. Hauling farm machinery is slow and dangerous.

The Mielkes also have a grain-drying and grain-storage complex that is used by seventy-five local farmers. When the construction of a golf course and several houses was proposed on a neighboring property recently, the Mielkes worried that if the golf course was built, golfers would complain about the noise of the grain dryer and the inconvenience of "corn lint" floating down on them. Such complaints could lead to a lawsuit that could put Trenton Mill Farms out of business. If the Mielkes were forced out of business, the acreage they rent could soon go out of farm use, and their former clients would have little alternative for grain drying and storage. Thus, the demise of a major farm operation could have a severe impact on the local farming industry. Should the Mielkes continue to invest in their farming operation or wait to see if the land next door is developed?

Farmland protection experts use the term *impermanence syndrome* to describe the vulnerability of farming to encroaching developments. Farmers are fiercely independent, but if they see a neighboring farmer sell his farm for big dollars to a housing subdivision or shopping mall, they may decide to do the same.

But for every acre converted to nonfarm uses, University of Delaware agricultural economist Gerald Vaughan warns that on average, an acre on another farm is likely to be idled; and other farmers reduce investment on two more acres. In other words, for every acre of farmland that is developed, neighboring farmers allow output to decline on three acres.

As farmland is developed, additional land on other farms often falls idle or is "undermanaged." How far and how fast this idling spreads into the countryside will vary, but a range of up to three miles is probable, and idling is accelerated by leapfrog development within the farming area. As development penetrates into farming areas, farmers tend to spend less on their farm buildings and equipment as they anticipate selling their land for development. The decision whether or not to buy a new $120,000 tractor or a $150,000 combine or put up a new dairy parlor or hog barn depends on a decision to stay in farming. As farmers decide to purchase less equipment and allow crop and livestock production to fall, the overall agricultural industry in the area weakens.

As the number of farms decreases, it becomes more difficult for the farm support businesses to survive; and once these businesses begin to disappear, the remaining farmers must spend more time and money traveling to fetch replacement parts, feed, seed, and machinery, not to mention finding shippers and processors of farm products. It simply becomes harder to farm efficiently.

The impermanence syndrome is not an all-or-nothing situation but a matter of degree, involving at least four factors. The more of these factors that exist at the same time, the greater the likelihood that farmland will be sold for development:

1. proximity to public sewer, water, and major roads
2. proximity to shopping and job centers
3. more farmland is rented rather than owner-operated
4. weak financial condition of the landowners

In judging the probable impact of proposed residential and commercial developments, planners, public officials, and concerned citizens should ascertain which of the above four factors exist and the likely chain reaction of change developments could bring in the local farming community.

Summary

Farms are a cultural tie to a time when most Americans worked and lived on farms. But we do not advocate protecting farms and farmland primarily for historic or cultural reasons. A farm must be able to pay its way as a business.

Farms come in many sizes, from family farms in the Midwest to huge cattle ranches in the West and California agribusinesses.

Federal farm programs have played an important role in supporting farm incomes and ensuring an adequate supply of food and fiber at affordable prices. The 1996 Farm Bill eliminated most crop subsidies, however, beginning in the year 2002. Thus, it is too early to tell what the impact of reduced subsidies will be on farm profitability.

For many farmers, land is their most valuable asset; farmland serves many financial purposes. As a result, farmers are wary of government regulations that affect what they can do with their land.

Decisions by farmers about whether to stay in farming or sell their land for development depend on a host of factors. Who owns the farmland, how much is rented, and the patterns of landownership will influence community planning possibilities and the long-term potential for farming. Community residents should support local farms by purchasing locally grown products whenever possible. Communities should also be careful in determining where development should go to avoid setting off a decline in farm investment that could become a self-fulfilling prophecy. A viable farming industry helps to keep local property taxes under control and to curb costly sprawl.

CHAPTER 5

Farmland Protection and the Federal Government

> The [farm]land being converted is the very best land. The impact is far more significant than the numbers would imply.[1]
> —Lloyd Wright
> Conservation Planning Division
> U.S. Department of Agriculture

> We are sensitive to the fact that farmland loss is a concern. But when we looked at it, we saw it wasn't a problem.[2]
> —Ken Krupa
> Economic Research Service
> U.S. Department of Agriculture

In March of 1991, we attended a conference in Washington, D.C., called "Saving the Land That Feeds America," sponsored by the American Farmland Trust. After keynote speaker James Moseley, assistant secretary of agriculture, delivered his speech, he was asked what the Bush administration's policy was on farmland protection. Moseley replied, "Next question, please."[3]

This episode reflects one side of the conflicting views on the loss of farmland within the U.S. Department of Agriculture. On the other hand, as early as 1975, the Department's Committee on Land Use recommended the maximum possible retention of agricultural lands.[4] But since then, the Department has not promoted a national vision for the future of America's farmland. Is farmland nothing more than private property? Isn't farmland also a strategic national resource? When and where is the loss of farmland likely to be a problem? Where should farmland be protected? Where should it be developed? What rate of loss of farmland is acceptable?

More than 100,000 people work for the U.S. Department of Agriculture. In fact, the USDA is so large that no one knows exactly how many employees it has. You would think that with a huge number of government workers earning their livelihoods from agriculture, there would be a federal policy and spending program to conserve and protect the nation's best farmland.

Well, there is something of a federal farmland policy, and a tiny amount of federal money has supported farmland protection efforts. But the federal government has nothing close to a coherent *strategy* to protect farmland. And there is no federal farmland policy that states and local governments are required to follow.[5] Instead, the federal government has viewed land-use matters as the domain of states, counties, and municipalities.

Yet the national government does influence land protection efforts through legal rulings by the Supreme Court (see chapter 3), tax policy (see chapter 12 on estate planning and transferring property), and farm lending that helps to finance farm operations and farm subsidy programs to bolster farmers' income (see chapter 4).

And when critics of farmland protection say that the free market should decide how land is used, they overlook the powerful effects of more than ninety federal spending programs on the location and cost of private development. In fact, lavish federal highway projects, federal grants to local governments for sewer and water projects, and the annual mortgage interest deduction for homeowners have subsidized the conversion of millions of acres of farmland over the past fifty years.

Since farmland preservation became a popular issue at the state level, the federal government has kept well out of the way. The federal response to states requesting assistance for farmland preservation has been painfully slow and meager, for fear of opening up the issue of federal involvement in local land-use planning.

There have been only a few federal programs that promote farmland protection, directly or indirectly: the Farmland Protection Policy Act of 1981; the Debt-Reduction-for-Easements program contained in the 1985 Farm Bill; and very limited assistance for state farmland preservation efforts through two separate programs, the 1990 Farms for the Future Act and a provision under the conservation title of the 1996 Farm Bill. To date, these programs have been little more than token attempts at farmland protection, and appropriations to the programs have been very small. In short, the federal government has responded feebly to states worried about permanent damage from farmland loss.

The Farmland Protection Policy Act of 1981

In 1979, the Department of Agriculture and the Council on Environmental Quality jointly funded the National Agricultural Lands Study (NALS). When the NALS report appeared in 1981, it caused a small uproar with the finding that between 1967 and 1977, the United States had lost three million acres of farmland each year to nonfarm uses. Although the NALS figures were later disputed, Congress responded by passing the Farmland Protection Policy Act as part of the 1981 Farm Bill (Public Law 97–98).

To reduce the loss of farmland to nonfarm uses, the Farmland Protection Policy Act put into place three measures. First, the law says:

> The purpose of this subtitle is to minimize the extent to which Federal programs contribute to the unnecessary and irreversible conversion of farmland to non-agricultural uses, and to assure that Federal programs are

administered in a manner that, to the extent practicable, will be compatible with State, unit of local government, and private programs and policies. (Subtitle I, Title XV, P.L. 97–98, Section 1540)

Each federal agency must identify and review any construction projects using federal funds that would result in the conversion of farmland.[6] Each agency must also determine if there are alternative locations or project designs that would conserve farmland. And the agency must make sure that, as much as possible, the projects are consistent with state and local government programs and policies and private efforts to protect farmland.

The Farmland Protection Policy Act does not require a federal agency to change a project just to avoid or minimize the impact of the conversion of farmland to nonfarm uses. But an agency has the discretion to decide whether to withhold funding from a project because of the conversion of farmland.

Since 1986, a federal agency must submit a Farmland Conversion Rating Form (AD-1006) to the local office of the Natural Resources Conservation Service when a federally funded project contributes directly or indirectly to the conversion of important farmland to nonfarm uses. Although the form serves only to report information, most of the completed forms have come from the Farm Service Agency and the Federal Highway Administration.

Each year, beginning in 1994, the Department of Agriculture must report to Congress on the impacts of federal programs and projects on farmland conversion. The AD-1006 forms are the basis for the annual report.

Somewhat surprisingly, the Farmland Protection Policy Act may not be used by an individual or group of individuals to challenge a federal project that may affect farmland. Yet in a state that has a state policy or program to protect farmland, the governor may file suit in federal district court to compel the federal agency to comply with the project review procedures. To date, however, no such legal suit has been filed.

Another shortcoming of the act is that it does not apply to farmland zoned for other uses, such as houses, businesses, and factories.

The second part of the Farmland Protection Policy Act calls for the Natural Resources Conservation Service to help state and local governments and nonprofit groups create farmland protection programs. The NRCS has mapped important farmlands in nearly every county in the nation to enable local governments to pinpoint their best farmlands.

The Land Evaluation and Site Assessment System

The Farmland Protection Policy Act also called for the creation of a farmland rating system to help federal agencies and local governments identify construction projects that would cause unacceptable conversions of farmland. Lloyd Wright of the NRCS spearheaded the creation of a land evaluation and site assessment (LESA) system.

The LESA system first rates the quality of land for farming and then rates the surrounding economic, social, and geographic features that indicate development pressures on the farm and farm viability. Both the land evaluation and the site assessment ratings include several factors and point scores. When the points for each factor are added up, they produce a total score for a farm (see tables 5.1 and 5.2).

The LESA system has the advantages of being objective, numerically based, and flexible, though there may be some trial and error involved in selecting point scores and the overall weighting of factors. The LESA system can serve as a guide for decision makers to target lower-quality farmland for future development and to protect highly productive farmland with long-term economic viability.

The information necessary to begin the land evaluation process can be obtained from a county soil survey. The evaluation may use one, two, or three soil ratings: land capability, important farmlands, and soil productivity. Land capability ratings indicate soil limitations and potential for crops, pasture, and development (see table 5.3). Generally, Class I and II soils are rated as prime farmland; Class III are soils of statewide importance; and certain Class IV soils are

TABLE 5.1 Land Evaluation Scores Based on Soil Productivity, Adapted from DeKalb County, Illinois

Soil Class	Corn Yield in Bushels per Acre Divided by the Highest Soil Class Yield	Ratio	Times 100	Land Evaluation Rating
I	160/160	1.00	100	100
II	140/160	0.88	100	88
III	120/160	0.75	100	75
IV	100/160	0.66	100	66
V	70/160	0.44	100	44
VI	50/160	0.31	100	31
VII	45/160	0.28	100	28
VIII	0/160	0	100	0

Sample 150-acre farm has 50 acres of Class I soils, 80 acres of Class II soils, and 20 acres of Class III soils:

50 acres × 100 rating = 5000
+
80 acres × 88 rating = 7040
+
20 acres × 75 rating = 1500
13540

13540 points divided by 150 acres =
Total Land Evaluation Score 90.26

Note: Within each soil class I–VIII there are often a variety of soils with different yields. The above table can be adjusted to include all soils within a soil class, such as one Class II soil with a yield of 145 bushels of corn per acre and another Class II soil with a yield of 135 bushels.

unique or of local importance. Soils in Classes V through VIII have very limited uses for farming.

Important farmlands consist of four levels of quality: the highest level is prime farmland; second is unique farmland that produces high-value crops such as orchard ground or cranberry bogs; third is farmland of statewide importance as determined by that particular state; and fourth is farmland of local importance.

Soil productivity rates the ability of different soils to produce a certain crop, such as the yield of corn in bushels per acre.

Of the three possible rating systems, land capability ratings are the most useful in evaluating how well the soils of a certain tract of land are suited to development in addition to farming. An important farmland rating is helpful mainly in public relations: "The farm was not rezoned for development because it contained a high percentage of prime farmland." Soil productivity is best used if you want to determine the quality of the land for farming purposes. Farmers

TABLE 5.2
Sample LESA System Site Assessment for a 150-Acre Farm, Adapted from McHenry County, Illinois

Site Assessment Factors	Weight Assigned	Sample Farm Points	Total Points Times Weight Assigned	Maximum Possible Points
Percentage of land in agriculture within 1.5-mile radius	2.0	9	18.0	20
Percentage of land in agriculture adjacent to the farm site	1.5	8	12.0	15
Percentage of farm site in agriculture	1.5	9	13.5	15
Percentage of farm site zoned for agriculture	2.0	10	20.0	20
Distance from a city or village	1.5	8	12.0	15
Distance to public sewer or water	1.5	5	7.5	15
Size of farm vs. average farm size in county	2.5	8	20.0	25
Road frontage of site	1.5	8	12.0	15
Farm support services available	1.5	8	12.0	15
Historic, cultural, and environmental features on farm site	1.0	6	6.0	15
Consistency with county plan	1.0	15	15.0	15
Consistency with municipal plan	1.0	15	15.0	15
Site Assessment Subtotal			163.0	200
Land Evaluation Subtotal			90.26	100
Total Points Possible				300
Total Points Scored			253.26	

Note: The land evaluation section rates soil productivity based on the yield of bushels of corn per acre, up to 100 points. The land being evaluated here has a high productivity rating.
Source: Adapted from *National Agricultural Land Evaluation and Site Assessment Handbook* (Washington, D.C.: U.S. Department of Agriculture, Soil Conservation Service, 1983).

TABLE 5.3 Land Capability Ratings

Soils Class	General Slope	Erosion Factor	Limitations
Class I	Slight	Slight	Few limitations that would restrict use
Class II	3–8%	Moderate	Some limitations; use conservation practices
Class III	8–15%	High	Many limitations; use special conservation practices
Class IV	15–25%	Severe	Many limitations; very careful management required
Class V			Very low productivity: pasture, range, wood land, wildlife uses
Class VI			Severe limitations; few crops, pasture, wood land, wildlife uses
Class VII			Very severe limitations; no crops, use only for range, pasture, wildlife
Class VIII			Most limited; use only for range, woodlands, wildlife, aesthetics

Note: In addition to the classes shown here, there are four subclasses that designate a particular soil limitation: e for erosion, w for wetness, i for internal soil problems, and c for climate. For example, a Class IIe soil would be at risk for erosion and a Class IIIw soil would be wet, indicating poor drainage or a high water table.
Source: USDA, Natural Resources Conservation Service.

in particular understand the value of a soil that can produce an average yield of 160 bushels of corn per acre.

A key feature to keep in mind about the LESA system is how farmland quality (land evaluation) is weighted against development potential (site assessment). In table 5.2, farmland quality is weighted one-third of the potential total of 300 points, and development potential makes up two-thirds. However, a different weighting on farmland quality, such as 150 points out of 300 total points, could lead planners and local officials to permit rather little development. On the other hand, if development potential receives a heavy weighting, such as 250 points out of 300 points, decisions could be made to allow considerable conversion of farmland to nonfarm uses.

The site factors, weightings, and total points listed in table 5.2 are not set in stone. Many places have included different development potential factors, such as distance to nonfarm zoning and the ability of the soils to support on-site septic systems. Allocating one-third of the overall LESA score just to soils makes sense in an area where prime soils are scarce; but in places with abundant prime soils, the land evaluation factors might include information on the quality of the farm operation, such as the gross annual income derived from the farm operation and the condition of the farm buildings.

One of the strengths of the LESA system is its flexibility. The factors, weightings, and maximum points can be changed as circumstances warrant. A LESA system involves testing and trial and error. It may be necessary to make adjustments. For example, thirty-eight counties in Pennsylvania use a modified LESA system in evaluating applications for the sale of development rights to farmland.

Although Delaware is the second smallest state, covering only 1.2 million acres, it has 550,000 acres in farm use and the state's farmers generate some

$820 million a year in product sales. The Delaware Department of Agriculture has created a LESA map for the entire state that indicates the current or potential use of land by section: the best farming regions, located mainly in the southern half of the state; cities or developed areas, found mostly in the northern half of the state; and areas that could go either way, some farming along with some development.

The LESA map serves two purposes. First, it is driving land planning by both the state and its three counties. It is especially useful for locating public infrastructure to service development and to keep infrastructure away from good farming areas. Second, it is being used to identify where the state should purchase development rights from farmland.

As of 1996, the LESA system has been used in over 30 states, and as many as 220 county and municipal governments have included a LESA system as part of their land planning or farmland protection efforts.

Debt-Reduction-for-Easements Program

Congress pieced together the 1985 Farm Bill at a time when a serious debt crisis was sweeping through the farms of the Midwest and the South. Interest rates reached 20 percent in the early 1980s as the Federal Reserve Board attempted to choke off double-digit rates of inflation. The high interest rates were especially burdensome for farmers who depended on borrowed money to buy land, equipment, livestock, seed, and fertilizer. The high interest rates also drove up the value of the dollar on world currency markets and made American farm commodity exports more expensive for potential buyers. The reduced demand for American crops, coupled with bountiful worldwide harvests, pushed down the price of soybeans, wheat, and corn, which in turn depressed farmland values. As farmland values fell, farmers found it more difficult to acquire loans because their farmland collateral was worth less. Thousands of farmers plunged into bankruptcy, even though in many cases, banks refrained from foreclosing on mortgages and forcing the sale of farms.

The Farm Service Agency acts as lender of last resort for farmers who are unable to obtain loans from commercial banks or the farm credit system. The agency has traditionally been an important source of financing for new and young farmers, who are usually greater credit risks. As of 1995, the Farm Service Agency had about 250,000 borrowers nationwide.

The debt-reduction-for-easement provision in the 1985 Farm Bill allows the Farm Service Agency to reduce the debt obligation of farmers who donate a conservation easement on their nonproductive land to the agency. This reduction in debt may or may not be sufficient to keep a farmer solvent.

According to the Land Trust Alliance, as of 1989 some 66,000 farmers had contacted the agency to have their debt reduced, but only 400 farmers wanted to be considered for the debt-reduction-for-easement program.

We are aware of only two debt-reduction-for-easement deals which were worked out in the early 1990s in Vermont. The Farm Service Agency staff and regulations have not been geared toward making debt-reduction-for-easements a popular program. This is unfortunate given the success of debt-for-nature-protection swaps that have occurred between industrialized countries and debtor countries in Latin America.

The 1985 Farm Bill also allowed the Farmers Home Administration to place conservation easements on land that had come into federal ownership through default on FmHA loans. The first of these easements were completed in Iowa in 1987 and applied to timberland, stream banks, and natural areas, as well as cropland. The 1990 Farm Bill authorized the secretary of the interior to acquire easements on wetlands that were part of properties held through foreclosure by the FmHA.

Farms for the Future Act

The Farms for the Future Act, passed in the 1990 Farm Bill, seemed to be a big break for the farmland preservation cause. It was the first time federal money was made available for direct assistance to state farmland preservation programs. Now defunct after just six years on the books, the Farms for the Future Act can only be seen as a rough attempt to put federal dollars toward farmland preservation.

The Farms for the Future Act was designed to lend federal money to states for the purchase of development rights to farmland. States would have been able to borrow up to $10 million a year for five years, by matching one dollar for every two dollars in loan money.

The Farms for the Future Act never got any further than a pilot project in Vermont, which borrowed $10.7 million in federal funds and purchased development rights to more than 9,000 acres of farmland between 1992 and 1995. Landowners and officials in Vermont rate the program a success, despite some awkward paperwork in borrowing money from private banks with a loan guarantee from Washington. Jim Libby of the Vermont Housing and Conservation Board called the program "very cumbersome", and predicted it would cost the federal government $1 million for Vermont to receive $600,000 in interest subsidies.[7]

The 1996 Farm Bill: Federal Grants for Buying Development Rights

To replace the lending approach of the Farms for the Future Act, the 1996 Farm Bill provided $35 million in federal grants over seven years. States and localities with farmland preservation programs can use the federal money to buy development rights (also known as conservation easements) to preserve valuable farmland. It was also hoped that new state and local farmland preservation programs would be started to take advantage of the federal money.

Section 388 of the Farm Bill authorized "a farmland protection program under which the Secretary shall purchase conservation easements or other interests in not less than 170,000 nor more than 340,000 acres of land with prime, unique or other productive soil that is subject to a pending offer from a state or local government for the purpose of protecting topsoil by limiting nonagricultural uses of the land." The funding for this program, however, is far from adequate to purchase conservation easements on over 170,000 acres of farmland.

In September, 1996, farmland preservation programs in 18 states received $14.5 million to help in the purchase of conservation easements on more than 55,000 acres. But as of this writing, Congress had budgeted only an additional $2 million for the program in fiscal 1997.

Despite short-term federal budget deficits, the federal funding role for farmland preservation is likely to expand within the next decade as the squeeze on farmland resources continues and more people bring farmland preservation to the attention of their representatives in Congress.

Federal Farmland Environmental Laws

Federal regulation of farming has driven many farmers to declare that the government should "get out of agriculture." Yet according to the U.S. Environmental Protection Agency, agriculture produces 64 percent of nonpoint source pollution in rivers and 57 percent in lakes. Soil erosion, feed lots, and runoff from fields on which pesticides, herbicides, and fertilizers have been applied all contribute to nonpoint source pollution. Soil sedimentation in waterways causes up to $2.2 billion a year in damage. In response, the federal government has taken some positive steps to protect farm soils and wetlands.

In 1977, Congress passed the Soil and Water Resources Conservation Act, which requires the USDA to issue a report on the nation's land and water resources every ten years.

This report then helps to shape national soil and water conservation programs. The *1987 Resources Conservation Act Appraisal* reported that 40 percent of America's cropland was suffering water and wind erosion in excess of five tons per acre per year—the rate at which topsoil can still be replenished by natural processes without a significant loss in productivity.

The 1985 Farm Bill established the Conservation Reserve Program, "conservation compliance" for owners of highly erodible cropland, and the Swampbuster and Sodbuster provisions. The Conservation Reserve Program has made payments to farmers to remove more than 36 million acres of highly erodible cropland from production for ten years. The Conservation Reserve contracts add up to nearly $20 *billion*. But ten congressional districts in the Midwest receive almost half of the payments. The 1996 Farm Bill authorized extending the Conservation Reserve Program, but it is expected that fewer than 30 million acres will be enrolled because of the high grain prices in 1996–97.

The Conservation Reserve Program has saved an estimated 670 million tons of soil a year, a savings of more than 10 tons of soil for each acre enrolled in the Conservation Reserve each year! Yet in 1992, according to the Natural Resources Inventory, 2.1 billion tons of cropland soil were lost to wind and water erosion, and 1 out of every 5 acres of cropland was eroding at a production-threatening rate of more than five tons per year. Still, the Conservation Reserve has resulted in less use of pesticides and herbicides, cleaner water, and more wildlife habitat. This protection could be temporary, however, if farmers bring much of that highly erodible land back into crop production.

The conservation compliance section of the 1985 Farm Bill required owners of some 120 million acres of highly erodible cropland to develop conservation plans by 1990 and implement those plans by 1995. Landowners who do not put conservation plans into action are not eligible to participate in federal farm programs. Conservation compliance has compelled farmers to adopt new practices, such as no-tillage or minimum-tillage farming, and to revisit proven techniques like contour cropping, terracing, and crop rotation.

The Swampbuster and Sodbuster provisions of the 1985 Farm Bill mean that farmers who plow up wetlands or highly erodible soils are not eligible for federal farm subsidies. In the lower forty-eight states, 117 million acres of wetlands—about 53 percent of the wetlands that have existed over the past 200 years—have been lost. In Illinois, Iowa, Missouri, and Kansas roughly nine-tenths of the original wetlands are gone, mostly to cropland.

Although protecting soil and water resources makes sense, the federal government has not been clear on what constitutes a wetland: wet soils, a quarter-acre bog, or land that floods periodically. The policy of "no net loss" of wetlands set down by the Bush administration was simply not well defined, however well intended. The goal was to create a new acre of wetland for each acre of wetland loss. But keeping track of wetland loss and identifying where to create new wetlands have proven to be difficult.

Between 1955 and 1975, an estimated 690,000 acres of wetlands were filled in each year, mostly for farm use. The 1990 Farm Bill created the Wetlands Reserve Program to restore and permanently protect 975,000 acres of wetlands and adjacent farmland by the year 2002. Roughly one-third of the wetlands acreage would be restored through cost-sharing agreements with landowners. Another third would be protected through the purchase of 30-year term conservation easements; and a final third preserved through permanent conservation easements. To date, the Department of Agriculture has purchased easements on about 315,000 acres of wetlands at an average cost of $600 per acre, or a total of nearly $200 million.[8] In addition, the Department has purchased easements on another 77,000 acres of wetlands, mostly in Iowa and Missouri, under the 1993 Emergency Wetlands Reserve Program, which was enacted in response to the devastating flooding of the Mississippi River and Missouri River systems.[9]

From 1982 to 1992, the annual rate of conversion of wetlands averaged 156,000 acres, with only one-fifth of the acreage involving farmland. While this progress is commendable, it is worth noting that federal spending on protecting wetlands has greatly exceeded federal funding to preserve productive farmland.

Summary

The federal government will probably continue to play a small direct role in promoting farmland protection. While federal farm subsidy and loan programs have a major influence on farm incomes, there is as yet no coordination between farmland protection policy and farm income policy.

The Federal Farmland Protection Policy Act has helped to raise awareness in the federal government about projects that convert farmland to nonfarm uses. Still, the Department of Agriculture needs to take a stronger role in working with other federal agencies to avoid the needless loss of high-quality farmland.

The land evaluation and site assessment system is a good start for federal assistance to state and local farmland protection efforts. But expanded federal funding for purchasing development rights to farmland could be crucial for starting new state farmland preservation efforts and maintaining existing state programs.

CHAPTER 6

State Farmland Protection Programs

> Taxes control the appearance of our communities.
> —Peter Wolf,
> *Land in America*

In the late 1950s, states began relieving farmers from having to pay taxes on the development value of their land. The purpose was to help farmers maintain a reasonable profit margin and to make the retention of farmland a greater possibility in the face of suburban encroachment. By the 1970s, every state had undertaken efforts to help farmers lighten the property tax burden. While these tax breaks may or may not actually help a farmer ward off development pressure, such programs were important forerunners for programs that would further protect farmland or even preserve it in perpetuity.

The states have generally recognized the importance of farmland and their agricultural industries, as well as the limited involvement of the federal government in protecting farmland, and thus the need for action at the state level. Some states have responded to the lack of local action by requiring local governments to undertake planning and zoning to protect land resources. And some states, notably Oregon and Vermont, have made grants to local governments to pay for improved land-use planning. In 1996, New York began offering grants to counties and municipalities to draft and implement farmland protection plans. In the early 1990s, Colorado and Michigan conducted state-level task forces to better understand the loss of farmland in terms of past and future trends and possible responses.

Farmland preservation programs have originated at both the local and state levels nationwide, but most local programs will accomplish substantially more when supported by state money and state program administration. A state role can help provide funding, guidelines, and legal standards for farmland protection efforts. Ideally, farmland protection is a cooperative effort between the state, its local governments, and private landowners.

States have four powers to influence farmland protection:

1. regulation: including zoning, subdivision ordinances, comprehensive plans, urban growth boundaries, and agricultural districts
2. spending: on roads, schools, sewer and water facilities, and utilities
3. taxation: income taxes, property taxes, and estate taxes

4. acquisition of interests in land: purchase of development rights and transfer of development rights

State farmland protection efforts have concentrated on three areas: 1) reducing property and estate taxes so that taxes do not force a sale of the farm; 2) slowing the rate of farmland loss; and 3) combining financial incentives with land-use controls.

All states offer favorable property tax programs for farmland, and several states appraise farmland at its use-value for estate tax purposes. Six states allow those inheriting a farm to defer estate taxes for five years and then pay off the estate tax bill over ten years.

Many states use agricultural zoning and agricultural districts to help slow the rate of farmland loss. More states are using the purchase of development rights to preserve land for farming, as discussed in chapter 9, and encouraging local governments to use transfer of development rights, as explained in chapter 10. In some states, a governor's executive order has helped to keep major state construction projects out of prime farming areas.

A few states have combined tax incentives with comprehensive planning and strong agricultural zoning. Oregon offers deferred taxation for farm property and requires local governments to adopt plans, agricultural zoning, and urban growth boundaries (see chapter 8). Wisconsin uses state income tax credits as an incentive for farmers to encourage local planning and zoning.

Table 6.1 shows a summary of the state farmland protection techniques. Most states have not done a good job of coordinating these techniques into a strategic package. For example, a property tax break for farmland is a public investment that needs to be protected; it makes little sense to provide tax breaks if the farmland can then be developed at any time. And the potential for abuse is enormous. Land speculators have often taken advantage of farm property tax breaks while holding land for eventual sale to developers.

If the current use of a property is farmland and it is taxed as farmland, then it is logical to zone it for farming as well.

You will notice that farm property tax relief and right-to-farm laws exist in every state. Some states have protection techniques "on the books" but not in practice. This is the case with purchase or transfer of development rights, which are allowed under law in forty-six states but are used in only nineteen. Other purchase of development rights programs, in Maine, New Hampshire, North Carolina, and Rhode Island, have become dormant through lack of funding. In North Carolina and Virginia two purchase of development rights programs have been implemented at the local level without state funding.

Twenty-six states authorize and encourage the use of agricultural zoning, though implementation must come at the county or municipal level. The specifics of agricultural zones may vary considerably among states and even within a state. For example, a zone with a 25-acre minimum lot size might be considered agricultural zoning in Pennsylvania but in California a 40-acre minimum lot size in Fresno County or even a 100-acre minimum lot size in Ventura County is considered agricultural zoning (see chapter 7 for an in-depth look at agricultural zoning).

TABLE 6.1
State Farmland Protection Programs

Key
CB = Circuit breaker
PDR = Purchase of development rights
DT = Deferred taxation
TDR = Transfer of development rights
PA = Preferential assessment
RA = Restrictive agreement

State	Property Tax Break	Right-to-Farm Law	Agr. Zoning	Agr. Districts	PDR or TDR	State Planning	Gov.'s Exec. Order
Alabama	DT	X					
Alaska	DT	X		X			
Arizona	PA	X		X			
Arkansas	PA	X					
California	RA	X	X	X	X	X	
Colorado	PA	X	X		X		
Connecticut	DT	X			X		X
Delaware	PA	X		X	X		X
Florida	PA	X			X	X	
Georgia	DT	X				X	
Hawaii	DT	X	X	X		X	
Idaho	PA	X	X	X			
Illinois	DT	X	X	X			X
Indiana	PA	X	X				
Iowa	PA	X	X	X			
Kansas	DT	X	X				
Kentucky	DT	X		X	X		X
Louisiana	PA	X					
Maine	DT	X	X		X	X	
Maryland	DT	X	X	X	X		X
Massachusetts	DT	X			X		
Michigan	CB	X	X	X	X		X
Minnesota	DT	X	X	X			X
Mississippi	PA	X					
Missouri	PA	X					
Montana	PA	X					
Nebraska	DT	X	X				
Nevada	DT	X					
New Hampshire	RA, DT	X			X		
New Jersey	DT	X	X	X	X	X	
New Mexico	PA	X					
New York	DT	X	X	X	X		
N. Carolina	DT	X		X	X		X
N. Dakota	PA	X	X				
Ohio	DT	X		X			X
Oklahoma	PA	X					
Oregon	DT	X	X			X	
Pennsylvania	DT, RA	X	X	X	X		X
Rhode Island	PA	X	X		X	X	
S. Carolina	PA	X					
S. Dakota	PA	X	X				
Tennessee	PA	X	X				
Texas	PA	X					
Utah	PA	X	X	X			
Vermont	RA	X			X	X	X
Virginia	PA	X	X	X	X		X
Washington	DT	X	X	X	X	X	X
W. Virginia	PA	X					
Wisconsin	CB	X	X	X		X	X
Wyoming	PA	X	X				
Totals	50	50	26	21	18	11	14

Sources: Lapping, Daniels, and Keller, *Rural Planning*, 166–167; *Farmland Preservation Report*, seriatim.

Agricultural districts are used in twenty-one states to provide some greater security for farmland owners against the intrusion and complaints of non-farmers. Some districts also offer property tax relief.

Eleven states have formal state land-use planning programs. These programs mostly require counties and cities to adopt comprehensive plans that meet state guidelines. Agricultural land protection is emphasized in Hawaii, Oregon, and Wisconsin and to a lesser degree in other states. State planning offers a set of standards for the location and type of development, and some states offer grants to localities to draft comprehensive plans.

Governors in fourteen states have issued executive orders that required state agencies to review projects that would result in the conversion of farmland to nonfarm uses (see appendix D). The executive order puts in place at the state level the same coordination and review by state agencies that federal agencies are supposed to follow under the Farmland Protection Policy Act of 1981.

Right-to-Farm Laws

Every state has a right-to-farm law that attempts to provide farmers with legally defensible protection from nuisance suits for standard farming practices. Ordinarily, the "nuisance doctrine" allows a landowner to file suit to stop a neighbor from using land in ways that detract from the landowner's enjoyment of his or her property. As people have moved out of cities and suburbs into the rural areas, conflicts have arisen between nonfarm residents seeking a rural haven and neighboring farming operations that generate odors, spray drift, noise late at night or early in the morning, and slow moving machinery.

PHOTO 6.1
Farming next to houses is not easy for farmers or residents.

As farmers become a political minority, it is not uncommon for local governments to pass, or attempt to pass, ordinances that restrict hours for operating farm machinery and regulations detailing when to spread manure and how much water can be used. The paperwork is time consuming and increasingly complex. These sorts of headaches can discourage farmers and speed their exit from farming.

Most newcomers to the rural-urban fringe do not understand that farming as practiced today is essentially an industrial process involving heavy machinery, powerful chemicals, and large concentrations of animals. In some cases it seems that the newcomers want farmland to look at without farmers farming the land!

State legislatures have passed right-to-farm laws to help protect farmers and agribusinesses from nuisance suits that would stop or limit normal farming practices. These normal farming practices must conform to state and federal laws and regulations. For example, if a farming practice results in manure runoff and water pollution, a neighbor could file suit to make the farmer stop that practice.

Most right-to-farm laws were passed between 1978 and 1984 and closely follow a North Carolina statute. Most include the concept of "first in time, first in right." That is, only farm operations that have been in existence for at least one year are guaranteed protection. Farmers start new enterprises, such as a hog facility, at their own risk. This may also be the case when farmers adopt new technologies.

Right-to-farm laws are largely untested.[1] One potential weakness of these laws is the matter of trespass.[2] Some courts have ruled that noise, dust, and odors may result in trespasses onto neighboring properties. A trespass is more serious than a nuisance, and a court is more likely to find in favor of the neighboring plaintiff. Also, the increase in large-scale animal confinement operations, such as 10,000 head of hogs or a million egglaying hens, may call for a re-thinking of right-to-farm laws and what constitutes a normal farming practice. In a well-publicized 1994 Vermont case, a judge denied a farmer's proposal to construct a 560-cow barn and large manure pit in an agricultural area because neighbors raised concerns about water pollution and offensive odors.[3]

Finally, some farmers are discovering that a right-to-farm law does not discourage a neighbor from filing a lawsuit. The farmer must then hire an attorney and mount a defense. Costs can rise into the tens of thousands of dollars. In 1995, Michigan revised its right-to-farm law to allow farmers to recover legal costs from plaintiffs who lose a nuisance suit. This way, nonfarmers will be discouraged from filing frivolous nuisance suits.

Farmers would be well advised to talk to their neighbors about their farm operations, especially if the farm produces livestock. Flies and the smell of manure often do not endear farmers to their neighbors. And farmers should be willing to try to mitigate the impact of activities that spill over from their farms onto neighboring properties.

Differential Assessment (Farm Property Tax Breaks)

Property taxes typically equal 10 to 20 percent of net farm income. But taxes on farmland vary among states, from an average of a few dollars per acre to over $10 an acre. On a 400-acre farm, that's a big difference. Among local governments, the range in tax burden is even greater. In rapidly growing communities, taxes on farmland can eat up more than half of net farm income and actually force the sale of the farm.

Several studies have shown that farmland generates more property tax revenues than it requires in public service costs (see chapter 3). Also, when farmland is taxed according to its highest and best use, it is taxed as a potential residential, commercial, or industrial site. Owners of farmland maybe unable to afford the tax burden and landowners are forced to sell some or all of their land, often for development. No politician enjoys being accused of compelling landowners to sell their land because of high property taxes.

Under differential assessment, farmland is valued for property tax purposes according to its current use in farming, not at its highest and best use for house lots or strip malls. The difference (differential) in assessed value between the highest and best use and the agricultural use multiplied by the local tax rate determines the size of the tax break. Obviously, the higher the local tax rate and the greater the differential between farm use-value and highest and best use, the larger the property tax savings to the farmland owner. In communities with considerable amounts of farmland, the savings to the farmer under differential assessment are generally small. But in growing communities, differential assessment can help a farmer stay in business.[4]

The farm use-value is determined from soil productivity ratings included in the Natural Resources Conservation Service soil surveys and from net income per acre. Commonly, state tax offices or agriculture departments estimate farmland value by computing annual net income per acre (or, in some states, rent per acre) capitalized by an interest rate that reflects the return on capital in other in-

---

BOX 6.1
How Differential Assessment Works

For a 200-acre farm

Property tax assessment based on highest and best use as residential building lots: $5,000 an acre, or $1,000,000

Property tax assessment based on current use as a farm: $1,250 an acre, or $250,000

Difference in assessments: $750,000

To farm owner with differential assessment of farm tax saving = $750,000 × local tax rate $.010 = $7,500 a year

vestments. Use-value assessment generally applies only to the farmland; farm buildings and farmhouses are taxed as commercial and residential property.

Preferential Assessment

Forty-eight states have enacted one of three types of differential assessment for farmland. Seventeen states offer *pure preferential assessment* (see table 6.1). Most states require farmland owners to apply for tax relief, showing a minimum acreage (usually 10 acres) and proof that the land is being actively farmed. However, there is no penalty to the landowner for withdrawing from the preferential assessment program or for converting farmland to a nonfarm use.

The problem with pure preferential assessment is that it may offer property tax breaks to land speculators and farmers acting as speculators who are waiting for the land to "ripen" in value before selling for development. This problem has been observed many times in urban fringe areas where there is heavy pressure to convert farmland to nonfarm uses. Florida, for example, spends $500 million a year on property tax reductions for owners of farmland. But the abuses by developers holding farmland have been widespread. In fact, Florida leads the nation in farmland loss, with 150,000 acres converted each year.

BOX 6.2
Estimating the Use-Value of Farmland

Soil productivity rating for 200-acre farm:

30 acres of Class I soil (productivity rating 130)
95 acres of Class II soil (productivity rating 120)
45 acres of Class III soil (productivity rating 100)
30 acres of Class IV soil (productivity rating 85)

Net returns per acre:

$160 for Class I soils
$140 for Class II soils
$100 for Class III soils
$80 for Class IV soils

Capitalization rate: 10%

Farmland value: Net returns per acre divided by 10% capitalization rate
    Class I soils: $1,600 an acre × 30 acres = $48,000
    Class II soils: $1,400 an acre × 95 acres = $133,000
    Class III soils: $1,000 an acre × 45 acres = $45,000
    Class IV soils: $800 an acre × 30 acres = $24,000
    TOTAL FARMLAND USE-VALUE = $250,000

A second problem with pure preferential assessment is that the tax breaks frequently go to "farmettes" and hobby farms, whose owners derive very little, if any, income from the farm operation. In these cases, preferential assessment simply subsidizes a rural lifestyle. A strong argument could be made to tighten up eligibility requirements. For example, to be eligible, a landowner must show minimum gross annual earnings of $25,000 from the farm. This standard would target *commercial* farm operations for property tax relief, not speculators or hobby farmers.

The consensus among land-use experts is that pure preferential assessment has reduced the tax burden on farmers, but it has not succeeded in keeping land in agricultural use when developers offer large sums for farmland. Also, farm property tax relief is expensive. In some states and counties, there are more acres enrolled in farm use-value taxation programs than there are acres of farmland. Preferential assessment has failed to dampen rising land values of farmland for nonfarm uses and has not curbed scattered development patterns that erode the farmland base.

Farm property tax breaks can be viewed as a public investment in private land, and taxpayers should be concerned about protecting their investment. Simply put, if land is receiving a farm use-value tax break, the land should be zoned as farmland. Otherwise, preferential assessment, without agricultural zoning to limit development potential, is a recipe for farmland conversion.

Deferred Taxation

Twenty-nine states offer *deferred taxation* to farmland owners. Deferred taxation combines preferential assessment with a rollback penalty to recapture some or all of the property tax benefits if the land is converted to a nonfarm use. That is, property taxation based on the farmland's highest and best use is deferred until the land is no longer used for farming.

Deferred taxation programs usually have stricter eligibility requirements than pure preferential assessment, but the requirements vary from state to state. Some states require only a minimum acreage; for example, Rhode Island has a 5-acre minimum and Connecticut a 25-acre minimum size. Other states require either a minimum acreage or a minimum level of gross farming income, such as 10 acres or $2,000 in Pennsylvania. Still other states have both acreage and income requirements, such as 10 acres and $10,000 in New York. A further common regulation is that the land must have been farmed for a certain number of years before entering the deferred taxation program. Maine, for instance, requires the land to have been farmed in one of the past two years and three of the past five.

The rollback penalty may recover all or a portion of the property taxes that would have been due if the land had been taxed at its highest and best use, plus an interest charge. In Pennsylvania, the rollback recaptures tax breaks up to seven years with a 6 percent penalty. Other rollbacks range in length from two years in Kentucky to fifteen years in Maine. Maryland levies a rollback penalty in the form of a land transfer tax based on a percentage of the fair-market value if the farmland is converted to a nonfarm use. The land transfer tax revenues then go into a fund to buy development rights to farmland.

The rollback penalty ensures that if participating land is taken out of farming, the landowner will not receive much or any property tax break. Yet the size of the penalty may be of little consideration if the land is to be sold for a large amount of money. Maryland, Nebraska, and Oregon require that land be zoned for agricultural use to be eligible for deferred taxation. This package of planning, zoning, and deferred taxation is likely to be more successful in retaining land in farming than reliance on property tax breaks alone. Through a comprehensive plan and on the zoning map, large blocks of farmland can be identified and given greater protection from development. The tax break then acts as a form of compensation for the greater restrictions, such as larger minimum lot sizes for house sites, resulting in fewer dwellings allowed. At the same time, the increased protection of the land serves to protect the public investment (tax break) in the farmland.

Restrictive Agreements

Four states—California, New Hampshire, Pennsylvania, and Vermont—use *restrictive agreements* to provide preferential assessment through a legally binding contract. Several individual townships in the Northeast have also used restrictive agreements. In the agreement, the landowner consents to restrict the use of the land for a specific period of time in return for preferential taxation or for a freeze on the actual amount of property taxes owed. Most contract periods run for 10 years, and a rollback penalty is imposed if the land is converted to a nonfarm use before the contract expires. But once the contract has expired, the landowner is free to change the land use without any penalty.

Restrictive agreements are voluntary, however, and farmers in rural-urban fringe areas are often reluctant to enter into contracts that limit their options to sell land. This problem has been evident under California's Land Conservation Act, better known as the Williamson Act. Landowners who enroll their land in Williamson Act contracts receive property tax breaks from year to year, but they must give notice if they intend to withdraw their land from the program. Once a landowner gives notice, he or she must then wait ten years before the land may be converted to a nonfarm use. In the ten-year waiting period, property assessments are allowed to rise to fair-market levels. Over half of California's farmland, some 15 million acres, is enrolled in Williamson Act contracts.

Shifting the Tax Burden

All differential assessment programs cause a shift in the local tax burden from owners of farmland to other property owners. Two researchers from the

University of Delaware have estimated that differential assessment programs in the northeastern United States have reduced local tax bases by a whopping $10 to $20 *billion*.[5] The shift in local tax burden can be especially felt in rural areas where farmland makes up a large proportion of the local tax base.

The state of California reimburses local governments for about one-third of the tax revenues lost from restrictive agreements, thus shifting some of the tax burden from local property owners to taxpayers statewide. New York also provides partial reimbursement. But in states without reimbursement provisions, local tax revenues may fall and taxes for nonfarm properties increase, prompting a search for commercial, industrial, and even residential development to bolster the local tax base.

For example, 210,287 acres were enrolled in farm use-value assessment in Loudoun County, Virginia, in 1989, which resulted in a loss of nearly $15 million in property tax revenues. But, as mentioned, farmland more than pays for itself in terms of taxes paid versus public services required, whereas, residential development costs more in public services than it generates in property tax revenue. In this sense, farm property tax breaks may be helpful in holding down overall property tax increases to pay for additional services for residential development.

How to Improve Differential Assessment Programs

If the three types of differential assessment are mostly ineffective in keeping land in farming, why are they so popular? The answer is that differential assessment has been extended to a wide array of landowners, from weekend hobby farmers to rural residential landowners to commercial farmers to imminent developers. Clearly, the lenient qualifying criteria have led to considerable abuse of farm property tax relief programs.

To improve differential assessment programs, two major changes could be made. The first step is to tighten eligibility standards for income and acreage; for example, a minimum of $25,000 in gross farm income and a 20-acre minimum farm size. These standards would continue to allow tax benefits to commercial farmers but would eliminate most land speculators and hobby farmers who earn the large majority of their income away from the farm and often have a solid ability to pay.

A second step is to link differential assessment with agricultural zoning, as is the case in Maryland and Oregon. It is more difficult to convert farmland if it is zoned for agricultural use, and it is less likely that the public will lose its investment in farm tax breaks.

A third step is general tax reform to replace the property tax with a local income tax, a local sales tax, or a combination of the two. A local income tax would go far toward eliminating the need for differential assessment. Two states, Michigan and Wisconsin, have shifted some of the farm property tax burden onto statewide income tax payers.

Circuit Breaker: Property Tax Relief through the State Income Tax

Since 1977, Wisconsin has offered farmers "circuit breaker" property tax relief through state income tax credits. A tax credit means that for every dollar of credit received, taxes owed are reduced by a dollar. Just as when an electric current triggers an overload and the circuit breaker mechanism shuts off the electricity, an income tax credit is triggered by a pre-set income figure.

The amount of the income tax credit depends on the level of property taxes and whether the county has agricultural zoning and an agricultural preservation plan. Tax incentives increase as farmland protection becomes stronger. For instance, a farmer qualifies for 70 percent of the credit in a county with only agricultural zoning and for the full tax credit in a county with both agricultural zoning and an agricultural preservation plan. Also, a farmer filing for the tax credit in a county with only a preservation plan must enter into a ten- to twenty-five year agreement requiring the land to remain in farming. If the farmland is rezoned or sold for a nonfarm use before the agreement expires, the landowner must pay back up to ten years of tax credits.

Thirty-two of 78 Wisconsin counties along with 391 towns and 36 cities and villages have exclusive agricultural zoning, which requires a 35-acre minimum parcel size to be considered a farm and prohibits construction of nonfarm buildings.

"The program was formulated to force farmers to force local officials to do planning," says Professor Dick Stauber of the University of Wisconsin.[6] And because farmers have supported the program, local officials have responded.

Wisconsin farmers have enrolled nearly 8 million acres in the circuit breaker program. In 1995, more than 23,900 farmers received $31.4 million in tax credits, equal to $1,314 per farmer, which lowered the average farm property tax bill by 35 percent. About 83 percent of the credits went to farms in exclusive agricultural zones, and 17 percent went to farms covered only by a county agricultural preservation plan.

A second property tax relief program in Wisconsin provides income tax credits equal to 10 percent of the first $10,000 of qualifying farmland, not including buildings. Virtually all farmers in Wisconsin receive the credit, which averaged $229 per farm, or a total of $14 million statewide in 1995.

Michigan offers circuit breaker tax relief through state income tax refunds when local property taxes exceed 7 percent of net farm income. To participate in the refund program (Public Act 116), landowners must enter into an agreement with the state not to develop their land for at least ten years. Landowners who do not renew their contracts must repay refunds over the last seven years. Landowners who withdraw from the program before the contract expires must pay back the entire amount of all refunds plus 6 percent interest. So far farmers have enrolled 4.5 million acres at an annual cost to the state of some $50 million.

The advantage of the circuit breaker approach is that it shifts some of the tax burden from farmland to income tax payers throughout a state. The program embodies the recognition that people pay taxes, land does not.

Property Tax Abatement

As an added incentive to get farmers to participate in the state easement-purchase program, some Maryland counties have offered property tax abatement.[7] Harford County, Maryland, has exempted farmers who sell an agricultural conservation easement from paying any property taxes. In addition, the county gives a 50 percent property tax break to farmers who form agricultural districts. Washington County, Maryland, offers a full tax abatement on farm buildings and land that are entered in agricultural districts. The Maryland Environmental Trust, a state-supported land trust, offers a fifteen-year property tax abatement for landowners who donate a conservation easement to the trust.

Since 1990, a number of townships in Connecticut have used a state law to reduce the property tax bills of local dairy farmers by one-half. The abatement is not necessarily permanent but was aimed at helping dairy farmers through a period of low milk prices.

Agricultural Districts

Twenty-one states allow farmers to form voluntary agricultural districts. To date, farmers have signed up more than 28 million acres in agricultural districts nationwide. Each district must have a specified minimum acreage, usually at least 500 acres, and must be approved by either the county or township government (see table 6.2). A farm property must be a certain minimum size (often 10 acres) to qualify for enrollment in a district. A landowner signs a petition to enroll land in a district for a certain period of time, often seven or eight years, but the landowner may remove land from the district without any penalty. And unlike zoning, agricultural districts do not place any restrictions on land use.

In a district, landowners may receive:

- property tax relief (either pure preferential assessment or deferred taxation)
- limits on sewer, water, and drainage taxes (Landowners must normally pay a fee on sewer and water lines that run through their property, based on the linear feet of pipe. For farmers not enrolled in an agricultural district and who have hundreds of feet of pipe running through their farms, the fees can be very expensive. In metropolitan Minneapolis, for example, these fees have forced several farmers to sell their land for development. An agricultural district keeps the fees from becoming a heavy burden on the farmer.)

- exemption from local nuisance ordinances that would restrict normal farming practices (which strengthens the right to farm), and possible alteration of local regulation of farming operations
- a state-level review of any proposed eminent domain action within the district
- limits on the extension of public sewer and water lines into the district (Also, state agencies must alter their policies to encourage commercial farming in the districts.) (See table 6.3.)
- limits on the annexation of farmland by a city
- eligibility to sell development rights to a county or state

Agricultural districts originated in California in 1965, and New York followed in 1971. (See appendix J for a list of agricultural district statutes by state.) The California districts cover about 15 million acres, and enrollment in a district is required for landowners who wish to enter into Williamson Act property tax contracts. The New York districts include preferential property taxation as well as some protection for farmers against encroachment from newcomers to the countryside, such as limits on assessments for sewer and water lines, limits on local regulation of farming, and state agency promotion of agriculture in the districts. About 8 million acres, or roughly half of New York's farmland and one-quarter of the entire state, have been placed in agricultural districts.

Pennsylvania farmers have enrolled over 2.7 million acres in agricultural districts, and Maryland has over 150,000 acres. Both of these states require farmland to be enrolled in an agricultural district in order for the landowners to be eligible to sell development rights to a county or the state.

Landowners in Delaware who enroll their farms in agricultural districts enjoy low property taxes and the added benefit of owing no estate or inheritance taxes on their farms. Land is enrolled in a district for a ten-year period,

TABLE 6.2
Factors to Consider in Creating an Agricultural District

State	Land Quality	Farm Viability	Current Land Use	Development Needs	Nearby Idle Land	Approved Conservation Plan
California	X					
Illinois		X	X	X	X	X
Iowa			X			
Kentucky	X	X				
Maryland		X		X		
Minnesota			X	X	X	
New Jersey	X	X	X	X	X	
New York		X	X	X	X	
N. Carolina	X		X			X
Ohio		X	X	X		
Pennsylvania	X	X		X		
Virginia	X	X	X	X	X	

Source: Bills, Nelson L., and Richard N. Boisvert, *The Agricultural District Approach* (Ithaca, N.Y.: Cornell University, Department of Agricultural Economics, 1988).

TABLE 6.3
Provisions of State Agricultural District Laws

Key
PDR = Purchase of development rights
DA = Development assessments
AN = annexation

State	Reduced Property Tax	Alter State Agency Policy	Alter Local Regulation	Limit Eminent Domain	Enable PDR	DA	Limits on Nuisance	AN
California	X			X				
Illinois		X	X			X	X	
Iowa		X						
Kentucky		X		X				X
Maryland			X		X			
Minnesota		X	X	X		X		X
New Jersey			X	X	X		X	
New York	X	X	X	X		X		
N. Carolina				X		X		
Ohio				X		X		
Pennsylvania		X	X	X	X		X	
Virginia		X	X	X		X		

Source: Bills, Nelson L., and Richard N. Boisvert. *The Agricultural District Approach* (Ithaca N.Y.: Cornell University, Department of Agricultural Economics, 1988).

and it can be renewed every ten years. To date, Delaware has sixty-nine districts covering 30,439 acres.

Most agricultural districts offer protection only for farm enterprises that existed at the time the district was created. In Michigan's agricultural districts, landowners may even change their operations, such as from dairy to hogs, without neighbors legally being able to stop them. In Delaware, a farmer in an agricultural district may recover court costs and legal fees when a neighbor files suit against the farmer's operation and the court finds in favor of the farmer. The threat of having to pay the farmer's court costs and legal fees has kept nonfarm neighbors from filing frivolous suits just to harass the farmer. Similarly, if a court in Iowa finds that a plaintiff has filed a frivolous nuisance suit against a farmer in an agricultural district, the plaintiff must pay all court costs and the farmer's attorney's fees.

Districts in Pennsylvania also provide some protection from eminent domain actions by government agencies. In one case, a local school board tried to condemn farmland in an agricultural district for a school site. The school board authorities went to the Pennsylvania department of agriculture for approval to condemn the land. The department denied the proposal, saying there were other suitable places for the new school outside of the agricultural district. The school board took the case to court and lost. The farmland remained farmland.

The Pennsylvania agricultural districts are created township by township. A key provision of these districts is that the township supervisors agree not to enact any nuisance ordinances that would restrict normal farming practices.

This provision strengthens the farmer's right to farm. Even though a neighbor might complain about normal farming practices, the neighbor would have no legal basis on which to force the farmer to cease normal farming practices. The Pennsylvania Farm Bureau has adopted a policy to encourage farmers to join agricultural districts (called agricultural security areas) primarily for the protection against eminent domain and the beefed-up right to farm.

Agricultural districts have been popular in rural areas where development pressure is low to moderate. But in areas close to urban and suburban development, fewer districts have been formed. Agricultural districts alone afford only limited protection from encroaching development. Even so, districts can be a good way for farmers to begin to band together as a commitment to stay in farming. Some Pennsylvania townships first adopted agricultural districts, saw a few farms preserved through the purchase-of-development-rights program, and then adopted strong agricultural zoning ordinances.

Agricultural districts can be used together with agricultural zoning or a purchase-of-development-rights program, or both. Agricultural zoning without the added protections of an agricultural district could leave farms vulnerable to eminent domain and legal suits over normal farming practices. A voluntary agricultural district combined with the regulatory power of an agricultural zone sends a powerful message to developers to look for other land to develop.

In communities and counties where agricultural zoning is not politically possible, the voluntary programs of agricultural districts and purchase-of-development rights might provide some protection for farmland. Several Pennsylvania counties without agricultural zoning have purchased development rights on about 60,000 acres of farmland. New Jersey counties have preserved over 24,000 acres of land enrolled in agricultural districts.

The risk in relying on an agricultural district with purchase of development rights is that land-use restrictions next to a preserved farm are loose. A neighboring farmer could easily withdraw from the district and sell his land for a housing development next to the preserved farm; this could make farming the preserved farm more difficult and perhaps drive it out of business. But if the agricultural district is at least a few thousand contiguous acres in size, then the preserved farms will tend to have some greater assurance and the likelihood of more preserved farms in the district increases.

Agricultural Economic Development

When most people think of economic development, they see factories with smokestacks, office parks, and civic centers. Agriculture is an industry that is much more than just the farms where food and fiber are produced. Farm products must be transported, processed, packaged, and distributed. Attracting and retaining these farm-related businesses are essential for maintaining a strong local agricultural industry and profitable farming. Agricultural economic development—the jobs, profits, incomes, and tax base that farming industries provide—is an important argument in favor of farmland protection.

Every state has economic development programs to promote agriculture. Most of these programs are operated through the state's department of agriculture. Specific economic development programs for agriculture vary among states, but may include one or more of the following:

- export promotion of crops, meat, and processed foods and promotion of locally grown food, such as Vermont's Seal of Quality and the New Jersey Fresh programs
- financing for value-added businesses that will process raw farm products into packaged foods
- loans for beginning or high-risk farmers (For example, the Vermont Economic Development Authority has made some $25 million in loans to 450 farmers. Under the Agricultural Improvement Act of 1992, states have the authority to sell tax-exempt aggie bonds to raise money for loans to first-time farmers. The maximum loan is $250,000 per borrower.)
- biotechnology and technology transfer to farms to increase the production of food and fiber
- grants for farmers to try new and different crops
- grants for farmers' markets
- marketing advice and a hotline for consumers to learn of farmers who sell produce directly to consumers
- a listing of farm bed-and-breakfast establishments for tourists
- a farm-link program to match beginning farmers with farmers looking to sell their farms

Inherent in these programs is the recognition that farming is a business and agriculture an important component of the state economy. If farming is not profitable, farmland protection programs ultimately will not be successful. Both farmland protection and economic development measures are needed to bolster the future prospects of agriculture as an industry.

Some local governments in states with farmland preservation programs have taken the initiative to establish programs to help their farmers bolster profits and stay in business. These localities have recognized that saving farmland is not enough when farming itself is only marginally profitable. In Hartford County, Maryland, for example, an agricultural planner was put in charge of working with farmers and conservationists to come up with a comprehensive program for helping local farmers stay profitable. Dozens of ideas were put into a plan, from streamlining local regulations to establishing a county value-added production center. The program was designed specifically because of the county's large investment in preserving farmland.

Summary

State governments have approved and adopted a variety of farmland protection measures. Most states emphasize voluntary incentives for farmers, such as differential assessment, right-to-farm laws, agricultural districts, and purchase of development rights. Half the states authorize local governments to use agricultural zoning to protect farmland. Eleven states have enacted statewide planning programs to direct and control development.

No one farmland protection technique can achieve all the protection goals. Rather, a coordinated array of voluntary incentives and land-use regulations is necessary.

Farmland protection efforts have worked best in those places where farmers have been supportive. This makes it easier for politicians to promote farmland protection policies and techniques. The general public is usually favorable toward farmland protection because of the open-space amenities that it provides. The key ingredient is that farmers need some assurance that if their land is going to be restricted for development, especially through zoning, they will be able to make a living on the farm. If farmers cannot make a living in farming, no amount of farmland protection efforts will be successful.

CHAPTER 7

Agricultural Zoning

> In time, most people come around to realize that agricultural zoning, if done properly, is for their best interest.
> —Pennsylvania township supervisor[1]

Zoning is the most widely applied land-use control in the United States. As we discussed in chapter 3, the main purpose of zoning is to separate land uses that might result in threats to public health, safety, or welfare or reduce a landowner's enjoyment of his or her property. For example, a cattle feedlot would probably be allowed in an agricultural zone but not in a residential zone where strong odors and flies from the feedlot could ruin a neighbor's barbecue and, more important, cattle manure could pollute residential wells.

Agricultural zoning makes good sense because it can reduce conflicts between farmers and nonfarmers. Many normal farming practices can cause irritants such as chemical sprays or the dust raised in plowing to spill over onto neighboring properties. And, nonfarm neighbors can cause problems for farmers, as well. When nonfarm residents move into rural areas, especially if they are new to country life, they often do not respect private property. They allow their dogs to roam onto neighboring farms where they may chase and harm livestock; they buy dirt bikes and all-terrain vehicles for their kids as if farm roads and fields were open to public use. Crop theft, litter, and vandalism are other headaches for farmers. Agricultural zoning provides some protection for farmers by limiting the number of nonfarm dwellings that can be built in the neighborhood.

Another purpose of zoning is to promote orderly growth of the community, which will help control the cost of public services and maintain a pleasing appearance. A local government may adopt zoning to implement the county or community comprehensive plan. If properly done, this plan should identify where development exists and where it should go in the future, as well as those areas that can and should continue to be farmed. Agricultural zoning typically discourages shopping malls, major housing subdivisions, and public land uses that would result in haphazard, widespread, and costly development in farming areas.

Unlike traditional zoning, agricultural zoning is used to protect valuable productive soils for both current and future generations. To be effective, an agricultural zone should apply to enough contiguous or nearly contiguous farms to allow efficient farming and to support the feed mills, hardware and seed stores, farm machinery dealers, and food processing and transportation businesses. This will help retain agriculture as an industry in the local economy.

Agricultural zoning is also important for other public values, such as retaining open space, protecting sensitive wetlands and wildlife habitat, and protecting water resources and air quality.

All of these purposes speak not only to the public interest in creating agricultural zones, but also to the benefits for farmers. These purposes are important to mention in generating political support for a community to adopt and support agricultural zoning. When it can be demonstrated that both farmland owners and the public stand to benefit, then agricultural zoning will be seen as a win-win technique for protecting farmland and the open space it provides.

Agricultural Zoning and Farmland Protection

Agricultural zoning is the most common land-use technique for limiting the development of farmland. Over five hundred counties and communities—mainly on the West Coast, in the Mid-Atlantic states, and in the upper Midwest—use some form of agricultural zoning. A few counties and townships created agricultural zones in the early 1960s, but most were created in the 1970s and 1980s. Over the past twenty years, agricultural zoning has become the first line of defense in most communities with successful farmland protection programs (see the Model Agricultural Zoning Ordinance in appendix A).

Most states authorize counties and municipalities to use zoning to protect farmland.[2] Hawaii and Oregon have enacted statewide zoning programs that emphasize agricultural zoning. Wisconsin has used agricultural zoning as an incentive for farm property tax relief. Agricultural zoning has created the foundation for leading purchase-of-development-rights programs in Marin County, California; Carroll County, Maryland; and Lancaster County, Pennsylvania. Agricultural zoning also enabled Montgomery County, Maryland, to protect almost 40,000 acres to date through a transfer-of-development-rights program.

Although protecting the farmland base is but one part of maintaining a viable farm economy, the farmland is one factor over which local governments have some control. Local governments can't influence national farm policy, for example, but through zoning, localities can provide a future for agriculture by protecting the land base that allows agriculture to happen. Agricultural zoning is attractive because it is inexpensive to put into place, and it can quickly protect thousands of acres. Also, zoning is flexible; it can be changed as circumstances and community needs and goals evolve over time.

PHOTO 7.1 Agricultural zoning provides protection for a farming neighborhood.

Legal Aspects of Agricultural Zoning

Elected county commissioners or township supervisors have the power to mandate that a farm be placed in an agricultural zone with or without the landowner's consent. This power can make agricultural zoning highly political. In general, land zoned for houses, businesses, and industrial plants is more valuable than land zoned for farming. Local officials are keenly aware that when they zone land, they are influencing land values and the potential wealth of landowners.

Zoning decisions may give rise to legal challenges. To be valid, agricultural zoning should meet six tests:

1. Agricultural zoning must serve a public purpose to justify the restrictions it imposes on a landowner. The Tenth Amendment of the Constitution allows the use of the "police power" of government to protect the public health, safety, and welfare. Zoning is an exercise of that police power, so agricultural zoning must be designed to achieve a clearly defined public purpose. There should be state legislation declaring that the protection of farmland is an important public goal. A state's zoning enabling act may expressly call for zoning to protect farmland. In addition, state laws on differential taxation for farmland, the right to farm, agricultural districts, and the purchase of development rights to farmland may contain language citing the need to protect farmland.
2. Agricultural zoning should be based on a carefully drafted comprehensive plan for the community or county. Agricultural zoning that does not agree with the comprehensive plan could be ruled invalid by a court.

3. Agricultural zoning must not result in the taking of private property. The Fifth Amendment to the Constitution guarantees a landowner's right to "just compensation" if the government takes his or her land. The Fifth Amendment clearly applies when government condemns land for a public purpose such as a road or a school. But if agricultural zoning is applied to actively farmed land, then it is not likely to be ruled a taking.
4. Agricultural zoning must also be reasonable. The U.S. Supreme Court has interpreted reasonable zoning to mean zoning that does not remove all of the economic use of the property;[3] this ruling supports agricultural zoning as long as farming is a viable use of the property.
5. Agricultural zoning cannot be used to exclude certain people from a community.
6. Agricultural zoning must be applied fairly and consistently among all landowners in the agricultural zone.

In some communities and counties effective agricultural zoning may not be politically possible because of a desire to avoid infringing on private property rights and profit potential. Often farmland is zoned at rural residential densities that do little to protect farmland. For example, in Massachusetts a 2-acre minimum lot size in agricultural areas is common. Orange County, North Carolina, employs a 5-acre minimum lot size in its farming areas. Although agricultural zoning has been found by a number of state courts to be legal,[4] many municipal and county attorneys are afraid of lawsuits from landowners and advise elected officials against adopting strong agricultural zoning.

We do not recommend the creation of agriculture-residential or agriculture-conservation zones with 1-acre to 5-acre minimum lot sizes. These are *rural residential* zones and should be located some distance from commercial farming areas. Lot sizes of 1 and 2 acres simply keep the land cheap for development: This zoning makes a large supply of building lots available, which leads to a buyer's market. Five acres is too large for a lot and too small to farm; it is hard for a homeowner to manage, often leading to problems ranging from thistle invasion to soil erosion. Lining residential lots along rural roads can result in hazardous driveway entrances.

Also, most rural building lots will use on-site well and septic systems. On-site septic systems vary greatly, from conventional leach fields to community on-lot systems to waste-water spray irrigation. The success of on-site systems depends on slope, soils, and depth to bedrock. Steep slopes, heavy clay or wet soils, and shallow soils are not suited to on-lot septic systems. A site should be tested by a licensed engineer. Some systems work, and others, such as the mound system, seem designed to fail. Because good-quality water is essential to many farm operations, especially for dairy and other livestock, the fewer residential waste-water discharges the better.

BOX 7.1
The Farmer's Perspective on Agricultural Zoning

Many farmers are unenthusiastic about agricultural zoning because it restricts the use of their land without offering compensation. The concern of farmers over private property rights is really a matter of wanting to have options. Most farmers have the majority of their wealth tied up in their land and view their land as both a retirement fund and an insurance policy. Especially if there is no family member to carry on the operation, a farmer may want to have the option to sell the farm for the highest dollar.

In farming areas on the urban fringe, many farmers do not see zoning as a guarantee against suburban sprawl. Zoning can be changed, and many local governments continue to favor development.

In places where agricultural zoning has won the support of farmers, elected officials, and the public, it has been durable. For farmers who intend to stay in farming and pass the farm to a family member, agricultural zoning can help provide protection by limiting exposure to conflicting residential and commercial development. At the same time, most farmers would like the flexibility to sell or give building lots to their children or sell off small amounts of mostly unproductive land to raise cash from time to time.

The key to starting and maintaining an effective agricultural zoning ordinance is to involve the farming community. After all, it is their land and they should have some assurance that they can live with the zoning restrictions. For example, Thurston County, south of Seattle, Washington, has seen the most rapid growth of any county in the state over the past decade. In 1995, county officials worked closely with farmers to lower the zoning density from one dwelling per 5 acres to densities of one dwelling per 20 acres and one dwelling per 40 acres on about 25 percent of the county's farmland. The county included a transfer-of-development-rights program (described in chapter 10) as a financial incentive for landowners in the new farm zones.

What Is in the Agricultural Zoning Ordinance?

The agricultural zoning ordinance has two parts, which work together: a set of written rules and standards, and a map showing the area that these rules and standards apply to.

The Agricultural Zone Map

The zoning map for the county or township should show the location of the agricultural zone in relation to the residential, commercial, and industrial zoning districts (see figure 7.1).

The area mapped for the agricultural zone should contain a "critical mass" of farmland to support the continuation of farming. As of 1996, a critical

FIGURE 7.1
Agricultural Zone (A) on the Community Zoning Map

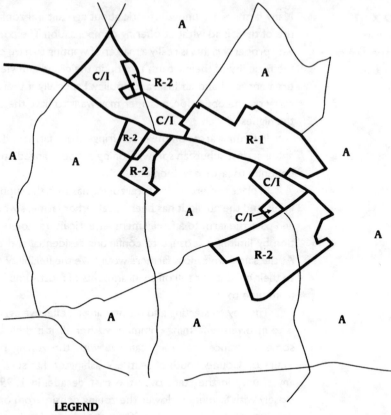

LEGEND
A = Agriculture
C/I = Commercial/Industrial
R-1 = Single Family
R-2 = Multi-Family

mass of farmland for a county is generally about 75,000 acres or agricultural production valued at $40 million a year. Counties with farm acreage or farm products below these levels will probably see their farming industry decline and change to nursery and greenhouse crops.

An important aspect of critical mass is that about 40 percent of America's farmland is rented. And rented farmland is crucial to many commercial farm operations. There is often substantial competition for rental ground among farmers even without competition from developers and people looking to purchase home sites. Agricultural zoning tends to stabilize the farmland base by making renting farmland a reasonable alternative for retired landowners and nonfarmers who hold farmable properties. Agricultural zoning makes it more likely that a farmer can negotiate a lease longer than on a year-to-year basis.

The Text of the Ordinance

The text of the agricultural zoning ordinance begins with statements about the purpose of the zone.

The purpose of agricultural zoning is to:

- protect good-quality agricultural soils from nonfarm development
- map out and protect a critical mass of farmland to encourage the continuation of crop and livestock production and to enable the farm support businesses—farm machinery dealers; feed and seed stores; food processors; vets; transportation, farm finance, and marketing companies—to remain profitable
- discourage land uses that would cause conflicts with farming operations[5]
- allow family farm-related businesses that generate extra income for the owners but are secondary to the farm operation
- designate minimum lot sizes, which determine the permissible subdivision of farmland for either farm or nonfarm uses
- establish setbacks for farm buildings from property lines to limit the spillover of odors, chemical sprays, and noise that could lead to complaints from neighbors
- provide some assurance that the public investment in farm property tax breaks and in the purchase of conservation easements on farmland will be protected

Permitted Uses, Special Exceptions, and Conditional Uses

After the purpose section, the zoning ordinance spells out permitted land uses (also known as uses permitted by right), land uses that require meeting certain standards and obtaining official approval, and land uses that are not permitted (see box 7.2).

Permitted land uses, or "outright permitted" uses, are agricultural, such as a dairy barn, or accessory uses such as a roadside stand for selling produce grown on the farm.

Uses allowed by special exception should affect only neighboring properties. These uses must meet standards to protect the right of neighbors to enjoy their property. A special exception is granted by a zoning hearing board (also known as the board of adjustment). A special exception use is generally considered a minor change to a property.

Conditional land uses typically affect an entire community and must meet certain conditions spelled out in the ordinance. For example, a community land fill involving several acres might be allowed as a conditional use in an agricultural zone. But first the landfill would have to meet safety standards and be located away from barns and houses. The elected governing body must hold a public hearing and give its approval before a conditional use is issued and the project built.

An agricultural zone should allow land uses that will not conflict with farming operations. At the same time, the more uses that are allowed in an agricultural zone, the more politically acceptable the zoning will be. For example, most agricultural zones allow a limited number of nonfarm dwellings and farm

> **BOX 7.2**
> Recommended Uses of an Agricultural Zone
>
> **Permitted Uses**
> - agriculture
> - forestry (including sawmills)
> - farm dwellings
> - production nurseries and greenhouses
> - wildlife refuges
> - fish hatcheries
>
> **Special Exceptions or Conditional Uses**
> - roadside stands for the sale of products grown on the farm
> - temporary housing for farm workers
> - new intensive agricultural uses, such as feedlots
> - family farm-related businesses, such as the sale of incidental farm supplies, repairs, and traditional arts and crafts, but limited in size: the farm-related business must be secondary to the farm operation
> - home occupations
> - bed-and-breakfast operations
> - accessory housing ("granny flats")
> - nonfarm dwellings in limited numbers
> - signs
> - elementary schools serving families in the agricultural area
> - churches
> - communication antennas
>
> **Uses Not Permitted**
> - landfills
> - quarries
> - sewage treatment plants
> - airports
> - golf courses
> - country clubs

support businesses. Generally speaking, agricultural zoning is best looked at as a tool to obtain the protection of farmland and farming by allowing some development.

Lot Areas and Subdivisions

The parts of the agricultural ordinance that most influence farmland protection are:

- the number of acres considered a farm and the amount of land required for any nonfarm uses in the farm zone
- the number of subdivisions of nonfarm lots and farm parcels to be created from the original farm (known as the parent tract)

The farm size in the zoning ordinance means the minimum number of acres that a landowner must have in order to build a "farm-related dwelling."

We advise that the farm size be only slightly smaller than the average farm size in the township or county. This will help prevent the subdivision of commercial farms into hobby farms in the agricultural zone. The smaller the minimum farm size allowed, the higher the price of farmland is likely to be because of competition from hobby farmers and urban refugees.

The model agricultural zoning ordinance in appendix A defines a farm as 25 or more acres, and nonfarm uses in the farm zone must be on lots between 1 and 2 acres in size. In places where big farms are common, the definition of a farm should be much larger. In Tulare and Fresno counties in California, minimum farm sizes of up to 80 acres are common. Colorado provides an example of a mistake to avoid. A successful ranch needs hundreds if not thousands of acres. But the state of Colorado allows 35-acre subdivisions by right; this standard has led to the carving up of homesites throughout the countryside and the needless loss of thousands of acres of ranch land.

The model ordinance in appendix A recommends that one nonfarm dwelling on a small building lot *or* one farm-related subdivision be allowed for every 50 acres. This discourages the creation of both hobby farms and several nonfarm dwellings, which could cause conflicts with neighboring farm operations.

Many communities employ minimum-lot-size zoning in farming areas (see table 7.1). Some of these so-called agricultural zones have small minimum lot sizes of 2, 3, or 5 acres that allow the land to be subdivided into parcels that are too small for commercial farming. Strong agricultural zones have minimum lot sizes of anywhere from one dwelling per 40 acres in Black Hawk County, Iowa, to one dwelling per 320 acres in the ranch country of Deschutes County, Oregon. Madeira County, California, has used a minimum lot size of 640 acres or *one square mile.*

Setbacks

The agricultural zoning ordinance will establish setbacks from property boundaries for any new farm buildings. Some ordinances have a 200-foot setback, which helps to protect neighbors from odors, chemical sprays, and noise and the farmer from complaints. An ordinance may include a farther setback of 350 to 500 feet for intensive farm uses such as feed lots or large hog or chicken facilities.

Setbacks may also be required for houses and other buildings on land next to an agricultural zone. These are discussed in the section on agricultural buffers later in this chapter.

Agricultural Nuisance Disclaimer

An agricultural nuisance notice in the zoning ordinance gives fair warning to people who are purchasing a house or business in an agricultural zone. An example of such a notice follows:

TABLE 7.1
Sample of County Agricultural Zones Using Minimum Lot Sizes

County	Minimum Lot Size in Acres on Which to Build a Dwelling	Main Type of Farming
Fresno, California	1 to 40	fruits, vegetables
Madera, California	1 to 640	cattle
Marin, California	1 to 60	dairy, cattle, sheep
Napa, California	1 to 40	grapes, wine
	1 to 160	hillside vineyards
Santa Barbara, California	1 to 40	fruits, vegetables
	1 to 100	cattle
Weld, Colorado	1 to 160	cattle
Ada, Idaho	1 to 80	cattle
DeKalb, Illinois	1 to 40	cattle, grains
McHenry, Illinois	1 to 40	grains, dairy, hogs
Black Hawk, Iowa	1 to 40	grains, hogs
Story, Iowa	1 to 40	grains, hogs
Woodford, Kentucky	1 to 30	horses
Baltimore, Maryland	1 to 50	horses, grains
Rock, Minnesota	1 to 80	cattle, grains, hogs
Waseca, Minnesota	1 to 160	grains, hogs
Deschutes, Oregon	1 to 320	cattle
Marion, Oregon	1 to 40	vegetables, grass
Juab, Utah	1 to 40	cattle
Skagit, Washington	1 to 40	dairy, nursery
Dane, Wisconsin	1 to 35	dairy

Source: Compiled by authors, and *1992 Census of Agriculture.*

Lands within the Agricultural District are used for commercial agricultural production. Owners, residents, and other users of this property may be subjected to inconvenience, discomfort, and the possibility of injury to property and health arising from normal and accepted agricultural practices and operations, including, but not limited to, noise, odors, dust, the operation of machinery of any kind, including aircraft, the storage and disposal of manure, the application of fertilizers, herbicides, and pesticides. Owners, residents, and users of this property should be prepared to accept these conditions and are hereby put on official notice that the Right-to-Farm Law of the State of _____ may bar them from obtaining a legal judgment against such normal agricultural operations.

Drafting the Agricultural Zoning Ordinance

The planning commission is responsible for drafting the agricultural zoning ordinance. Because the agricultural zoning ordinance has legal effect, people trained in planning or law should draft or review the ordinance. The planning commission may want to hire a consultant to work with landowners and the commission. The planning commission may want to obtain agricultural zoning ordinances from other communities for guidance. We do not encourage copying an ordinance from another community, however, because other communities may have different land-use patterns and types of agriculture.

The planning commission must recommend the zoning ordinance to the elected governing body, which has the option to approve, reject, or amend the text or the map of the ordinance. In most states, only the governing body can legally approve the creation of a zoning ordinance or any changes to the ordinance.

Types of Agricultural Zoning

In drafting the agricultural zoning ordinance, a county or municipality may choose one of several types of agricultural zoning (see figure 7.2). One type of zoning may be more appropriate than another, depending on: 1) the number and size of farms; 2) the number and size of parcels in the agricultural area; 3) the limits the community wants to place on the number of nonfarm dwellings and nonfarm uses in farming areas; and 4) the local political and legal realities.

The two general types of agricultural zoning are exclusive farm zoning and nonexclusive farm zoning. In an exclusive farm zone, landowners may construct only farm buildings or farm-related housing. Few jurisdictions outside of Hawaii and Wisconsin have used truly exclusive farm zones.[6] Exclusive farm zones make sense when farming is the dominant land use, the farmland is in large contiguous blocks, and there are few nonfarm dwellings or other nonfarm buildings in the area.

Hawaii introduced exclusive agricultural zones as part of its 1961 state land-use plan. This type of agricultural zoning avoids the problem of leapfrog and buckshot development that can fragment farmland. Development is kept at a distance and conflicts between farms and nonfarm neighbors are kept to a minimum. Also, in the ag zone, the market value and the agricultural value of a piece of land tend to be close, reflecting the absence of nonfarm alternatives. This tends to keep down property taxes and makes farm expansion more likely. Farmland may be rezoned and converted to nonfarm uses, but it must touch on an urban or rural residential zone to do so.

In the 1960s, Hawaii's sugar cane, pineapples, and cattle were valuable products. Then came the tourist boom, and as the commitment to farmland protection waned, the value of some farmland for development soared to over $100,000 an acre. Between 1982 and 1987, Hawaii led all states in the percentage of farmland lost. This fact is a lesson of both the flexibility and the impermanence of agricultural zoning. Ag zoning is a political process, and as local elected officials change, so can zoning.

Wisconsin began a Farmland Preservation Program in 1977 that uses the carrot of property tax relief for farmland owners to get counties to implement exclusive agricultural zones. Farmers in counties with exclusive agricultural zoning may enter agreements not to develop their land for ten to fifteen years in

FIGURE 7.2
Types of Agricultural Zoning

Exclusive Agricultural Zoning	Nonexclusive
No construction of nonfarm dwellings allowed. Rarely used. Risks government "taking" private property rights without compensation.	Some nonfarm development allowed. Goal is to balance some nonfarm development with farmland protection. Generally avoids the government "taking" of private property.

Two Types of Nonexclusive Agricultural Zoning

Large Minimum Lot Size	Area-Based Allocation
Refers to minimum size on which a new dwelling may be built. Size varies among counties and municipalities. Lot sizes of 5 and 10 acres encourage the creation of "rural estates." Size should reflect minimum viable farm size, such as 50-acre minimum lot size in Baltimore County, Maryland, to 320-acre minimum lot size in the rangeland of Deschutes County, Oregon.	Number of nonfarm dwellings allowed depends on size of farm. Nonfarm dwellings must be built on small lots, usually under 2 acres. Greater flexibility in siting nonfarm dwellings. Less farmland used.

Two Types of Area-Based Allocation

Fixed Area	Sliding Scale
The landowner may build one nonfarm dwelling on a small lot for every certain number of acres, such as one building lot of up to 2 acres for every 25 acres in West Hempfield Township, Lancaster County, Pennsylvania.	The number of nonfarm houses per area decreases as the size of the farm increases. Thus, smaller parcels with limited farming potential are allowed to be developed at a somewhat higher density than larger commercial farm parcels.

Agricultural Buffer Zones

Large Minimum Lot Size	Cluster Development
The minimum lot size will be smaller than in a normal agricultural zone, usually of 5 to 10 acres. The buffer may be intended to last several years and eventually be developed more intensively as a nearby town expands and public sewer and water become available.	A grouping of the permitted nonfarm dwellings on part of a property with a minimum percentage maintained as open space. The open space should be used to buffer neighboring farms. Cluster developments are often designed to be more of a long-term buffer than a large minimum lot size buffer.

return for an income tax credit of up to $4,200 a year. The county zoning ordinances are reviewed and certified by the state Land Conservation Board, and agricultural zones require a minimum parcel size of 35 acres to build a farm dwelling. Since 1982, exclusive agricultural zoning has been put in place on 6.7 million acres in 32 of the state's 72 counties, 391 townships, and 36 cities and villages.[7]

Nonexclusive Agricultural Zoning: Several Options

Local governments can choose from a number of nonexclusive agricultural zones. The goal is to balance the long-term protection of farmland with some nonfarm development. The main issue is how much nonfarm development and subdivision of land can be allowed without disrupting the stability of the farming community.

Large Minimum Lot Size

The most common type of agricultural zoning is called *minimum-lot-size* zoning, which says that a farm cannot be broken into parcels below a certain size for farming purposes. For example, several Oregon counties, some California counties, and a few Minnesota counties use an 80-acre minimum lot size; this means that an 80-acre farm cannot be subdivided, nor could a 150-acre farm be split in half. A landowner would need at least 160 acres to create two 80-acre farm parcels.

The idea is that if a minimum lot size is large enough, it should be too expensive for residential buyers. Also, a large minimum lot size should help keep farmland in blocks big enough to farm profitably, either individually or as a group. A large minimum lot size discourages the intrusion of nonfarm uses into a farming area, and the farmland is not broken up by numerous house sites. Nonfarm dwellings may be permitted but usually on small lots, on less productive ground, and in locations where they will not interfere with commercial farming operations.

Deciding on the right minimum lot size is a delicate balance between effective land protection and legal and political acceptance. The courts in each state have the power to determine if a minimum lot size is too large. The Illinois Supreme Court upheld McHenry County's 160-acre minimum lot size; Pennsylvania courts have upheld a 50-acre minimum lot size; but the courts in the New England states and Ohio tend to frown on any minimum lot size larger than 5 acres.[8]

Because zoning is flexible, the minimum lot size can be changed over time as circumstances change. Between 1990 and 1992, McHenry County was the fastest-growing county in Illinois. Taxes soared and traffic congestion brought gridlock. The price of farmland doubled because of its potential for residential lots. In 1994, the county reduced the minimum lot size in the agricultural zone to 40 acres.[9]

Most agricultural zoning ordinances that use minimum lot sizes also allow several nonfarm uses such as schools, churches, and sometimes even golf courses. These uses tend to attract traffic and housing developments and increase conflicts between farmers and nonfarm neighbors. Whenever possible, these nonfarm "growth magnets" should be located in or next to existing villages or built-up areas. In many regions, golf courses are accompanied by development of large estate homes along the greens.

Sliding-Scale Agricultural Zoning

The number of lots allowed can also be determined through a sliding-scale approach. In sliding-scale zoning, the number of development units allowed is based on parcel size with allowed density (not number of rights) decreasing per acre as the size of the parcel increases. Smaller tracts will have a higher percentage of land developed than will larger tracts. The sliding-scale approach reflects a recognition that there are small tracts of land in the agricultural zone that are difficult to farm profitably and have basically passed out of the farmland market into the residential land market. Tracts below 25 acres, for instance, are allowed an extra dwelling. But larger tracts that can be farmed for a living are allowed fewer houses, because these houses could cause conflicts with the neighboring farm.

In Clarke County, Virginia, a sliding-scale allocation allows a range of density from one dwelling unit for a lot of record of less than an acre to one dwelling unit for every 25 acres for a 100-acre parcel (see table 7.2). As parcel sizes increase over 100 acres, the allowed density greatly decreases. For example, a landowner with 500 acres is allowed eleven dwelling units, which is a density allowance of one unit per 45.5 acres.

Fixed Area-Based Allocation

In many localities, density allowances in rural areas are set by fixing a number of building rights according to area, or number of acres owned. For example, one building right per 25 acres owned would allow four lots to be created on a 100-acre farm. Along with the density allowance, a maximum lot size may also be adopted, such as a 2-acre maximum, to allow for adequate septic area. Some communities further restrict the number of dwellings allowed according to environmental constraints or availability of public services, but others have increased the number of dwellings allowed to appease landowners when the zoning was adopted. For example, Harford County, Maryland, in 1977, changed its rural zoning from one dwelling per 3 acres to one dwelling per 10 acres, but a generous grandfather clause added additional "family conveyances" without even requiring that the lots be created for use by family members. These

TABLE 7.2
Sliding-Scale Agricultural Zoning, Clarke County, Virginia

Size of Farm	Number of Dwelling Units Allowed
0–14.99 acres	1
15–39.99 acres	2
40–79.99 acres	3
80–129.99 acres	4
130–179.99 acres	5
180–229.99 acres	6
230–279.99 acres	7
280–329.99 acres	8

Note: Minimum lot size is 1 acre, maximum lot size 2 acres. Adopted in 1980.

7. AGRICULTURAL ZONING

additional building rights, in effect, diminished the one-dwelling-per-10-acres zone to about one dwelling per 7 acres, a density that invites an overpopulated agricultural zone despite the county's active farmland preservation program.

Minimum-lot-size zoning can be combined with a maximum lot size for any nonfarm dwellings built. This keeps the farmland from being broken up into large residential estate lots. For example, several townships in Lancaster County, Pennsylvania, allow one building lot per 25 acres and the building lot can be no more than 2 acres. So if a farmer owns 100 acres, part of the land can be subdivided into four building lots of up to 2 acres each, or a total of 8 acres. The farmer would then have 92 acres left for farming. (See figure 7.3.)

Agricultural Buffer Zones

There are two types of agricultural buffer ordinances. One type refers to the siting of nonfarm dwellings on building lots subdivided off a farm. The second type regulates how much development is allowed next to a farm.

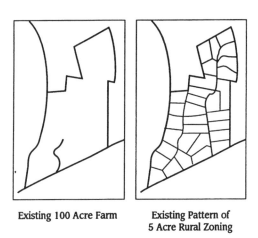
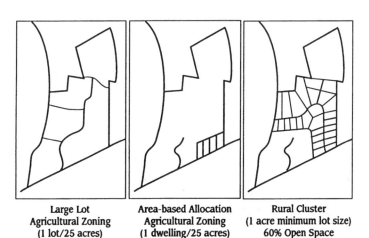

FIGURE 7.3
Agricultural Zones and Permitted Development

> **BOX 7.3**
> Agricultural Zoning with Mandatory Conservation Easements
>
> The countryside around Marin County, California, features open expanses of hilly grassland, which are grazed for dairy and beef operations. The scenic beauty of western Marin also contributes to an important tourist industry. Especially breathtaking is the drop of the grasslands into the ocean along Tomales Bay. In this coastal farming area, the agricultural zoning requires one dwelling per 60 acres, and only 5 percent of a farm or ranch can be developed. The remainder of the property must have a permanent conservation easement placed on it.
>
> For example, say a 600-acre ranch in the area was to be developed according to the agricultural zoning ordinance. No more than ten houses could be built on no more than 30 acres. The remaining 570 acres would then come under a permanent conservation easement, restricting the use to farming and open space.

The purpose of a buffer is well explained in the buffer ordinance of Stanislaus County, California: "Urbanization and the proliferation of rural residences throughout the county has led to increased conflicts over agricultural operations.... By separating incompatible uses, a buffer minimizes the impacts of development on surrounding agricultural operations and decreases the likelihood of conflict."[10]

Sussex County, Delaware, requires a buffer of land between farmland enrolled in an agricultural district and new development. A new house must be placed at least 50 feet away from the boundary, and major subdivisions must have a 30-foot-wide forested buffer designed by a state forester. In addition, deeds for new homes built within 300 feet of a farm in the agricultural district must contain restriction clauses acknowledging adjacent agricultural uses. This lessens the ability of a nonfarm neighbor to win a nuisance suit against a farmer who employs normal farming practices.

Fremont County, Idaho, has come up with an innovation called *resource easements* to reduce if not eliminate complaints from nonfarmers who move into agricultural areas. Owners of new nonfarm parcels must record an easement with the county recorder of deeds that recognizes that farming activities may conflict with residential life. The easement goes on the landowner's deed and runs with the land, that is, it applies to future landowners. And the easement must be officially recorded before any new home construction can begin. The easement takes away the nonfarmer's legal basis for making a case against normal farming practices.

Since 1991, Stanislaus County, California, has required realtors and sellers to disclose to potential home buyers in farming areas that they will be exposed to the effects of commercial agriculture. In 1992, the county supervisors enacted an ordinance requiring a buffer negotiation between a developer and a farmer as a condition of approval for a proposed development. According to county planner Leslie Hopper, "It's up to the project applicant and the farmer to hammer out the specifics; then it is written into the conditions of approval."[11] The Stanislaus County ordinance allows for a broad range of uses for buffer areas, including open space and recreational uses, industrial uses, and cemeteries.

The state of Maine prohibits any "inconsistent development" within 100 feet of a registered farm tract. This includes nonfarm residences, school buildings and play areas, wells, drinking-water springs, commercial food outlets, and campgrounds and picnic areas.

A few townships in Lancaster County, Pennsylvania, prohibit the planting of trees or shrubs on residential land within 30 feet of a farm property. This will keep shade from covering neighboring farm ground and reducing crop yields. But if planted correctly, a screen of trees and plants can reduce the noise, dust, and chemical spray drift from a farm onto neighboring properties.

Locating Nonfarm Dwellings

Some agricultural zoning ordinances have design requirements for siting nonfarm dwellings that are built on lots subdivided off a farm (see figure 7.2). These dwellings must be built on lower-quality soils or in places where they will cause the least interference with farming operations. An ordinance may include a provision that requires nonfarm dwellings to be clustered together. A bonus of additional dwellings could be allowed to encourage the clustering of nonfarm houses.

Cluster Zoning

Both large minimum lot sizes and cluster developments have been used as buffers next to farming areas. The large minimum lot size of from 3 to 10 acres can limit the number of nonfarm dwellings and can provide opportunities for hobby farming. But a large minimum lot size can also create an awkward pattern that eats up the land in the buffer zone fairly quickly (see figure 7.2). On the other hand, if public sewer and water are not available and are not expected to be for many years, a large minimum lot size can more safely accommodate on-site septic and well systems than a cluster development.

Many localities require 2-acre minimum lot sizes because that is believed to be necessary if an on-site septic system is used. This would allow the landowner to identify a primary leach field and a back-up field. Sometime during the life of the septic system, the leach field might have to be moved.

Pennsylvania's Department of Natural Resources has a standard that says if a landowner wants to build a house with an on-site septic system, the landowner must first have the groundwater tested. If the groundwater shows more than ten parts per million (ppm) of nitrate, an on-site septic system cannot be used. The ten ppm standard was set by the U.S. Environmental Protection Agency to protect against health problems such as "blue baby." Pennsylvania is the only state that actively enforces the ten ppm standard. The standard has meant that areas with heavy use of manure, ammonium nitrate fertilizer (the norm for corn), or existing septic systems have become largely undevelopable without public sewer service. This has helped curb residential leapfrogging throughout the countryside.

Cluster development has the advantage of being able to site buildings away from environmentally fragile areas, rather than in the "cookie cutter" pattern associated with the large minimum lot size. But an important issue is the density that is allowed with the cluster (see figure 7.2). For example, a standard of one dwelling per 2 acres can result in overly intense development that can conflict with adjacent farms and cause sewage and water quality problems. A standard of one dwelling to 5 or 10 acres is more likely to result in an effective buffer and not overtax groundwater supplies.

What happens to the open land in a cluster development is a matter of some debate. According to Bob Wagner of the American Farmland Trust, "There's no way you can guarantee that the remaining open space will be usable for farmers. Cluster development should be a last-ditch effort for use in urbanizing areas."[12]

Maryland counties with active farmland protection programs and subdivision regulations that allow clustering are experiencing mixed results. According to farmland preservation administrator Donna Mennitto of Howard County, cluster development was never envisioned as a farmland preservation tool. She feels it can work as a buffer, but the remaining open-space land will probably not be farmed. A further problem is that the county's health department opposes the use of shared community on-lot septic systems, which would allow more land to stay undeveloped. "It means that all the good percable land is where the lots go," says Mennitto.[13]

Baltimore County, Maryland, allows clustering at a rate of one dwelling per 5 acres on watershed resource lands within the agricultural zone. To date, legal challenges by neighbors have kept any cluster developments from being built. John Bernstein, executive director of the Valleys Planning Council, points out that the cluster ordinance works in favor of development. Without a cluster ordinance, a developer might be able to put only one house on the 5 acres with road frontage out of a 20-acre lot. But under the cluster ordinance, if the same parcel had 15 acres of unusable land behind the 5-acre lot, the developer would be allowed to put four houses on the property. That simply puts more houses in the countryside, Bernstein argues. The county may repeal its cluster ordinance.

Carroll County, Maryland, has cluster zoning that allows lots on every farm. County farmland preservation administrator Bill Powel predicts that clusters of from three to six lots will occur on every farm. One effect of cluster development, he notes, is the fragmenting of farmland, which may create a psychology of impermanence wherein farmers cease to invest in their farms as they anticipate the sale of their land for development.

Queen Anne's County, Maryland, has zoned about 90 percent of its land-area either agriculture at one dwelling per 8 acres or "countryside" at one dwelling per 5 acres. Clustering is optional, and lots with on-site septic must be at least 20,000 square feet, a little under half an acre. Since 1987, the county has approved over 430 applications for more than 1,100 lots. Most clusters contain two to four houses. County planning director Steve Kaii-Ziegler says that there hasn't been a big problem yet with clusters, but he is aware of new residents in the countryside becoming frustrated with nearby farming operations.

Calvert County, Maryland, has made clustering mandatory since early 1993 in conjunction with the county's transfer-of-development-rights program. Parcels in the county's agricultural "sending" area must remain 80 percent open space. But a parcel in the designated growth "receiving" area is required to remain only 50 percent open space. "We have noticed most development has been going into the receiving area," said Greg Bowen, who manages the county's farmland preservation program. "We had a voluntary cluster for ten years. It didn't work."[14]

Lancaster County, Pennsylvania, has removed cluster zoning in rural areas from its growth-management plan. The county planning commission was concerned that a rural cluster ordinance would encourage putting new residents in the countryside and would cause conflicts between farmers and nonfarmers.

Even cluster zoning advocate Randall Arendt of the Natural Lands Trust warns that clustering should be seen not as a farmland preservation technique but as a way to save some rural character and open space. According to Arendt:

> [cluster zoning] is definitely a second best technique if not third best. If you want suburbia, have a suburban zoning density. If you want to remain agricultural, have an ag zoning density, which would begin at one [dwelling] per 25 acres. Communities where open space zoning is appropriate are those with one [dwelling] per 3 acres. [But they] should realize they will not remain rural.[15]

Cluster zoning has been popular in states that do not expressly allow agricultural zoning and in counties and municipalities where political opposition has thwarted agricultural zoning.

In general, cluster zoning makes sense in a suburban setting where there is little farming and the goal is to preserve some open space. But in farming areas, cluster zoning can lead to serious conflicts between farmers and nonfarm neighbors. The danger is that excessive use of clustering in farming areas can simply produce clustered sprawl. Clustering homes in an agricultural district doesn't curb development or restrict growth there, it just changes the way it looks. Clustering merely tinkers with the form of development, and cluster advocates overlook the fact that the function of clustering is often to put more houses in the countryside.

How to Create and Support Agricultural Zoning

Local governments can legally create agricultural zones, and, once formed, the zones need the support of elected officials, landowners, and concerned citizens. The keys to success are education and communication between farmers and local officials. Any proposed change in land regulation brings initial opposition. Open discussion can lead to understanding of community goals and landowners' needs. The result can be agricultural zoning that satisfies both.

It may take time for a community to come around to adopt agricultural zoning. Education takes time, and people need time to discuss and evaluate new ideas. The time factor may be frustrating to those who feel that the community urgently needs agricultural zoning and has only a short window of opportunity to put it in place before development overruns the community. Yet there is often nothing like a crisis, such as a major proposed development, to generate a widespread call for action.

The first step in creating an agricultural zone is to involve the farmland owners. Organize a meeting of farmers and nonfarmers. Show them what is likely to happen to the farmland with and without agricultural zoning. A very useful way to illustrate these two outcomes is through a build-out scenario.

Take the existing zoning ordinance and map of the farming area. The current ordinance may be a rural zone, allowing one house per acre or 2 acres. Then draw on the zoning map all of the houses that are permitted to be built. Everyone will be amazed. You will hear comments such as "That's not what I want" and "But that would never happen."

Next, take the soils maps from the county soil survey and the tax parcel maps showing the location of farms to identify an area to protect. Ask the farmers how many nonfarm dwellings they can farm near for every 100 acres. The number of nonfarm dwellings per 100 acres will give you a basis for the zoning. For example, if farmers say they could reasonably farm near four nonfarm dwellings per 100 acres, that would suggest a zoning standard of one nonfarm dwelling per 25 acres.

Take the map of the farming area and draw in all of the houses that would be allowed under stricter zoning; one house per 10 acres, one per 25 acres, one per 50 acres, and so on. See which of these scenarios the farmers and nonfarmers agree on. Is there a consensus? Don't expect complete agreement, but there should be one scenario on which a majority agree.

Finally, examine the different types of agricultural zoning and discuss specific uses that should or should not be allowed in the zone. Again, a consensus should emerge.

Be careful not to put the farmers on the defensive. Do not make agricultural zoning solely a matter of government imposing restrictions. Agricultural zoning must be something that most people in the community want and that farmers can live with. Farmers should be encouraged to state their needs and what restrictions on land development they can tolerate. Nonfarmers should say what they want to see happen to the farmland and what support they are willing to give the farmers.

If agreement can be reached, a group of farmers and nonfarmers should approach the local or county planning commission or elected officials to institute agricultural zoning. The group should be prepared to show that agricultural land and the agricultural industry are valuable, with soils maps, a map of a proposed agricultural zoning district, and a map of active farms. Your county extension agent or conservation district can help you find the information for these maps.

Get a copy of the community's comprehensive plan from the planning commission. Be prepared to show how agricultural zoning fits the goals and objectives of the plan.

You may also want to have examples of agricultural zoning ordinances from other counties or municipalities. These examples can give your planning commission direction in drafting an ordinance for your community.

You might suggest some general rules for the agricultural zone:

1. The agricultural zone or zones should cover large contiguous areas. Some nonfarm development may be allowed within the agricultural zone through the special-exception or conditional-use process. A contiguous area for farming will tend to keep nonfarm development away and keep down conflicts.
2. The splitting of land into nonfarm house lots and smaller farm tracts should be limited. The fragmenting of the farmland base harms the long-term health of the local farm economy.
3. Design standards could require nonfarm development to be located on lower-quality soils and in places where there will be little interference with farming operations.
4. A nuisance notice could warn newcomers to the agricultural zone that farming has priority and may inconvenience nonfarmers.
5. Setbacks from farm property lines should apply to both new farm buildings and new housing developments.
6. Farm-based businesses should be provided for. A farm-based business ordinance permits certain on-farm enterprises and agricultural support businesses, which can provide jobs for farm family members and supplemental income, and can improve the efficiency of farming in the area. Farm-based businesses must remain an accessory use, secondary to the farming operation, and should not interfere with adjacent farms or cause nuisances for nearby residents.
7. The agricultural zoning map should be carefully drawn so that landowners can see whether their land is included in the proposed zone.

The planning commission will make a recommendation on the zoning proposal to the elected officials. The elected officials make the legally binding decision whether to adopt the agricultural zone. If the zone is rejected, you may need to organize a political campaign and work to elect candidates who support the creation of an agricultural zone. Don't give up!

Monitoring Development Proposals and Changes to the Zones

In communities with agricultural zoning, citizens and landowners alike must monitor proposed and actual changes to the agricultural zones. A good idea is to ask elected officials to appoint a farmland protection committee to keep track of changes to the farmland.

Local zoning decisions are made on a case-by-case basis, often without a sense of how each decision affects remaining farmers. Also, the lack of coordination among townships or counties can hinder attempts to protect a *regional* critical mass of farms and farmland. But in townships, counties, and states where agricultural zoning has won the support of farmers, most of the farmland will probably be protected for at least ten to twenty-five years.

Zoning is meant to be flexible in response to changing circumstances. The local government may amend the zoning text, for example, changing the minimum lot size for a nonfarm dwelling from two acres to one, or the zoning map may be altered to increase or decrease the amount of land included in the agricultural zone.

Any changes to the zoning text or map must first go through the planning commission and then the elected governing body. Proponents should be sure to obtain and review the agendas for the planning commission meetings and be prepared to attend the planning commission meetings in numbers. Early input on proposed changes to the agricultural zones is often the most effective approach.

Zoning decisions are made by elected officials who often have a bias in favor of growth. They reason that turning farmland into a factory site, shopping center, or housing development will broaden the local property tax base. But in the case of housing, as mentioned earlier, several studies have shown that it costs more to provide public services to houses than the houses generate in local taxes.

Some local officials may feel that development is inevitable because of increasing population. Development pressure is a powerful force in changing zoning maps and ordinances. Local farmers, who have owned their land for many years and who are looking to retire, make very sympathetic figures when they petition local officials to rezone their land from agriculture to residential or commercial use.

Elected officials may feel compelled to rezone farmland to nonfarm uses when:

- there are no heirs to take over the farm
- the farm goes bankrupt
- there is access to public sewer and water, making development easy
- land prices for nonfarm uses are high
- existing farmers are not willing to buy land for farming

Rezoning land from one use to another is a fairly common and accepted practice. But rezoning can become a problem if it creates a "spot zone," which could be ruled illegal by the courts. A spot zoning occurs when a piece of land in the middle of one zone is rezoned to another land use that does not serve a public purpose. For example, a property owner might propose a rezoning to allow fifty houses in the middle of an agricultural zone. The property owner

> **BOX 7.4**
> A Point System Approach to Rezoning Decisions
>
> Since 1986, Tulare County, California, has relied on a point system of fifteen factors to determine whether to allow farmland to be rezoned to another use. The factors include soil quality, existing parcel size, surrounding land uses and parcel sizes, level of groundwater, access to paved roads, whether it's in a flood-prone area, and whether irrigation water is available. If a proposed rezoning scores seventeen or more points out of a possible thirty-two points, it is denied. A score of eleven points and under means that a rezoning should be granted, which leaves a gray area of twelve to sixteen points.
>
> Between 1986 and 1994, thirty-one rezonings were submitted, and twenty-two were approved. Twelve proposed rezonings scored eleven or fewer points and were approved. Five proposals scored seventeen or more points and were denied. Fourteen proposals rated from twelve to sixteen points, and ten of those were approved. Overall, Tulare County has seen a decline in the number of rezoning requests since implementing the points-based system.
>
> The advantage of a points-based system is that it is fair, objective, and easy to understand. The Tulare system recognizes that there are some cases, however, in which a judgment call by local officials is necessary.

would stand to benefit from the rezoning, but the public goal of protecting farmland would be harmed.

In judging whether a rezoning is valid, the planning commission and elected officials should consider:

- compatibility with neighboring properties
- public benefits
- compliance with the local comprehensive plan

A rezoning is usually allowed if:

- there is little or no injury to nearby properties;
- the property is located on the edge of the zone;
- the rezoning will have benefits to the neighborhood, such as a rezoning to allow a farm machinery dealer as a commercial use in an agricultural zone.

Whatever the planning commission or elected officials decide, they should make findings of fact to explain their decisions.

Keeping Records of Changes to the Agricultural Zone

Most communities with agricultural zoning have not kept good records of the number of rezonings or the number of acres rezoned out of agricultural zones or the number of nonfarm dwellings built in agricultural zones. The local zoning officer or zoning administrator is the public official responsible for enforcing the zoning ordinance. The zoning officer or the planning commission should be directed to keep records of changes to the amount of land in the agricultural zone. Also, when a rezoning out of agriculture is allowed, public officials should make findings of fact to support their decision.

Tax maps, in conjunction with zoning maps, can show nonfarm development in agricultural areas.

With more communities and counties acquiring geographic information systems, keeping track of the location and amount of land zoned for agriculture should be rather easy to do. This information should in turn be readily available to farm groups and citizen groups with an interest in farmland protection. The more information these groups have, the better they can assist in monitoring efforts and comment upon proposed rezonings.

Case Study: Agricultural Zoning in Oregon

Agricultural zoning is a key component of the state of Oregon's pioneering land-use program, which began in 1973.[16] Because agriculture is one of the state's leading industries, a main feature of the Oregon program is the protection of farmland. Oregon's farms cover roughly 17 million acres and earned nearly $3 billion in gross sales in 1992. About 2.5 million acres of farmland are located in the fertile Willamette Valley—between Portland and Eugene and between the Cascade and Coastal ranges—where almost three-quarters of the state's 2.8 million people live. Most of Oregon's high-quality farmland lies within commuting distance of major urban areas.

The Oregon farmland protection program features six techniques:

1. The state government has reviewed and approved county comprehensive plans that comply with state goals, including a farmland protection goal (Goal 3), which carries the force of law.
2. Each county must identify its agricultural lands.
3. Each county is required to zone agricultural lands into exclusive farm use (EFU) zones, with minimum lot sizes ranging from 20 to 40 acres in western Oregon and 80 to 320 acres in the rangeland of eastern Oregon. (Currently, about 16 million acres have been placed in EFU zones, covering about one-quarter of the state.)
4. Farmland owners in EFU zones receive property tax deferral and the benefit of a right-to-farm law.
5. Some nonfarm uses are allowed in EFU zones, such as schools, churches, and home occupations, but nonfarm single-family dwellings are allowed only if they will not interfere with farm operations and are located on lower-quality soils. Counties are encouraged to designate rural residential zones to accommodate growth away from commercial farming areas.
6. Counties and cities have jointly created urban growth boundaries (UGBs), which designate the limit of urban service extensions, especially public sewer and water.

Oregon's Performance Record

Oregon's record in farmland protection is mostly positive. One of the shortcomings is that not all of the county plans were approved until 1986, thirteen years after the initial legislation. In the meantime, there was ample opportunity for scattered housing development in the countryside that ran counter to farmland protection. Some counties, particularly the rural counties outside of the Willamette Valley, tried to ignore, if not blunt their responsibilities in, the statewide planning effort. The lessons here are that implementation of a state program must be fairly rapid, and that the state must provide ample resources for monitoring and enforcement of land-use controls at the local level.

The Oregon program looks good according to the standard of total acres in farming over time. Between 1982 and 1987, the amount of land in farming actually increased throughout the state. From 1987 to 1992, 200,000 acres of farmland (just over 1 percent of all land in farms) were converted to other uses, and commercial farm revenues jumped from $1.79 billion to $2.24 billion.[17]

A 1986 study found that a number of farms created between 1978 and 1982 were in reality hobby farms of fewer than 50 acres and produced less than $10,000 a year in gross farm sales.[18] Information compiled by the Oregon Department of Land Conservation and Development (DLCD) for 1993–94 showed that a total of 1,113 "farm" and "nonfarm" dwelling permits were granted in EFU zones, even though very few new commercial farms were created.

This finding is consistent with prior data that indicated that dwelling permits increased substantially from 1987 to 1989 and with the fact that 37 percent of the new farms created in EFU zones in 1990 did not produce any farm income.[19]

In 1993, the Oregon legislature revamped the state's agricultural zones. People owning small "substandard" lots (below the minimum lot size) in EFU zones were allowed to build a nonfarm residence on them. Counties were required to adopt 80- or 160-acre minimum lot sizes unless they could show that lower sizes were viable. Also, in the highly productive Willamette Valley, a landowner was required to gross at least $80,000 a year from farming in order to be able to build a farmhouse.

An almost forgotten element of the Oregon program is the rural residential zone, with 3- to 5-acre minimum lot sizes, appropriate to hobby farming and rural ranchettes. Over 350,000 acres are zoned for rural residential use in the Willamette Valley alone. However, little if any coordinated effort has been made to encourage development in those areas and away from EFU zones.

If the proliferation of nonfarm rural residences and hobby farms is not curbed, Oregon runs a very real risk of endangering commercial farming operations. Commercial farms could be increasingly forced to compete with buyers who often have a greater ability to pay for land. The greater number of nonfarmers in the countryside could lead to more land-use conflicts, such as complaints about farming activities and interference from trespass and vandalism.

On the one hand, Oregon's use of a coordinated set of farmland protection techniques has provided greater stability to the farmland base and has aided in increased agricultural production. On the other hand, more direction, monitoring, and enforcement is needed from the state of Oregon to help counties comply with urban growth boundaries, channel rural development into rural residential zones, and apply agricultural zoning to pursue the goal of protecting farmland in large blocks.

Farmland Protection Packages

Agricultural zoning works best as part of a package of farmland protection techniques. Differential assessment of farm property ideally should apply only to farms in agricultural zones. Otherwise, farmers could receive property tax breaks and then easily turn around and sell their land for development. In Oregon, a farmer must have land included in an agricultural zone to be eligible for property tax deferral. The agricultural zone helps to protect the public investment in farm property tax breaks.

Agricultural zoning should be in place on farmland before the landowner can be eligible to sell development rights or transfer development rights. Agricultural zoning will tend to keep down the cost of development rights, enabling more farms to be preserved. There is always a danger that a preserved farm can act as a magnet for development, which defeats the purpose of preserving the farm. Agricultural zoning can provide some protection to farmers who have sold or donated their development rights by limiting neighboring development. And agricultural zoning can limit development until funds are available to purchase the development rights to important farmland.

In Montgomery County, Maryland, the nation's leading transfer-of-development-rights program was made possible by the downzoning of 78,000 acres to agricultural at a density of one building lot per 25 acres. Montgomery County has protected almost 40,000 acres with transfer of development rights and another 7,000 acres through the purchase of development rights to farmland.

Summary

Agricultural zoning plays a key role in many local farmland protection programs and in some state growth-management efforts. The purposes of agricultural zoning are to protect productive soils, to separate farm operations from conflicting nonfarm uses, and to strike a balance between restricting and allowing the development of farmland.

A local government should be careful in using agricultural zoning because of the possibility of legal challenges. Agricultural zoning must be based on the local comprehensive plan. Also, the zoning must be reasonable, consistent, and not result in a taking of private property.

There are several types of agricultural zoning to choose from, and these vary according to the kinds and density of nonfarm development allowed in the agricultural zone. Generally speaking, the less nonfarm development allowed, the more effective farmland protection will be.

Some counties and communities are trying to use zoning to create buffers next to agricultural areas. Buffers make good sense, given that the smells, noise, dust, and sprays from farming operations can spill over onto neighboring properties.

Agricultural zoning requires constant vigilance on the part of elected officials and the public to make it effective. Agricultural zoning relies on the judgment of local government officials and the willingness of the public to challenge proposed developments in farming areas.

Because agricultural zoning is not permanent, techniques such as the purchase or transfer of development rights can be helpful in providing long-term protection of farmland. Agricultural zoning also works well with agricultural districts, preferential property tax assessment, and urban growth boundaries.

CHAPTER 8

Controlling Sprawl

URBAN GROWTH BOUNDARIES AND URBAN SERVICE AREAS

> It is precisely here, at the edge of development, that sensible and informed farmland and open space preservation policies must be inaugurated.
> —Peter Wolf
> *Land in America*

> Urban growth boundaries shall be established to identify and separate urbanizable land from rural land.
> —Oregon Department of Land Conservation and Development

Farmland protection efforts have most often succeeded in areas located some distance from development. There, farmers can see a future for farming and often feel they can live with a combination of incentives and land-use controls designed to encourage farming and limit nonfarm development. States have authorized the use of agricultural zoning, agricultural districts, differential taxation, and right-to-farm laws in response to land-use conflicts between farmers and newcomers within the countryside. But at the edge, where a city or suburb meets the countryside, these incentives and land-use controls have not succeeded in protecting farmland for more than a few years. Once the price of farmland rises far above what a farmer will pay for it, development is usually only a matter of time.

A key issue where farms are close to cities or suburbs is the public control of infrastructure spending and location. Sewer and water lines in particular are powerful inducements for residential and commercial development. A number of local governments and the state of Oregon have done an impressive job of closely coordinating infrastructure planning and land protection goals as part of a well-conceived strategy to retain farmland and open space in growing urban areas. Careful infrastructure planning helps to control the costs of community services; makes for more compact, cheaper-to-service development; and reduces sprawl onto open farmland.

The Rise of Sprawl

The American Dream has long included a single-family detached home with a lawn and a garage. The friendly, tree-lined street is safe from crime. The residents enjoy fresh air and clean water. The neighborhood is located within easy access of work, shopping, and schools. Property taxes are low, the public schools good, and public services efficient. There is a sense of community, of shared aspirations and concerns, of belonging to an identifiable place.

In search of this aspect of the dream, Americans have been leaving urban centers in droves and locating farther out into the countryside. This pull is matched by the decline of living conditions in the cities, which push people to leave for the suburbs. Since World War II, millions of acres of farmland and natural lands have been lost, fragmented, or severely pressured because of increasingly mobile populations and a dramatic shift in the location of new houses, office parks, and shopping malls. In many places, a well-defined separation between agricultural and residential areas has given way to fingers of high density commercial and residential development along major roads and houses on 1- and 2-acre lots dotting the countryside. Not only has suburban development occurred in ever expanding rings around urban centers, but when suburbs grow, they often expand onto land ideally suited for farming.

As an example of population shift, consider Detroit: between 1955 and 1990, nearly one million people left Detroit. Where did they go? Look no further than the suburbs. The number of suburbanites in southeast Michigan is now three times greater than the population of Detroit.

The number of households has increased nationwide because of more single-parent families, more retirees, and just more people. This in turn has meant a growing need for housing. Meanwhile, suburban and ex-urban house lots have continued to increase in size, resulting in a lower density and a greater consumption of land per person. Some communities now require a minimum of 2, 5, or 10 acres per house lot.

Spread-out development is expensive to service. A 1992 report by the Urban Land Institute estimates that sprawling residential development may cost from 40 to 400 percent more to service than compact housing.[1] Property taxes and utility rates must rise to pay for extending sewer and water lines, building new schools, and purchasing school buses to carry widely scattered students. For example, between 1970 and 1990 the population in northeast Illinois increased by only 4.1 percent, but the amount of land consumed by development jumped an estimated 45 percent. A study by the Northeastern Illinois Planning Commission asked, "How can costs not rise if public services and infrastructure must be stretched over 45 percent more territory to serve virtually the same sized population?"[2]

Land fragmentation is a major problem in urban fringe areas. The division and sale of land, even without intense development, can rapidly reduce the viability of land for farming or forest management, or the environmental benefits of wildlife habitat and water quality. And once farmland is fragmented or converted to other uses, the loss of that farmland is usually permanent.

Between 1970 and 1990, 25,000 shopping centers were built in America, mainly in the suburbs. As more people moved to the suburbs, the jobs followed. Office parks, often featuring a campus-style setting of walkways and green grass, became popular. Golf courses dotted the landscape, offering outdoor

BOX 8.1
The Causes of Sprawl

- Increasing population and number of households
- Migration from urban centers to newer suburban housing
- New suburban housing on large lots
- Commercial developments in suburban areas: shopping malls and office parks
- Golf courses
- Increasing dependence on the automobile: more vehicle miles traveled, more cars per family, and new road construction
- Increased telecommuting
- Isolation of communities in the inner cities and old suburbs
- Lack of regional planning or tax-base sharing
- Perception that new suburbs are safer
- Perception that suburbs have lower costs for businesses because of property tax incentives to attract more tax base
- Perception that suburbs have lower home ownership costs

BOX 8.2
The Costs of Sprawl

To taxpayers:
- more highways
- social problems
- environmental problems

To businesses:
- lower quality of life for employees
- higher direct business costs and taxes
- higher labor costs
- abandoned investment in older inner cities

To residents of new suburbs:
- cost of automobiles
- higher commuting times and costs
- cost of new suburban infrastructure

To farmers:
- loss of prime farmland
- loss of rental ground
- production losses from pollution, vandalism, complaints, and nuisance ordinances
- decline of farming communities
- loss of farm support businesses
- long-term uncertainty: the impermanence syndrome and disinvestment in farming

recreation. And more roads were built, enabling more Americans to commute to work between suburbs rather than just from a suburb to the center city. And today, many Americans work in one county and live in another.

An obvious cause of sprawl is the automobile. People can live farther away from where they work and shop. But further causes of sprawl are the decisions made by dozens of local governments in a region. These governments compete for development because they believe it will increase their property tax base and provide jobs. They do not coordinate their planning efforts, and development spreads from one community to the next.

Compact Development: The Cure for Sprawl

A three-year Congressional study concluded in 1980 found that federal infrastructure grants and mortgage subsidies had resulted in wasteful land consumption. The purpose of the study, called "Compact Cities: Energy Saving Strategies for the Eighties," was to recommend ways for the federal government to help states and localities curb sprawl and create incentives for development within growth centers, thus saving energy and rejuvenating core communities. But memories of the 1970s energy crisis were already fading and pro-growth politicians frowned on recommendations in the report that states exercise more oversight in planning and zoning and actually mandate growth boundaries.

In the 1990s, there has been a resurgence of interest in more compact, sustainable, space-saving development. This concept, called New Urbanism, emphasizes neighborhoods, town centers, a mix of land uses, public transit, and pedestrian access to shops, schools, and recreation. If a community wants to curb urban and suburban sprawl, new development must be contained as much as possible. At the same time, new development must be attractive and affordable. Home designs that include porches and balconies liven up streets. Low-rise, compact development of four to twelve dwelling units per acre is cheaper to service; new sewer and water lines and roads do not have to be extended into the countryside. New, compact development can make use of existing infrastructure, recycle older buildings, and fill in open lots that have been fragmented through leapfrog development. Compact development lends itself to traditional pedestrian-oriented designs that are socially and aesthetically attractive. Developers can mix houses and shops so people do not have to get in a car and drive to do their shopping. The compact development fits in well with mass transit (called transit-oriented development or TOD by architect Peter Calthorpe). The result is considerable energy and transportation cost savings, as well as improved air quality.

The Urban Growth Boundary

The tool that has been put in place to promote more compact development is the urban growth boundary or urban service area. The purpose of a growth boundary or service area is:

> To contain urban development within planned urban areas where basic services, such as sewers, water facilities, and police and fire protection, can be economically provided.[3]

And:

> To provide for an orderly and efficient transition from rural to urban land-use.[4] (Oregon's Statewide Planning Goal 14)

An urban growth boundary comes to life through an agreement between a city and county (or city and surrounding township) in which an area of county land adjacent to the city is designated for urban-density development. The growth boundary is shown on a map, such as in figure 8.1. Within the boundary of the growth area and city, there should be enough buildable land to accommodate development and population growth for twenty years. This long-term supply of land avoids artificial shortages that could drive up land prices. On the other hand, the growth boundary should not be drawn so loosely as to designate too much for development and create sprawl.

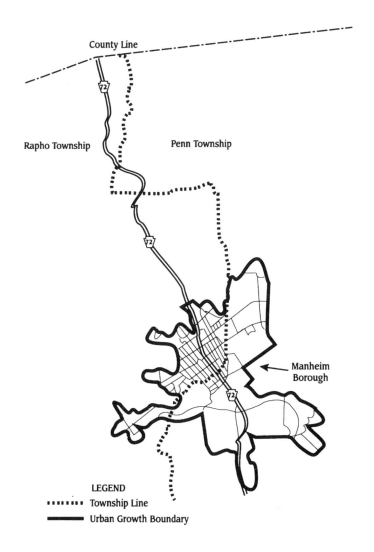

FIGURE 8.1
Urban Growth Boundary Example, Lancaster County, Pennsylvania

Before drawing the boundary, the city and county should make three studies:

1. A projection of population growth, housing needs, and land needs for residential, commercial, industrial, and public spaces and buildings. This is the kind of projection and determination of land needs made in a community comprehensive plan.
2. An inventory of public facilities, their capacity, and the projected needs. This is simply a capital improvements program (see chapter 3).
3. An estimate of a twenty-year supply of buildable land, taking into consideration topography, land needs, availability of public facilities, and a "market factor" of 10 to 15 percent additional land. The market factor is a margin of safety to make sure that land supplies are not constrained.

Then the city and county should amend their comprehensive plans and zoning maps to show the location of the urban growth boundary. Finally, the city and county should enter into a formal urban growth boundary agreement (see appendix F).

Once the boundary is drawn, the city and county agree in writing that urban services, especially public sewer and water lines, will not be extended beyond the Growth Boundary—though the boundary itself may change over time. Without public sewer and water lines snaking out into the countryside, sprawl onto farmland is curbed. The provision of infrastructure is efficient and cost effective.

The county must implement adequate agricultural zoning or conservation zoning on lands outside of the growth boundary. This will protect not only farmland but also water supplies, wildlife habitat, and sensitive rural lands. Agricultural zoning outside the growth boundary can ensure that large residential and commercial developments do not simply leapfrog over the growth boundary, and that the agricultural area will not be overrun with hobby farms and ranchettes.

For example, the purpose of the agricultural zoning ordinance of Utah County, Utah, is "to maintain a greenbelt around urban centers, as a means of cleansing the atmosphere and preserving a quality living environment."

The city and county should review the UGB at least every five years and extend the boundary to add more land if needed. Within the growth boundary, zoning should encourage urban-density growth. Because developers want certainty in where they can develop, commercial and residential building projects will occur within the growth boundary where the land is zoned appropriately and public services are available.

Some farmers and land speculators at the urban edge may be concerned that an urban growth boundary will prevent them from cashing in their land for major developments. But a growth boundary is not designed to be a permanent, unchanging limit to growth. Where a growth boundary will expand, and when, is not definite. But a growth boundary can influence the location and timing of future development so that urban growth occurs in an orderly, gradual, and phased pattern.

The Track Record

European countries have used the urban growth boundary concept since before World War II. Perhaps the best example is the greenbelt area of 1.2 million acres (about 2,000 square miles) west of London, begun through an act of Parliament in 1938. Most of the greenbelt is private land, and there are gradually growing villages included within the greenbelt. The purpose of the greenbelt is to prevent urban sprawl and to keep nearby towns from merging.

Germans have separated land into two main categories: land to be included within cities and villages, and land in the countryside for farming and forestry. This practice creates a visual edge to the cities and villages; strip development does not stretch on for miles along major highways. You don't see suburbs with huge shopping malls, because most of the shopping still occurs in the city center. At the same time, valuable farmland is conserved and a sense of order, community, and tradition are maintained.

The first American urban service area was put into place around the city of Lexington, Kentucky, in 1958. Lexington and Fayette County agreed on where to provide capital improvements that would encourage urban development. Much of the Lexington–Fayette County service area coincides with a major highway. Outside the service area is the famous Kentucky Bluegrass Country, of horse farms and large estates. In 1990, Fayette was Kentucky's leading agricultural county, with farm product sales of over $130 million.

Fayette County's 1988 comprehensive plan update stated, "Capital improvements shall not be extended outside the Urban Service Area in such a way that rural areas are subjected to urban development pressure." The county set agricultural zoning at a 10-acre minimum lot size.

Over the years, housing development at high density has sprung up just inside the boundary. In 1995, the Lexington–Fayette Urban County Planning Commission approved expanding the urban service area by 5,700 acres.

Oregon: The Nation's Leader in Urban Growth Boundaries

The best known and most extensive experience with urban growth boundaries has taken place in Oregon. Enacted in 1973, Goal 14 of the Oregon State Land-Use Program legally requires the state's 36 counties and 242 cities to work together to create urban growth boundaries (see appendix F for an example of a UGB agreement). Although it took thirteen years for all of the UGBs to be put into place, growth boundaries have been an important tool in curbing urban sprawl and in complementing the zoning of millions of acres of countryside for farm and forestry uses.

PHOTO 8.1
An urban growth boundary in Oregon runs along the road and separates farmland (right) from development (left). (Photo courtesy of Kevin Kasowski, 1000 Friends of Oregon).

To date, about 2 million acres of Oregon's 28 million acres of privately owned land have been included inside UGBs.[5] The state government's role has given consistency to the UGB process and forced local governments to cooperate with each other. The state government also requires local governments to:

- draft public facilities plans—especially for sewer and water—to ensure that land inside growth boundaries will be developed at urban densities (Goal 11)
- provide housing for all income levels (Goal 10)

Inside the boundaries, development occurs at a higher density than usual, and developers receive approvals more quickly and predictably. Developers recognize that they can meet market demand inside the growth boundary. Environmentalists accept the urban growth because they can see the rural lands being protected and air and water resources conserved.

In 1990, the Oregon Department of Land Conservation and Development undertook a major study of the performance of the urban growth boundary in curbing sprawl. The results were somewhat alarming. Of the four case studies, only the Portland Metropolitan Service District (MSD) had successfully contained growth, though the MSD contains several thousand acres of buildable land. The Portland MSD was formed in 1979 over an area of 365 square miles. In the early 1980s, land within the MSD was up-zoned to accommodate as many as 300,000 homes, up from 129,000. The average house lot size dropped from 13,000 to 8,800 square feet, while multifamily housing increased from 30 percent to just over half of all new housing construction. This greater density won strong support from the building industry.

Between 1985 and 1989, only 5 percent of the new homes were located outside of the Portland MSD. In the Bend area, 57 percent of new homes were built outside of the growth boundary; beyond the Brookings UGB, 37 percent; and outside the Medford UGB, 24 percent.

A land-use watchdog group, 1000 Friends of Oregon, has compiled a Map of Shame indicating a sizable amount of development just beyond the urban growth boundaries in the Willamette Valley. The threat to Oregon's farmland appears to come not from urban sprawl but rather from suburbanites building houses with on-site septic and well systems amid farming areas.

The conclusion to be drawn is that local jurisdictions have not done an adequate job of enforcing zoning outside of urban growth boundaries. The result is continued traffic congestion, higher public service costs, more development with on-lot septic and water systems, and more low-density housing development on farmland. Better monitoring and enforcement appear to be needed. Another suggestion would be to use a buffer zone of 10- to 20-acre minimum lot sizes just outside of the UGB to cushion the impact between farms and residential development. The lot sizes could then be lowered to enable more intensive, urban development as the growth boundary is extended.[6]

On the positive side, Oregon's UGBs have resulted in greater certainty in the land-use planning and development process. And there has not been a shortage of buildable space. The amount of affordable housing has increased within the UGBs, in part because developers in the Portland metro area can build up to ten units per acre. Also, more land has been zoned for industrial purposes. The UGB process clearly has not impeded economic growth.

Growth Boundaries Reach the East

In Lancaster County, Pennsylvania, several townships and boroughs (villages) have cooperated to form a dozen urban growth boundaries. The county planning commission made projections of population growth over the next twenty years and then computed how much land would be needed to accommodate that expected population. The county then shared the projections of population and land needs with the townships and boroughs, along with maps of suggested urban growth boundaries. After some negotiation, most of the townships and boroughs agreed to adopt the growth boundaries.

Although under Pennsylvania law these boundaries are not legally binding between the township and borough and county, the boundaries have been incorporated by townships and boroughs in their comprehensive plans and zoning ordinances. In other states, urban growth boundaries could be formed through intergovernmental agreements between a county and city.

Lancaster County is the only jurisdiction in the nation that is intentionally purchasing development rights to farmland to create at least parts of urban growth boundaries (see figure 8.2). The land is preserved for farming and open-space uses. In the countryside behind the line of preserved farms, the land is zoned for agriculture at one building lot per 25 acres. This means that leapfrog development over the UGB will probably not occur, except for a few scattered nonfarm residences.

Figure 8.2 shows eighteen farms that were preserved around the village of Maytown over nine years at an average cost of about $3,500 an acre. An area to the southeast was not preserved because it is planned for residential and commercial development. Maytown may expand in that direction over time.

The county has agreed not to purchase development rights to farmland within existing or proposed urban growth boundaries. So far, the development community in Lancaster County has gone along with the urban growth boundary concept. Scott Jackson, executive vice president of the county Building Industry Association, publicly proclaimed, "As long as there are places where we can logically extend our business, we have no qualms."[7]

The Spread of the UGB Concept

A number of states are using or beginning to use the UGB concept. Nationwide, cities annex farmland for urban development hundreds of times each year.

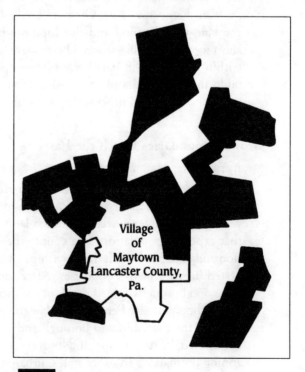

FIGURE 8.2
Using the Purchase of Development Rights to Create a Growth Boundary

8. CONTROLLING SPRAWL

Since the mid 1960s, California counties have created Local Agency Formation Commissions (LAFCOs) to settle disputes over the annexation of county lands by cities. The purpose of a LAFCO is to discourage urban sprawl arising from the premature extension of urban services and to encourage orderly, sensible city boundaries.

Some cities may be tempted to annex commercial properties that are not close by because of the desire for an increased tax base. This practice could lead to leapfrog development and eventual sprawl.

The LAFCO process allows cities to identify "spheres of influence" on county land; the cities have planning control over the sphere areas, which will eventually be developed. In essence, the LAFCOs establish an urban growth boundary separating urban and rural lands for the short to medium term.

A LAFCO must follow the state policy that:

> Development or use of land for other than open space uses shall be guided away from existing prime agricultural lands in open space use toward areas containing non-prime agricultural lands, unless that action would not promote the planned, orderly, efficient development of an area.
>
> Development of existing vacant or non-prime agricultural lands within the agency limits should be encouraged before any proposal is approved for urbanization outside the agency limits.[8]

The Yolo County LAFCO policy adopted in 1994 recognizes that "agriculture is a vital and essential part of the Yolo County economy and environment. Agriculture shapes the way Yolo County residents and visitors view themselves and the quality of their lives. Accordingly, boundary changes for urban development should only be proposed, evaluated, and approved in a manner which, to the fullest extent feasible, is consistent with the continuing growth and vitality of agriculture within the county."[9]

In 1975, Tulare County took the normal LAFCO approach and followed the Oregon model by adopting a formal urban growth boundary for twenty years. In 1992, most cities in Sonoma County agreed through the LAFCO process not to annex those farmlands and open lands that act as "community separators or greenbelts." Then in 1996, Sonoma County voters passed ballot measures that created urban growth boundaries around four urbanizing areas, creating the nation's first true network of greenbelts and further restricting development potential outside the growth boundaries.

Since the 1950s, Boulder, Colorado, has followed a "blue line," which does not allow urban services above a certain altitude. The lack of services in effect limits development and protects the unspoiled views of the Rocky Mountains.

Since the 1970s, Baltimore County, Maryland, has included an urban/rural demarcation line in the county master plan. Sewer and water service has not been extended beyond the line into rural areas. Most of the rural lands are zoned for agriculture at a density of one dwelling per 50 acres, and conservation easements have been purchased on about 15,000 acres.

The city of Virginia Beach, Virginia, has drawn a "green line" in the southern part of the city to separate the urbanizing area from 20,000 acres of farmland and open space. The city has kept public sewer and water from crossing the green line. The absence of these urban services enabled the city to target the area outside the green line for protection through a purchase-of-development-rights program, beginning in 1995.

The state of Washington encourages urban growth boundaries in its 1990 Growth Management Act: "To prevent sprawl by defining urban growth areas and providing open space and low-density rural development at the perimeter of urban areas" and "to prohibit development that requires or encourages urbanization of lands not designated for urban uses in the comprehensive plan."

Thurston County, Washington, has entered into a Memorandum of Understanding with three cities, Lacey, Olympia (the state capital), and Tumwater, to place development in areas served by municipal utilities.

Florida's 1985 Growth Management Act requires that urban services (sewer, water, and roads) be in place before development can occur. This policy of "concurrency" has resulted in local governments setting urban service limits, which act as short-term urban growth boundaries to accommodate urban development.

Vermont's Act 200 of 1988 recommends that development be concentrated in "growth centers," as does the New Jersey State Plan of 1992. An evaluation of the New Jersey State Plan by Rutgers University concluded that the growth center strategy would save the state $1.7 billion over twenty years. Although development would consume an estimated 78,000 acres of farmland by the year 2010 under the State Plan, without the plan 108,000 acres would be consumed.[10]

An important benefit of urban growth boundaries is they compel local governments to plan together. A city and a county can undertake a joint comprehensive plan to identify areas of the county that should be protected and where the city should grow. This sort of forethought and cooperation will conserve farmland and save taxpayers millions of dollars in the long run.

Summary

Urban growth boundaries, urban service areas, and village growth boundaries hold considerable promise as tools to organize the location of urban-type development. These boundaries can be effective in controlling the location of infrastructure, limiting costly sprawl, and helping to protect greenbelts of farmland and open space around cities. The urban growth boundary process ties together regional land-use planning, capital improvements programs, economic development, and phased growth to combat sprawl. Regional planning between a city and county or village and township is the sort of cooperation that is long overdue in America.

CHAPTER 9

The Purchase of Development Rights

> One fellow told me, "You're trying to rule the next generation from the grave." I told him, "That's exactly what I'm trying to do." My home farm has a deed restriction. It's the farm where I was born and raised, and I made up my mind it would never be developed.[1]
>
> —Harold Trimble
> Pennsylvania farmer

The purchase of development rights (PDR) has become an increasingly popular farmland protection method for state and local governments, primarily because it is voluntary and nonregulatory.[2] Eighteen states have enacted PDR programs, and forty-six states have passed legislation to allow state or local governments to acquire development rights to private property.[3]

In America, as stated previously, when you own a piece of land, what you own is a bundle of rights that go with the land. These include water rights, air rights, the right to sell the land, the right to pass it along to heirs, the right to use the land, and the right to develop it. Any one of these rights can be separated from the bundle and sold, donated, or otherwise encumbered.

When landowners sell development rights to their local or state government, they give up only the right to develop their land. They retain all other rights and responsibilities that go with landownership, such as the right to sell the property and liability for property taxes. Despite the government's investment in the land, it remains private property and no public access is allowed. But landowners must practice good stewardship, and the property will occasionally be visited and inspected by the agency holding the development rights.

Since the late 1970s, most local governments have been less inclined to use zoning powers to restrict the use of land. Some state and local governments, mostly in the northeastern United States, have opted instead to spend millions of dollars in an attempt to stop development one random parcel at a time (see tables 9.1 and 9.2). By contrast, effective agricultural zoning can protect whole contiguous farm communities quickly and at a small cost.

But in a farm community of, say, a five-mile radius, when a few farmers sell their development rights, they can influence and inspire others. Eventually, several landowners can create a block of contiguous preserved farms that can mean

TABLE 9.1
Purchase of Development Rights by State to August 1996

State	Year PDR Begun	Acres/Farms Preserved	Funds Authorized/Spent
Connecticut	1978	25,194/165	$82.2/$75.4 million
Delaware	1995	8,600/31	$26/$11.3 million
Maine	1990	307/1	$.38/$.38 million
Maryland	1977	117,319/809	$133.2/$125.1 million
Massachusetts	1977	35,907/398	$111.6/$86.1 million
New Hampshire	1979	2,787/33	$4.9/$4.9 million
New Jersey	1983	27,924/189	$195/$88.5 million
Pennsylvania	1989	80,069/645	$180/$145.1 million
Rhode Island	1982	2,428/30	$14/$14 million
Vermont	1987	45,511/140	$31.3/$26.3 million
Total		345,746/2,410	$735.08/$564.78 million

Note: Kentucky (1994) and California (1995) have enacted PDR programs into law, and both are expected to make their first easement purchases in 1997. Michigan and New York began programs in early 1997.
Source: Farmland Preservation Report and American Farmland Trust.

PHOTO 9.1 The 221-acre Harold Trimble farm, Drumore, Pennsylvania, preserved through the sale of development rights.

the difference in the community succumbing to sprawl or remaining agricultural. In counties where PDR has been practiced for ten or more years, regions of preserved farms have begun to emerge and are stabilizing local agricultural economies. Because the land is restricted in perpetuity, the result is substantial—with the cost of buying the easement paying for itself in less development to service, an active farming economy, and a beautiful landscape year after year, decade after decade.

Dairy farmer Jeff Boissoneault sold development rights on his 150 acres, which added to a neighborhood block of 5,000 acres of preserved farmland in northwest Vermont. He looks at a PDR program as giving farmers the opportunity to compete for land. "Farmers can't compete with 1-acre house lots sold

TABLE 9.2
Leading Locally Supported PDR Programs as of Febuary 1997

This table shows the nation's leading purchase of development rights programs operated and funded by local governments. Local programs may also receive state funding. These are the top ranking local programs according to number of acres preserved. Dozens of other localities administer state programs and may also have comparatively high acreage totals, but do not operate or fund local programs. State programs in Connecticut, Massachusetts, Vermont, and Delaware are not administered at the local level. California, Michigan, Kentucky, and New York have started state programs, but were not yet fully active at the time of this survey; thus, few localities in those states have programs.

Locality	Acres Preserved
Carroll Co., MD	24,068
Harford Co., MD	19,519
Howard Co., MD	17,200
Baltimore Co., MD	11,716
Frederick Co., MD	9,784
Lancaster Co., PA	23,500
York Co., PA	9,418
Berks, Co., PA	8,357
Adams Co., PA	5,630
Chester Co., PA	5,584
Lehigh Co., PA	5,417
Burlington Co., NJ	9,621
Hunterdon Co., NJ	5,219
Salem Co., NJ	4,960
Monmouth Co., NJ	4,486
Town of Pittsford, NY	1,200
Suffolk Co., NY	7,641
Sonoma Co., CA	21,000
Marin Co., CA	25,504
Peninsula Twp., MI	2,259
Routt Co., CO	0
City of Virginia Beach, VA	1,923
Town of Dunn, WI	0
Thurston Co., WA	0

Source: *Farmland Preservation Report,* Vol. 7, No. 4, Feb. 1997.

for $35,000," said Boissoneault. "Removing the development rights makes land more affordable."[4]

The purchase of development rights was first used in the early 1970s in Suffolk County, New York, which covers the eastern half of Long Island. Over the years, that PDR program has succeeded in preserving almost 7,000 acres, but at an average cost of more than $5,000 an acre. In some cases, the value of the development rights has reached $20,000 an acre, equal to as much as 85 percent of the fair-market value of the property.[5]

Despite the potential high cost of buying development rights and because local governments had failed to move to protect their farmland resources, states stepped in to establish programs to purchase development rights. In 1977, Maryland and Massachusetts enacted PDR programs, and other northeastern states soon followed. Legislators supported PDR programs because:

- They avoided the public outcry over property rights and the "takings" issue that lawmakers knew downzonings would generate.
- They were politically expedient: the programs could be seen as cost effective because they would result in truly permanent farmland preservation that was legally binding.
- They could curb sprawl, something the public was beginning to scorn more articulately.
- They were justifiable on economic development grounds because they pumped money into the local farm economy and stabilized the land base for future farm use.
- They would help keep some farmland parcels more affordable for the next generation of farmers because the speculative value would be removed.

While there were—and are—some farmers who resent a government program that tames inflated land values, most have supported the establishment of PDR programs. But their support is often accompanied by a fervent belief in a farmer's right to sell his land for development if need be. These conflicting opinions have resulted in only a lukewarm advocacy of farmland preservation on the part of farm organizations. Without support from the general public—based mostly on the preservation of scenic open space and "rural character"—PDR programs might not exist.[6]

From the landowner's standpoint, whether to sell development rights is a very personal decision, involving individual financial situations and the owner's vision for the future of the land.

Selling development rights is a way to obtain cash from the land, a way out of being land rich and cash poor without giving up any part of the farm or the family's way of life. Landowners typically use the payment to pay off debts, invest in the farm, or set up a retirement fund. A farm with no development rights remaining may be assessed lower for property tax purposes. Also, the sale of development rights reduces the value of the farm for estate tax purposes, making it easier to pass the farm to the next generation. Estate planning and farm transfer using the sale of development rights are discussed in chapter 12. Finally, the PDR program can strengthen a landowner's confidence in the future of farming as more farms in the vicinity sell development rights.

How It Works

Once a government sets up a purchase-of-development-rights program, the actual purchase of development rights to a farm typically takes a year from application to settlement.

First, a farmer applies to the county or state advisory board, giving information on number of acres, type of farm operation, crop yields, mortgages, deed reference, and whether there is a soil conservation plan.

Second, the county or state board reviews the application according to minimum standards for size, quality of the farm, and development pressure. The board then ranks the farm among other applications.

Third, an appraiser is hired to estimate the value of the development rights, the difference between the fair-market value and the agricultural value. That is,

$$\text{Fair-Market Value} - \text{Agricultural Value} = \text{Easement Value}$$

Next, the board will show the appraisal to the landowner and make an offer. Some negotiations may occur, and a landowner may want to obtain a second appraisal as a bargaining chip. The offer may not be for the entire appraised easement value if the appraised value is high or the board wants to stretch its money to preserve more farmland. In Maryland and some counties, the landowner makes a "bid" in a competitive process, indicating that he or she will accept a price for the development rights below the appraised value. The larger this discount, the greater the chances the landowner will have in selling the development rights to the government.

Most Maryland farmers, for example, are currently making bids of between $1,800 and $2,200 an acre for the sale of development rights; and the offers go to bidders who accept the largest discounts below appraised value, regardless of location, proximity to another preserved farm, or ranking scores for soil quality and development pressure. These are issues of contention among both farmers and PDR program administrators.

The average per acre cost of development rights varies considerably. In areas close to major cities, such as King County, Washington, and Montgomery County, Pennsylvania, the cost of development rights has averaged more than $4,000 an acre. In Vermont, a rural state, the cost has been below $1,000 an acre.

If the board and the farmer agree on the price, the board will conduct a title search to determine the legal owner of the property, any mortgages or liens, and a legal metes-and-bounds description of the property. All mortgages and liens must be subordinated or satisfied prior to settlement. A subordination means that if a mortgage holder were to foreclose on the mortgage, the easement would not be removed.

The board and the landowner sign a *deed of easement*, which is placed on the landowner's deed to the property. The deed of easement limits the use of the land to farming and open space, usually in perpetuity (i.e., forever). Some easements last only for a certain number of years specified in the deed of easement; but we do not recommend buying easements in less than perpetuity (see appendix C). The deed of easement runs with the land; if the land is sold or passed on to heirs, the easement remains on the property.

The easement typically prohibits new houses, except one additional house for the owner, the owner's children, or a farm worker. Good stewardship requirements prohibit the dumping of garbage or the removal of soil. Normal agricultural practices and structures are permitted as long as they comply with state and federal statutes.

The government agency holding the easement has the responsibility to monitor the property to ensure that the landowner is complying with the terms of the easement. A staff person from the agency visits the farm perhaps once a year, files a monitoring report, and sends a copy to the landowner. If any violations are found, for example, that the landowner has allowed old tires to be dumped on the land, the holder of the easement can compel the landowner—by legal action, if necessary—to remedy the problem.

The Use of PDR Payments

Farmers most often use their easement payment to pay off debts, set up a retirement fund, or reinvest in the farm operation. In a study by the American Farmland Trust, 46 percent of farmers surveyed reported that at least part of their easement income went to purchase equipment, livestock, and land or to construct farm buildings. A very large percentage of those purchases were made locally, which means that taxpayer dollars are being reinvested in the community, a fact that gives the PDR program added political strength.

Landowners' Questions

Landowners may have many questions about the sale of development rights. Conversations that PDR administrators or attorneys have with farmers who are making an initial inquiry about PDR will usually focus on the following examples:

Why should a landowner sell development rights?

For those who love their land, who see it as something other than a commodity, who value the privacy and gratification they get from land ownership, and who care about conservation, the sale of development rights is an excellent opportunity to be rewarded for having those personal values.

Yet selling development rights can be puzzling, even for the most conservation minded. A Maryland attorney tells the following story to help his clients understand the opportunity PDR provides:

> A farmer came to my office to discuss estate planning and said, "Why should I sell my development rights for $2,000 an acre when I can sell 2-acre lots for $50,000 each?"
>
> I looked the farmer in the eye and offered him a hundred dollars not to put his face through my office window. I asked the farmer if he would accept the payment, seeing that he didn't want to put his face through the

window anyway. The farmer said he would take the hundred dollars.

I then asked, if he didn't want to develop his farm, why wouldn't he take payment not to do something he wasn't going to do in the first place? The farmer got my point (but not the hundred dollars)![7]

Some farmers caution that the PDR is a one-time payment and afterwards the land is restricted to farming and open space. Farmland is not likely to appreciate in value as rapidly as other real estate. And the economics of farming appear uncertain, requiring more capital and innovation for farms to be profitable. For these reasons, farmland protection experts advise landowners to look at the sale of their development rights as seriously as they would consider selling the entire farm.

What if a farmer retires and the land is no longer farmed?

Even though an easement has been placed on the land to preserve it for farming, the government agency holding the easement cannot force a landowner to farm. There may be times when traditional farming does not occur on the land. The land can simply continue as open space. There is concern in some states and counties that when a landowner sells a preserved farm, the next owner may want to use the property as a rural estate rather than an operating farm. But because the easement runs with the land, the next owner may not develop the property for a nonfarm use.

Keeping the land *undeveloped and available* for agriculture is an important purpose of PDR programs. Some public officials worry that the government's interest in the land is devalued if farming is not occurring, but the land is still saving public dollars by not being developed, the soils will benefit from rest, and the government's interest is in perpetuity—a very long time. Chances are that the land will be put back into farming sometime in the future.

What if a preserved farm becomes surrounded by development? Can a landowner buy back the development rights?

All state programs, with the exception of New Jersey, have escape clauses that allow a landowner, in extreme circumstances, to repurchase the development rights. The burden is on the landowner to prove that farming can no longer be possible without economic loss and considerable conflict with nonfarm neighbors.

The process for buying back development rights can be daunting. In Massachusetts, for example, no landowner of a preserved farm, after examining the procedures for a buy-back, has attempted to pursue it. First, the commissioner of agriculture must determine that farming on the property is no longer economical or functionally possible. Then the legislature must agree by a two-thirds vote that a public purpose would be served by relinquishing the easement. Finally, the landowner would have to pay for the development rights at the current market value based on an appraisal.

In Maryland and Pennsylvania, the buy-back process is similar, but a landowner cannot apply until after the easement is twenty-five years old. As of this writing, neither program is yet that old.

Can land be taken by eminent domain if protected by an easement?

Yes. Through the power of eminent domain, a government can condemn private property, including land protected by an easement, but only for a public purpose. The government must pay fair-market compensation to the landowner and to the holder of the conservation easement. The conservation easement restriction is then removed.

How does selling development rights affect my local property taxes?

The result may vary from state to state. In some states, the landowner can inform the local government that the development rights have been sold and therefore the assessed value of the property should be reduced. In other states, there may be no change at first, and the landowner will be wise to enroll the property in the state's differential assessment program for farmland.

Rising property taxes are a big concern. Although the assessed value on a preserved farm may be controlled, the millage rate can always rise, resulting in a higher tax bill.

In this situation, local governments can provide a tax incentive to encourage landowners to sell development rights and to enable those with preserved farms to remain on the land. In Harford County, Maryland, for example, the local property tax is completely excused for preserved farmland. This tax break has been a strong incentive for landowners to participate in the county's PDR program.

Will a farmer lose any ability to secure loans after selling the development rights?

The experience around the country is that farmers have been able to obtain financing after selling off the development rights. Lenders understand that even without the development rights the farm has a value as a farming asset. And the farm can still be sold for farming. Whether a bank or farm credit agency makes a loan depends on the ability of the farmer to pay off the loan based on the income from the farm. In many cases, farmers use the money received from the sale of development rights to buy down debt or invest in the farm operation. These moves enhance the farmer's ability to obtain future loans.

The claim made by some farm organizations that maintaining high development values for farmland on the urban fringe is the only way farmers can secure loans has little validity, according to Darvin Boyd, senior vice president and agri-finance director of CoreStates Bank, one of Pennsylvania's biggest ag lenders. "My response is that land values . . . continue to have a significant value for agricultural purposes, and although such land used for building purposes would have a higher value, there comes a point where you have to look at the larger picture. You have to look at what value is reasonable to lend to an agricultural operation," Boyd explains.[8]

Moreover, farmland that has been preserved by an easement has retained a good agricultural value. For example, in Lancaster County, Pennsylvania, more than fifty preserved farms have sold for between $4,000 and $6,000 an acre, compared to the average price of $5,000 an acre. In some cases, a preserved farm attracted a higher price because it was located among several preserved farms, and the threat of conflict with housing developments was remote.[9]

Setting up a PDR Program

There are seven basic steps to setting up a purchase-of-development-rights program and acquiring development rights:

1. First, state legislation enables the state and/or local governments to purchase development rights to farmland (agricultural conservation easements). (Forty-seven states have passed enabling legislation. Make sure your state legislation defines a conservation easement as "an interest in real estate.")
2. State legislation and/or local ordinances appropriate funds and establish an administrative agency to operate the program.
3. State and/or local agency drafts PDR program regulations and guidelines, including a method to target farms and farmland for preservation.
4. The agency solicits and then receives applications to sell development rights and ranks them according to regulations and guidelines.
5. An independent appraiser conducts an appraisal of the development rights value on the selected farms.
6. The agency negotiates the conservation easement price with the landowner and purchases the development rights.
7. The agency holds the conservation easement and monitors and enforces the terms of the easement.

The use of public money to purchase development rights to farmland requires state legislation and a resolution by a county if the PDR program is to be administered at the county level. The New England states do not have county government and have operated PDR programs through state agencies, usually the departments of agriculture.

In states with county governments, such as Maryland and Pennsylvania, counties administer the PDR program together with the state department of agriculture. The advantage of the county-state programs is that funds for the PDR program are appropriated not only by the state but by the counties as well. The counties are able to budget such expenses better than individual townships or cities. Also, the county-state approach provides for some local control of the PDR program. Each county has its own PDR program guidelines, which generally meet state standards, and the county can either purchase easements itself or apply for state funds to purchase easements.

How to Identify the Land You Want to Protect

Farmland targeted for protection should become apparent from the comprehensive planning process. The key factors are:

- soil quality
- an area of contiguous farmland

- the size of the farms
- the location of nonfarm development and public infrastructure

Maps showing these features should be available from the county planning office or county extension office.

The next step is to set up an application process. In Pennsylvania and Maryland, only farmland enrolled in an agricultural security area or an agricultural district is eligible for the sale of development rights. Also, the property must be above a certain size. The state of Pennsylvania has a 50-acre minimum for a farm not next to a preserved farm and a minimum of 10 acres for one adjacent to a preserved farm. The state of Maryland has a 100-acre minimum, except for land next to a preserved farm.

Because funds for purchasing development rights are limited, it is necessary to have a system to rank the applications (see appendix G). The ranking system should reflect a strategy. An obvious strategy is to create contiguous blocks of preserved farmland. This will help keep residential and commercial development from encroaching on a core area of preserved farms. Another strategy is to preserve those farms that have the best soils and have potential for surviving far into the future. Most PDR programs are designed to protect farmland away from built-up areas and thus removed from heightened development expectations. On farms at some distance from public sewer and water, development rights will be less expensive, and there will be a greater chance of putting together contiguous blocks of preserved farmland.

But some programs, such as in Lancaster County, Pennsylvania, favor preserving farmland under development pressure because of the urgency and the possibility of using preserved farms to block the spread of development into the countryside.

Marin County, California, adopted a comprehensive plan that divided the county into three distinct areas. On the western edge, the federally controlled Point Reyes National Seashore and Golden Gate National Recreation Area include nearly all the land along the Pacific Ocean and are off limits to development. On the eastern side is a designated growth corridor straddling U.S. Route 101. In the middle is grazing land of beef and dairy ranches zoned for agriculture. This is where the Marin Agricultural Land Trust has purchased development rights to over 25,000 acres.

PDR program administrators and their boards must decide which of these strategies or what blend of strategies they wish to pursue. Farms that will achieve the goals of the program should receive the most points in the ranking system. Creating a ranking system often involves trial and error. It may be necessary to experiment, but the system shouldn't be changed every year; that will only confuse landowners. The ranking system determines: which applications will be accepted to have appraisals done, and the order in which the appraisals will occur. The ranking system does not determine the value of the development rights. Only the appraisal does that.

The land evaluation and site assessment system, discussed in chapter 5, can be modified to rank easement sale applications. The LESA system evaluates each farm according to objective numerical values for several factors and comes up with a final score. These scores indicate the priority of the farms for preservation and determine which farms will be considered for appraisals of easement value and the order for conducting the appraisals of easement value. For example, a farm that scores below 50 points will not be accepted and no appraisal will be conducted. A farm that scores 88 points will be appraised for easement value before a farm that scores 65 points.

Appraisals

The next step is to hire a professional, independent appraiser to estimate the value of the development rights. An appraisal is an educated guess of the value of a property at a certain point in time. The appraisal of the development rights is actually a double appraisal of the fair-market value of the property and the restricted agricultural value; the difference between these values is the development rights value. To estimate the fair-market value, an appraiser will most often look at the sale of comparable properties and then make adjustments to those sales to arrive at a value for which the subject property would sell on the market today.

The appraiser must also consider zoning restrictions, proximity to public sewer and water, road frontage, and soils in estimating the fair-market value.

The appraiser may use a comparable sales approach to estimate the agricultural value if preserved farms have sold in the general vicinity. If there have been few or no sales of preserved farms, an appraiser may use local farmland rental values or crop values capitalized by an interest rate to determine a pure agricultural value of the subject property.

In some PDR programs the appraiser has the option to include the value of farm buildings; the farm buildings would presumably be demolished if the property were sold for development, and the purpose of the PDR program is to preserve the land not the buildings.

Finally, the appraiser estimates the development rights value as the difference between the fair-market value of the property and the agricultural value.

In most PDR programs, the landowner gets to see the appraisal along with a written offer from the county or state agency to purchase the development rights. There is no requirement that a landowner receive 100 percent of the appraised value of the development rights. Like any real estate transaction, the sale of development rights is a negotiated process. However, a county or state agency may not offer more than the appraised value of the development rights unless the landowner conducts a second appraisal that shows a higher value.

Appraisals of the value of development rights are expensive, typically ranging from $1,000 to $2,500 per appraisal. Appraisals also take time. The appraiser must visit the subject farm, identify comparable sales, adjust those sales to determine the fair-market and agricultural values, and finally arrive at the development rights value. The appraiser then compiles a rather lengthy report.

Appraisals are based on property sales, and because the prices of those sales may fluctuate over time, value of the development rights may vary as well.

Points-Based Appraisal Formulas

Most counties and states are required by law to use appraisals before public funds can be spent on development rights. But if the law allows it, a local government may want to consider devising its own formula for determining the value of development rights.

Montgomery County, Maryland, has led the way on this innovation. In 1989, when the county began its own local PDR program, administrators devised a formula that assigns points and dollar values for features important to farmland preservation (see table 9.3). The formula considers farm size, soil quality, road frontage, and distance to public sewer and water.

A staff person from the county can punch the dollar values and features into a calculator and come up with an estimate of the development rights value within a few minutes. This "kitchen table" method gives farmers a quick estimate of the value of their development rights. And the county avoids paying up to $2,000 for a standard appraisal.

The advantages of using the appraisal formula are substantial. First, it allows far greater speed to settlement—usually six months, whereas the state pro-

TABLE 9.3
Points-Based Appraisal for the Purchase of Development Rights, Montgomery County, Maryland[a]

Quality Factor	Property Data	Dollar Value
Base price per acre		$700
Farm acreage	250 acres	$350
(1% of base price per acre for every 5 acres)		
Soils		
Class I (3 × percentage of acreage)	20 acres	$168
Class II (2 × percentage of acreage)	50 acres	$280
Class III (1 × percentage of acreage)	150 acres	$420
Other (0 points)	30 acres	$0
Stewardship		
(10% of base price for NRCS Soil Conservation Plan)	Yes	$70
Road frontage		
(1% of base price for every 50 feet up to 5,000 feet)	3,500 feet	$490
Distance from Edge of Agricultural Reserve Zone		
(100% of base price if within 1/2 mile)	Yes	$700
Gross annual farm product sales (25% of base price if more than $5,000 per year)	Yes	$175
Total per acre price		$3,353
Total price for 250 acres		$838,250

[a]From James E. Peters, 1990. "Saving Farmland: How Well Have We Done?" *Planning* 56, no. 9 (1990):12–17.

gram, which uses standard appraisals, can take a year. Second, a formula can be modified to reflect changes in county policy. Third, the formula approach eliminates disputes over the value determined; farmers understand that if the value seems low to them, it is because county policy places a low value on a particular farm, not because the appraisal is questionable. If farmers are dissatisfied with the formula value, the Montgomery County program allows for a standard appraisal to be conducted. The county will pay up to 25 percent more than the formula-derived value, based on the value determined by the standard appraisal.

By using a formula, a local government gets to decide which farms are preserved, based on its own values and criteria, rather than on the criteria in standard appraisals, which focus on development pressure. For example, if a main strategy of a PDR program is to preserve the best farms, the points-based appraisal could be adjusted to give an extra reward for farms with good-quality soil and large acreage. This is a major departure from the Maryland state program, in which development rights values and purchase priorities are determined by standard appraisals and the farmer's offer of a discount in the competitive bidding process.

Development Rights Payment Options: Minimizing the Landowner's Tax Burden

Whenever money changes hands, there are important tax considerations for the landowner. A landowner should always consult with an accountant or tax advisor before deciding among payment options, if any, offered by the PDR program. Some programs offer two options: a lump-sum payment or installment payments plus interest.

State and federal governments tax the development rights payment as a capital gain; state tax rates vary, but the federal rate is currently 28 percent. Because the payment is treated as a capital gain, a landowner can use the basis in the land and buildings (the original price plus improvements minus depreciation) to offset the development rights payment. For example, if a lump-sum payment is $200,000 and the basis is $50,000, then the taxable portion of the payment is $150,000.

The lump-sum payment is most attractive to a young farmer who has recently purchased the farm and has a high basis and a large debt load.

For example, say that Charlie Johnson bought his farm for $450,000 two years ago and has taken only a little depreciation for income tax purposes. The PDR payment is $200,000. Because Charlie's basis is greater than the PDR payment, he can reduce his basis by the amount of the payment and owe no federal capital gains tax. The payment can help Charlie buy down his debt load and increase the profitability of his operation.

A second option is to offer the landowner installment payments plus interest on the balance outstanding. For instance, Pennsylvania counties can offer installments over five years. That is, the landowner can choose to take the money in two, three, four, or five calendar years. Spreading the payments can help the landowner reduce taxes owed, especially if the landowner has losses to carry forward. Installments may also be attractive if the landowner is looking to retire; the payments can provide a stream of income for some years.

The interest rate on the installments is negotiated between the county or state and the landowner. One note of caution is that the interest rate selected should not be greater than the comparable rate on a U.S. Treasury note for a similar period. A higher rate on the payments could trigger additional capital gains taxes. The interest portion of the installment payments is taxed as ordinary income.

The Securitized Installment Purchase Agreement

A refinement of the installment payments option is the securitized installment purchase agreement. The brainchild of investment banker Daniel P. O'Connell, the securitized installment purchase agreement effectively turns an easement contract into a municipal bond. The landowner receives tax-exempt interest payments each year over the life of the contract, and at the end of the contract the principal amount of the easement is paid to the landowner. In the process, the landowner defers capital gains taxes on the easement sale price until the principal is paid. (See table 9.4.)

The landowner can sell the contract at any time on the municipal bond market. If the contract is sold before it expires, capital gains taxes become due.

The securitized installment purchase agreement is attractive to landowners who: 1) have little basis in their property and would otherwise pay up to one-third of the easement price in federal and state capital gains taxes; 2) have been offered a large payment for the development rights ($200,000 or more); 3) wish to set up a retirement stream of income; and 4) would rather preserve their land than sell it for development.

It would be possible to add a further twist by offering the landowner a partial cash payment at settlement with a securitized installment purchase agreement. For example, John and Kathy Hershey have been offered $400,000 for

TABLE 9.4
Securitized Installment Purchase Agreement Programs

Location	Farms Preserved	Acres Preserved	Cost	Discount	Term
Howard County, MD	73	8,000	$47m	40–50%	30yr
Harford County, MD	44	9,600	$20m	30–40%	20yr
Mercer County, NJ	2	300	$ 3m	20%	30yr
Virginia Beach, VA	N/A	10–15,000(est.)	$40m(est.)	0%	25yr
Peninsula Township, MI	N/A	9,000(est.)	$20m	0%	15yr
State of Pennsylvania	N/A				

Source: Evergreen Capital Advisors.
N/A = Planned, but not yet in use.

the development rights on their farm. They have a $100,000 mortgage outstanding, which they would like to pay off. Otherwise, the Hersheys would like to set up a retirement income and continue to live on the farm as long as possible. The Hersheys take $100,000 in cash at settlement to pay off the mortgage. Gains taxes are owed on the cash, depending on the Hersheys' basis in the farm. The Hersheys take the remaining $300,000 in an installment purchase agreement that pays $18,000 a year in *tax free* interest over a period of twenty years. At the end of the twenty years, the Hersheys receive a check for $300,000 for the remaining principal, at which time capital gains taxes apply.

Howard County's Pioneering Program

The securitized installment purchase agreement was originally created for Howard County, Maryland, a rapidly urbanizing county between Baltimore and Washington, D.C. The county wanted to purchase development rights on up to 20,000 acres of farmland. The estimated cost of the development rights was $5,000 to $6,000 an acre. In 1986, the county felt that development rights needed to be purchased as soon as possible because of escalating property values.

The county was advised to purchase thirty-year federal zero-coupon bonds to fund the development-rights-purchase program. A zero-coupon bond requires a small downpayment relative to the face value of the bond and produces no annual interest; instead, the bond pays a lump sum when it matures.

The zero-coupon bonds enabled Howard County to purchase the principal payments for the development rights for a fraction of what they would have cost up front. The county also placed a payment cap of $6,600 per acre and offered landowners 50 to 60 percent of the appraised easement value because of the tax advantages of tax-exempt income and the deferral of capital gains taxes. Landowners could also claim the difference between the appraised value of the easement and the actual offer as a charitable deduction for income tax purposes.[10]

The next matter was how to make the annual interest payments on the installments. The county took in about $3 million a year in real estate transfer taxes. This money was earmarked to pay the interest on the installment purchase agreements. To comply with federal regulations, the interest paid to the landowners could not exceed the interest on the zero-coupon bonds.

To date, seventy-three agreements have been settled covering 8,000 acres. Ten agreements have been resold in the municipal bond market, often at a premium price above the face value of the installment agreement.

In 1993, Harford County, Maryland, began offering the installment purchase agreement, and over one hundred landowners applied for the program. By mid 1995, the county had increased its preserved farmland by 9,600 acres and climbed from tenth to seventh place in the national ranking of counties that have preserved farmland.

The installment purchase agreement has also been used in Virginia Beach,

Virginia, and in Peninsula Township, Michigan. Virginia Beach is the largest city in Virginia, with well over 300,000 residents. It is a popular tourist spot along the Atlantic Ocean. The southern 20,000 acres of Virginia Beach have a very high water table and are mainly in farming use. Virginia Beach has protected this region with a "green line" that restricts the extension of public sewer and water lines. Peninsula Township, Michigan, includes the two-mile-wide, fourteen-mile-long Old Mission Peninsula with its world famous cherry orchards. In both Virginia Beach and Peninsula Township, the local governments identified an amount of farmland they wanted to preserve as quickly as possible: 15,000 acres in Virginia Beach and 9,200 acres in Peninsula Township. Virginia Beach is funding its program from a 1.5 millage increase in local property taxes and a tax on cellular phones! Peninsula Township is using a hike in property taxes to pay for the program. The township officials were concerned that property taxes would go up as fast or faster if farmland was not preserved. After setting up the installment purchase program, Peninsula Township received a $1 million grant from the state of Michigan Natural Resources Trust Fund to help with easement purchases.

Pennsylvania enacted legislation in 1994 that allows counties to use their money and state funds in an installment purchase agreement. To date, no development rights have been purchased this way; however, several counties are looking into the option.

Using the Development Rights Payment in a Like-Kind Exchange

A like-kind exchange is a fairly common real estate tool to enable a trade of properties to occur while minimizing capital gains tax liability. (See examples in box 9.1.) For example, if you wanted to purchase a farm but the farm owner wanted to avoid a high capital gains tax on the sale, you could purchase a second farm and then trade farms with the owner of the first farm; capital gains tax for the owner of the first farm would be deferred until the sale of the farm that he or she traded for.

In 1992, the Internal Revenue Service issued two private letter rulings allowing the use of a development rights payment in a tax-deferred, like-kind exchange under Section 1031 of the Internal Revenue Code.[11] This means that a landowner, instead of receiving a cash payment for the development rights, can use a "qualified intermediary" (such as an attorney, bank, or real estate agent who has not represented the landowner within the past two years) to put the development rights money toward purchasing real estate. The real estate purchased can be additional farmland or other "like-kind" real estate for business, trade, or investment purposes, such as an apartment building. Up to three properties may be acquired, but the fair-market value of the additional real estate may not exceed 200 percent of the easement value.

Capital gains taxes that would normally be due on the sale of development rights are deferred under a like-kind exchange. Basis (the original price paid for

BOX 9.1
Like-Kind Exchange with Development Rights Payment

Development rights being purchased on 100-acre farm (Property A) for $300,000
Fair-market value of Property A: $600,000
Basis in farm: $20,000
Taxable capital gain: $280,000 = $300,000 − $20,000

Example 1: Full Exchange

Value of Exchange Property B: $400,000
Owner of Property A trades the $300,000 easement payment through a qualified intermediary for $400,000 Property B. Easement goes on Property A and the owner must take out a $100,000 mortgage on Property B.
Basis in Property A = $12,000
Basis in Property B = $100,000 + $8,000

To figure the allocation of basis, take the ratio of the fair-market values of Property B and Property "A": $400,000/$600,000 or 2/3. This means that the ratio of the basis in the two properties must also be 2/3. Property A retains a basis of 3/5 times $20,000, or $12,000. Two-fifths of the basis in Property A is carried over to Property B: 2/5 times $20,000, or $8,000. There is an additional cost basis of $100,000 in Property B, which represents the extra cash above the easement payment that the owner of Property A had to pay to acquire Property B.

Basis must also be allocated between land and buildings. The basis allocated to buildings can be depreciated. A capable accountant or tax attorney can be consulted for particulars.

Example 2: Partial Exchange

Value of Exchange Property B: $200,000
Owner of Property A trades the $200,000 of the easement payment through a qualified intermediary for Property B. Easement goes on Property A, and the owner must pay capital gains taxes on $92,500 of the easement payment not used in the like-kind exchange.
Basis in Property A = $7,500, after $7,500 in basis used against
$100,000 taxable gain.
Basis in Property B = $5,000

To figure the allocation of basis, take the ratio of the fair-market values of Property B to Property A: $200,000/$600,000 or 1/3. This means that the ratio of the basis in the two properties must also be 1/3. Property A retains a basis of 3/4 times $20,000, or $15,000. One-quarter of the basis in Property A is carried over to Property B: 1/4 times $20,000, or $5,000. Of the $15,000 in basis remaining in Property A, only $7,500 may be used to offset the $100,000 taxable gain. Property A ends up with a basis of $7,500.

Again, the basis in Property A and Property B must be allocated between land and buildings.

the farm plus improvements minus depreciation) is allocated between the farm property to be preserved (Property A) and the exchange property acquired (Property B). If either the preserved farm or the exchange property is sold, federal gains taxes will be due. If the farm owner holds the replacement property (Property B) at death, the heirs receive a "stepped-up basis" in the farm and the gains tax that has been deferred as a result of the exchange is avoided entirely. Estate taxes may still apply.

This ruling is especially helpful to landowners who have owned their farms for many years and have little basis left in the farm and so would face a sizable gains tax bill if they sold an easement to preserve their farms. Through a like-kind exchange, a long-time landowner could acquire additional farmland for a family member or purchase an investment property that would provide a nest egg for retirement.

Over fifty sellers of development rights in eight Pennsylvania counties have used the development rights payment in a like-kind exchange. Several development rights purchases in Vermont and New Jersey have also involved a like-kind exchange. Most exchanges have been for additional farmland and some have been to acquire rental properties.

Preservation organizations in other states should be careful to make sure that their state's statutes define development rights or a conservation easement as an interest in real property. A further caveat is that a private letter ruling cannot be cited as precedent by another individual or group. A letter ruling does indicate, however, how the IRS is likely to view similar situations. To be on the safe side, landowners selling development rights to preservation organizations in other states should seek a private letter ruling that confirms the IRS position in the Pennsylvania rulings.

There are other things to keep in mind about like-kind exchanges. First, it is important to note that the government agency purchasing the easement on Property A does not show up in the chain of title on exchange Property B. Hence, potential liability from the unforeseen condition of Property B, such as a hazardous waste dump, is minimized. Second, only Property A is being preserved by an easement, not Property B. If Property B is additional farmland, it could be eligible for preservation, depending on state and local eligibility criteria. Third, from the date the deed of easement is settled on Property A, the landowner has 45 days in which to "identify" the exchange property (Property B) and 180 days in which to acquire the exchange property. Fourth, and most important, a successful tax-deferred exchange can be accomplished only with careful legal documentation and adherence to IRS regulations. The assistance of an attorney trained in like-kind exchanges is essential.

How to Pay for PDR Programs

Preserving farmland is a long-term commitment for governments, taxpayers, and landowners. A government and taxpayers must have the commitment to purchase development rights over several years and to fund an adequate staff to implement the PDR program and monitor the preserved farms far into the future. The preservation of a few hundred acres will not have much benefit for growth management or the farm economy. The nation's best PDR programs are those with dedicated funding sources that keep a certain level of funds coming in, preserve a certain minimum number of acres per year, and, therefore, maintain farmers' faith in the program.

Farmers who sell their development rights have some assurance that other farmers will also preserve their farms. Over time, a critical mass of farms can be preserved, ensuring enough business to sustain farm support operations and hence a working farm economy.

State-level PDR programs have been funded in three ways: bonds, general appropriations, and real estate transfer tax revenue. But state-level funds alone are not enough; county or municipal funds are necessary as well. The PDR programs of Maryland and New Jersey require counties to contribute their own funds in order to receive state money.

County-level PDR programs operating independently of state programs have used general obligation bonds, local real estate transfer taxes, sales taxes, and other dedicated taxes to fund preservation programs.

Bonds

It makes sense to use a method of long-term funding for a long-term program. That's why bonds are a familiar approach to funding land preservation. Voter approval is often required to sell general obligation bonds, and is often obtained.[12] In 1979, on the third try, voters in King County, Washington, passed a $50 million bond issue to fund its PDR program. In 1988, California voters passed Proposition 70, which earmarked $776 million in bonds to purchase natural areas, open space, and development rights to farmland. About

---

BOX 9.2
What the Voters and Politicians Will Support

Before a purchase-of-development rights program can be implemented, there must be a way to pay for it. This requires the support of politicians and voters. A good way to win the support of the public is to conduct several meetings about what a PDR program is, how it works, what it costs, and how it is expected to help the community. A public opinion poll will show how much support there is for a PDR program at different levels of cost. For example, the poll might say, "How much extra would you be willing to pay in property taxes each year to support a purchase-of-development-rights program? Please choose one of the following:
a) $0   b) $10   c) $25   d) $50   e) $100"

If taxpayers and politicians believe that a purchase-of-development-rights program is a good idea, they must be presented with an acceptable way to finance the program. There are several ways to finance a PDR program; which one is best depends on the taxpayers and politicians in the county or state.

$63 million of the bond money was targeted to purchasing development rights in eight counties.

New Jersey voters have repeatedly approved state bond measures aimed at preserving land, including $200 million for purchasing development rights on farmland.

A 1990 nonbinding referendum asked the voters of Chester County, Pennsylvania, if they would support a $50 million bond issue for farmland preservation and acquiring park land. *Ninety percent were in favor.* The following year Chester County began selling bonds and raised $12.5 million to finance its PDR program.

Even townships have approved bond issues for buying development rights to farmland. In 1987, South Windsor, Connecticut, passed a $3 million bond with a tenth of the funds for farmland preservation. In July 1996, Pittsford Township outside of Rochester, New York, approved raising $9.9 million to buy the development rights to about 1,100 acres of farmland, an estimated cost of $9,000 an acre. In the process of updating the township's comprehensive plan, a fiscal model projected the cost of preserving farmland at less than $1,400 per household over twenty years. But letting the farmland be developed into houses would cost $5,000 per household.[13]

The advantage of bonds is that a county or state can raise a large amount of money fairly quickly and purchase development rights before nonfarm development eats severely into farming areas. General obligation bonds are tax exempt and are attractive to investors in high tax brackets. The disadvantage of bonds is that the annual interest payments increase the overall cost of the PDR program. Most general obligation bonds pay interest for twenty years. Most bonds are paid for out of property taxes, through an increase in the millage rate. Other sources of funds to pay off the bonds could be real estate transfer taxes or a local sales tax.

In 1987, a statewide referendum in Pennsylvania asked voters to approve the sale of $100 million in bonds for the purchase of development rights to farmland. A coalition of over forty farm, civic, and business groups pushed hard for the referendum, which passed by a two to one margin. Even the city council of Philadelphia endorsed the referendum, paraphrasing William Jennings Bryan that if the farms were lost, weeds would grow in the streets of cities.

The $100 million in bonds was spent by 1994, and in its place the legislature passed a two-cent-a-pack tax on cigarettes, which generates $22 million a year. But with campaigns to discourage smoking, Pennsylvania is looking for more reliable ways to fund the state PDR program.

Some counties and municipalities frown on selling bonds, fearing they convey a "credit card mentality" and could lower the county's credit rating and hence increase the cost of borrowing. But Howard County, Maryland, was able to improve its credit rating to AAA, the highest rating, because the county convinced Fitch's rating service that selling bonds for farmland preservation meant that the county would not need to sell as many for development infrastructure.

Pay As You Go

Some counties prefer a pay-as-you-go approach to financing the purchase of development rights. The PDR program can be an expense item in the county's annual budget. This way the county pays for the program out of annual property tax revenues. In a county with over 100,000 residents, the cost to any one property owner is likely to be only a few dollars a year in property taxes. In a county with increasing land values, property tax revenues are likely to rise as well. The other advantage of the pay-as-you-go approach is that it costs less than borrowing money through bonds. The main drawback is that, unlike the situation with bonds, a county may be able to raise little money for buying development rights up front. Also, bonds commit the county to the PDR program, whereas with a pay-as-you-go method, elected officials can decide in any year to end the PDR program. That was the fate of the PDR program in Forsyth County, North Carolina, which had preserved over 1,300 acres. The county has dropped the program staff and the monitoring of farms under easement.

Bonds and Pay-As-You-Go

A combination of bonds and pay-as-you-go can provide a happy medium between the different funding methods. The sale of bonds can get a PDR program off to a flying start. And the pay-as-you-go approach makes the PDR program an item in the county's annual budget, providing a continuing stream of funding for easement purchases.

Real Estate Transfer Taxes

The voters of Harford County, Maryland, in 1992 approved a local purchase-of-development-rights program that strongly suggested a local real estate transfer tax would be the dedicated funding source. The tax was set at 1 percent of the sales price, and half of the $4 million in annual revenue is used to fund the county's installment purchase PDR program.

In Howard County, a transfer tax of $1/2$ of 1 percent was levied to fund the local installment purchase program.

These two dedicated funding sources are secure because the revenue is required to cover the cost of the installment purchase agreements. It could be disastrous to a county's financial performance rating to remove or divert a revenue stream dedicated to bond debt.

But if transfer tax revenues are not tied to bonds, the transfer revenues could be up for grabs, even if they are "dedicated." The Maryland state program was forced to give up its $20 million in transfer tax funds to shore up the state general fund during the state's budget crisis of 1990. The Maryland program did not purchase any development rights for two years and struggled with minimal funding for years afterward. Had the transfer tax revenue been needed to pay bond debt, the governor might have had to look elsewhere for quick cash.

The Maryland program, while operated by the Department of Agriculture, receives most of its funds through Program Open Space of the Department of Natural Resources, funded through the state's 5 percent real estate transfer tax. In addition, the farmland preservation program receives funds from a separate agricultural land conversion tax: the state of Maryland imposes a 5 percent tax on the sale price of farmland that will be going out of farm use and will no longer qualify for farm property tax assessment.

Sales Taxes

In 1993, voters in Sonoma County, California, north of San Francisco, passed a local dedicated sales tax of ¼ of 1 percent for the next twenty years to pay for the purchase of development rights to farmland and open space. The sales tax was attractive because Sonoma County is a major wine-producing area with a high volume of tourists, who help fund the land protection program with their purchases.

The sales tax has been generating about $20 million a year even though it applies at a rate of twenty-five cents for every $100 of taxable purchases. In the first three years, the sales tax money has enabled Sonoma County to preserve over 21,000 acres.

Developer Exactions

There are two examples of California cities, Carlsbad and Fairfield, that have negotiated deals with developers to either pay money for purchasing development rights to farmland or actually place easements on farmland in return for development approval.

Unless a local government is working with a developer who supports farmland preservation, this method can pose a difficult legal problem. The U.S. Supreme Court has recently ruled that "exactions on development must address a problem created or contributed to by the development, or the cumulative effects of development."[14] The court requires a direct link between the exaction and the problems caused by the development. Such a link may be made in the case of the cost of community services. As mentioned previously, studies have shown that farmland more than pays for itself and residential development uses more in services than it generates in property tax revenues. An argument could be made that if a developer proposes to build houses on farmland, the developer should help pay to ensure that other farmland cannot be developed. Any such exaction should be spelled out in the community's zoning ordinance. Still, there is no guarantee of how the courts would look at such an exaction.

Special Taxation Districts

Three California counties—Marin, San Joaquin, and Solano—have used special taxation districts, in which landowners and home buyers pay an extra tax, to fund farmland preservation. The tax acts much like an impact fee, in that it applies when certain farmland or open space near the urban fringe is subdivided for development or is built upon. The tax on undeveloped parcels is $25 a year over twenty-five years; on developed parcels it is $85 a year. The proceeds from the tax are used to pay off revenue bonds, which are sold to raise money to buy development rights.

Money: The Key Ingredient

Money for PDR programs remains a central and ongoing problem for state and local governments. Maryland's efforts received a blow in 1990 when the governor raided $20 million from the agricultural conversion tax fund that had been earmarked for farmland preservation throughout the state (see figure 9.1). Responding to the state government's funding shortfall, Carroll and Frederick counties created option-purchase programs, which they call Critical Farms programs. The county purchases 75 percent of an appraised easement and holds an option to complete the purchase unless the state can make an offer for the entire easement. Carroll County has used the Critical Farms program as a way to make emergency purchases of easements.

Although $35 million in federal funds for buying development rights was authorized in the 1996 Farm Bill, that money might be stretched over seven years and across as many as fifteen states.

It will be increasingly important for farmland preservation advocates to build political strength to sustain funding in state programs. At the end of 1995, for example, Pennsylvania had nearly one thousand landowners waiting to sell their development rights. Although the commonwealth and individual counties have appropriated an average of $25 million a year, landowner interest in the PDR program far exceeds the available funds. The danger is that some farmers may not be able to hold on long enough to sell their development rights. There is no substitute for a steady, reliable, and hefty source of funding.

Using PDR with Other Farmland Protection Tools

The nation's best farmland preservation programs combine PDR with growth-management techniques that work effectively in a particular community (see chapter 13). Without additional techniques to protect the investment of development rights purchases, the community may end up losing that investment through default.

The danger occurs when a purchase-of-development-rights program is not backed up with effective agricultural zoning, and building rights are too numerous and therefore land values too high to make the purchase of many development rights financially possible. If a PDR program is weakened by inappropriate zoning, easements may be purchased on only a few farms that then become isolated among developed properties. In a region experiencing residential growth, these stand-alone preserved farms can attract developers to adjacent

FIGURE 9.1
Farmland Preservation Versus Farmland Conversion in Maryland

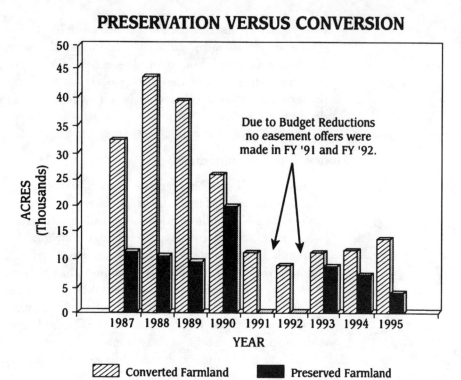

PRESERVATION VERSUS CONVERSION

Due to Budget Reductions no easement offers were made in FY '91 and FY '92.

Converted Farmland   Preserved Farmland

NOTE: Acres of preserved farmland in the above chart only reflects that which was preserved by the Maryland Agricultural Land Preservation Foundation. There may have been more farmland acres preserved through various county programs, local land trusts and other state programs.

	1987	1988	1989	1990	1991	1992	1993	1994	1995	TOTAL
CONVERTED FARMLAND	32524	44269	39801	26079	11070	8719	11132	11430	13743	198767
PRESERVED FARMLAND	11091	10366	9301	19768	0	0	8358	6805	3708	69397
DIFFERENCE	-21433	-339.3	-30500	-21433	-11070	-8719	-2774	-4625	-10035	-129370

land where lots can be marketed as possessing permanent views. But a preserved farm, like any other farm, does not need more nonfarm neighbors. Eventually, farming on the isolated preserved farm could come to an end, resulting in a rural estate that contributes little or nothing to the local agricultural economy. And the purpose of the PDR program would be defeated.

While agricultural zoning can be thought of as the foundation on which PDR programs should be built (see chapter 7 for a full discussion), other growth-management tools can be helpful in protecting easement investments.

Agricultural districts offer an extra layer of protection to landowners who have sold off their development rights. Depending on the state, the agricultural district gives the landowner use-value property tax assessment, exemption from sewer and water assessments, greater protection against eminent domain actions by government agencies, and some protection against nuisance ordinances that would restrict normal farming practices. A preserved farm needs to be able to operate as a farm with as little interference as possible from nonfarm neighbors and local governments.

Differential assessment programs are crucial to owners of preserved farms. If the development potential of a farm has been limited to agriculture, the landowner will not be able to sell off building lots to meet property tax expenses. The sale of the development rights should mean that the assessed value of the farm is reduced. However, the millage rate may still increase. On the other hand, landowners must be careful in how they use the development rights payment. They should put some money away to meet future property tax needs.

Urban growth boundaries can work very well with PDR programs to curb development. Like zoning, an urban growth boundary can be changed, but by preserving farms along a boundary, it can be made more or less permanent. It is possible to extend public sewer and water lines through a preserved farm or for a city to annex a preserved farm, but these actions are not likely to happen.

Summary

The purchase of development rights to farmland will continue to increase in popularity because, despite the cost, the public feels it gets something for its money: the preservation of open farmland without other techniques in use. The purchase of development rights appeals to landowners because it is voluntary and provides a cash payment in return for land-use restrictions. By preserving farmland, governments can take an active role in channeling growth away from prime farming areas.

It may be difficult for a purchase-of-development-rights program to preserve a critical mass of farmland. Other techniques should provide incentives (such as differential property tax assessment) and regulations (such as agricultural zoning) that encourage farming and limit the encroachment of nonfarm development. A purchase-of-developments-rights program may be the first step in getting a community to recognize the need for better planning, zoning, and support for local growers of food and fiber.

Taxpayers and elected officials must take a long-term view and continue to fund PDR programs if farmland preservation is to have a chance of success. Often, a PDR program must be in existence for ten to fifteen years before it has a significant impact.

A conservation easement does not guarantee that a farm will remain a farm because it cannot require that land be actively farmed; the land may revert to open space, especially in suburban communities. The economics of farming in a region and nationwide will determine whether land stays in farming.

CHAPTER
10

The Transfer of Development Rights

Space adrift.
—John Costonis[1]

In chapter 9, we discussed how development rights can be severed from a landowner's "bundle of rights" and sold to a local or state government for the purpose of preserving the land. But development rights can also be sold to private developers who transfer those rights to develop real estate in another location. This technique, known as the transfer of development rights (TDR) is an innovative way to accommodate both preservation interests and development interests. A TDR program allows potential building space to "drift" from one parcel to another.

Unlike most community comprehensive plans, TDR requires much more certainty of where development will happen and where it will not. A good TDR program will be able to determine, from the start, how much development will occur and even what it will look like, that is, whether it will be conventional single family houses on quarter-acre lots or more compact housing in a neo-traditional village style that is pedestrian-oriented and pleasant to see from the road.

Hundreds of counties and municipalities across the United States have transfer-of-development-rights programs on the books, if not in practice. TDRs have been used to protect historic sites, wetlands, and scenic areas, as well as farmland. While TDR has not been used extensively for any of these purposes, in the last two decades it has been used more often for farmland preservation.

Next to establishing effective agricultural zoning on the urban fringe and the political struggles that involves, TDR is the most difficult farmland preservation technique to establish. And, once adopted, it is the least likely to be used effectively because of the large amounts of land involved and because few localities devote the time or expertise to do the necessary community-wide planning.

Nonetheless, some communities believe that purchasing development rights is too expensive and that zoning alone will not enable the community to achieve its farmland protection goals. Public funds are indeed limited, especially for buying development rights close to urban areas. And private land trusts are hard pressed to preserve a critical mass of farmland through the donation or purchase of conservation easements. Agricultural zoning only partly restricts a landowner's ability to develop the land, and because zoning is a political process, it might be changed to allow more houses and shopping malls to be built.

Many communities have examined the transfer of development rights as a possible technique for farmland preservation because:

- the private sector buys the rights, saving the public from a large expense
- it compensates landowners for land-use restrictions and provides open space to the public
- it helps to keep farmland prices affordable for agricultural uses
- it can encourage development in designated growth areas where there is adequate public infrastructure
- it can promote good community growth management

In this book, we cannot offer a detailed guide to starting a TDR program, but we will explain the main features, evaluate the leading programs, and point out the shortcomings.[2]

The Basic Ideas

The transfer of development rights does more than preserve farmland: it changes the way development occurs in a community. TDR can create greater certainty about the future because, in designing an effective TDR plan, public officials and citizens—not just landowners—are actually deciding where and how much development will occur. This is why when a community is contemplating a TDR program, both professional expertise and participation by all the stakeholders are essential.

The stakeholders are the general public and all the affected parties: landowners, farmers, conservationists, developers, homeowners, and elected officials.

When all the stakeholders gather for the purpose of creating a TDR program, it is a good idea to have the meetings led by a professional facilitator.

The right to transfer development rights is not ordinarily part of the bundle of rights that comes with land ownership. A state government must enact specific legislation to enable a local government to legalize the sending of a building right from one parcel to another.[3]

In a purchase-of-development-rights program, a government agency or land trust purchases the development rights from a particular property, which is then restricted to agriculture or open space, and the development rights are "retired" through the deed of easement. The transfer of development rights involves not only buying the development rights and preserving the land, but then

FIGURE 10.1
Transferring Development Rights from One Property to Another

A. Landowner Sells TDRs to a Developer

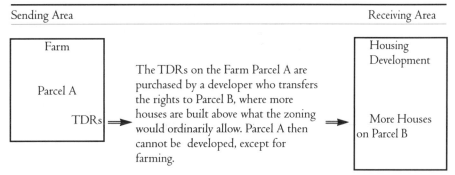

B. Using a Local Government TDR Bank to Transfer Development Rights

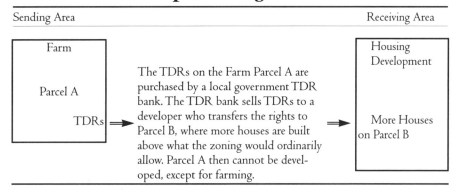

moving those rights from the restricted property to enable additional development (known as a "density bonus") on another property.

The transfer of development rights usually involves, first, purchase of development rights by private developers from individual landowners. A government agency, known as a TDR bank, may act as a broker between TDR sellers and developer-buyers to maintain an active market (see figure 10.1). The TDR program allows the buyer of a TDR to build more houses on a certain parcel of land than the normal zoning regulations would allow. This is the incentive for private developers to buy TDRs.

A local government creates a market in development rights by defining "sending" areas designated for preservation and issuing transferable development rights "tickets" to landowners in the sending areas. Landowners receive transferable development rights based on the number of acres owned, such as one right per 5 acres. The transferable development rights cannot be used to develop land in the sending area. Instead, the government identifies "receiving" areas and requires that developers who wish to build at increased densities in the receiving areas first purchase a set number of transferable development rights tickets from the landowners in the sending areas.

The Four Main Elements of a TDR Program

The transfer of development rights provides ample opportunity to be creative, and there are often important differences among TDR programs. But each successful TDR program has four basic elements:

1. a designated preservation zone, usually called the sending area from which development potential is sent away or transferred
2. a designated growth area, usually called the receiving area, to which development rights are transferred
3. a pool of development rights (from the sending area) that are legally severable from the land
4. a procedure by which development rights are transferred from one property to another.

The *sending area* consists of land the locality has decided should be protected from development. It can be one contiguous area or several smaller areas within a locality, where farmland or lands valued for their environmental or scenic attributes predominate. A sending area can also be a historic preservation district or a rural village with adjacent farmlands.

Private landowners usually have at least some development rights attached to their land. The landowners in the sending area must be compensated for their development rights or allowed to develop their property according to the underlying zoning.

And, without an adequate receiving area, landowners in the sending area would have trouble finding a buyer for their development rights. Under a TDR program that mandated that sending-area landowners sell TDRs, the absence of a receiving area for growth could be grounds for a lawsuit. Under a voluntary TDR program, the lack of a receiving area would result in development occurring in the sending area just as before and no land preservation.

In identifying a sending area, a community should document the land quality, as well as the environmental and human-made resources, and the reasons for protecting them. Then the community should create a map that displays these resources. The map will be an important tool to assure that the TDR program is legally defensible and soundly conceived.

The *receiving area* actually requires the most attention while drafting a TDR plan. A TDR program will work only if there is space for additional dwelling units on parcels already slated for development. Developers must have an incentive to purchase development rights. To create the incentive, the government creates additional building capacity within certain zoning districts. The extra capacity on a site will be approved only after the developer purchases a certain number of development rights from landowners in the sending area.

Receiving areas should generally provide for about 30 to 50 percent more building units than the actual number of transferable rights would allow. This will allow for competition among landowners to sell land and among developers to purchase developable land; for the possibility that the amount of developable land has been overestimated; for additional transferable rights that might be generated through appeals; and for properties in the receiving area whose owners may not want to sell for development.[4]

Ideally, the receiving areas will allow for both major mixed-use developments and smaller single-family housing projects. Obviously, it is best to have receiving areas with access to public water and sewer service and other amenities.

Identifying receiving areas is a sensitive task. But the task should be entered into with open minds. For example, a receiving area does not have to be one contiguous mass of developable land. It simply needs to be placed where additional development will fit in with what is already there. A receiving area can be placed in one particular location, or it can be scattered throughout the community's developed areas in small clusters. But wherever the receiving areas are, they should be consistent with the community comprehensive plan, future land-use map, zoning, and capital improvements program.

One of the first specific questions about TDRs is how to determine how many building rights are in the sending area? Put another way, how does the community determine how much new development could potentially occur?

In Montgomery County, Maryland, a suburban area north of Washington, D.C., that has operated a TDR program since 1982, the county created a sending area first by downzoning 78,000 acres of predominantly agricultural land from one building right per 5 acres to one per 25 acres. Then, the county government gave landowners in the sending area one transferable development right for every 5 acres owned. The county also identified receiving areas in which a developer could build an extra dwelling unit per acre by purchasing one transferable development right.

Montgomery County thus provides a ready model for determining the number of building rights: downzone the sending area, decreasing the number of rights under current zoning but at the same time increasing the number of rights *if they are transferred*.

This sounds easy enough, but obviously many politicians get cold feet when they hear the "d" word (downzoning). They are afraid landowners in the sending area will claim (erroneously) that their property rights are being stolen, and the media and the public will believe them. This is likely the reason the Montgomery County TDR model has not been widely used.

Determining the amount of new development to occur is critical to the success of a TDR program and to the livability of the community. This task will establish how much land must be set aside as receiving areas and how many development rights a landowner is to have.

There are at least two ways to approach this task. The locality can first determine how much new development it can ultimately accept and then decide how this new development, broken down into development rights, can be allocated among landowners in the sending areas. Or, in the second method, the locality can establish a method of allocating development rights as a policy decision. This method would determine the amount of development to occur by adding up the development rights that already exist in the sending area. This second approach is used by localities that decide to let existing zoning determine the number of rights each landowner will have to transfer. This approach can cause problems if TDR is voluntary and existing zoning in the sending area is too lenient, such as a building right for every 3 acres owned. Landowners in the sending area then have the luxury of choosing among: 1) selling many TDRs; 2) selling some TDRs *and* developing parts of their land at fairly high densities; and 3) simply developing their land at a rather high density. These latter two options in effect defeat the land preservation purpose of the sending area.

If TDR is mandatory (meaning that a landowner cannot develop his or her land but can only sell off TDRs), under zoning that allows one building right per 10 acres, a farmer who owns 100 acres could have ten building rights to transfer to compensate for being disallowed to create ten building lots on his or her farm.

Thus, if there is lenient zoning in the sending area, the community may be unable to establish a sufficiently large receiving area to accommodate the large number of TDRs. This may force the community to open up the sending area to development or to designate whole new receiving areas for development.

The point is that a TDR program should not be used without other land-use controls. It needs to be accompanied by effective agricultural zoning so that the development rights to be transferred have a better chance of being accommodated in a properly planned and zoned receiving area.

At the same time, developers must be able to profit from purchasing development rights in the sending area and then using them to build at a higher density in the receiving area. The prices of development rights are determined by developers' bids and landowners' asking prices, just as in a private market. Initial TDR prices in Montgomery County were about $600 an acre, but more recently they have averaged about $3,000 an acre, with some selling in the range of $5,000 to $10,000 an acre. In the Pinelands of New Jersey, TDRs have recently sold for slightly over $500 an acre, there being far less development pressure than in suburban Maryland.

Necessary Conditions for a Successful TDR Program

The transfer of development rights works best in towns and counties where farmland is clearly separated from existing development and properties planned for development. This enables the local government to create distinct sending and receiving zones. If housing developments are already scattered throughout the farming area, a viable sending zone will be difficult to identify. This is often the case in suburban communities.

Also, an active real estate market and a growing population must exist to ensure that developers will be willing to purchase TDRs in order to build more intensively in the receiving areas. TDRs have not worked well in purely rural areas because there is not enough population growth or demand for new housing.

The support of politicians, the public, landowners, and developers is essential for a successful TDR program. The politicians and the public must

BOX 10.1
Features of an Effective TDR Program[a]

Many local governments throughout the United States have either adopted a TDR program or are considering one. TDR can be an effective land preservation and growth management tool. But as with any tool, certain conditions must exist for it to work properly.

1. *Simplicity:* The TDR system must be easy for landowners and the public to understand. The system must be simple to administer. A common criticism of TDR programs is that they are too complicated.
2. *Growth management:* A county or municipality must use a solid comprehensive plan and tight zoning ordinances to support a TDR program. Sending areas that allow too much development will fail to protect farmland or entice landowners into selling TDRs. Receiving areas must include zoning incentives for developers to use TDRs and build at higher but acceptable densities. The ultimate purpose of a TDR program is to create more efficient growth patterns.
3. *Growth pressure:* There must be an expectation of long-term growth to make landowners in the sending areas confident that they will eventually be able to sell their TDRs to enable more development in the growth areas. TDR programs are not effective in entirely rural areas because there is not enough demand from developers for TDRs.
4. *Adequate incentives:* The farmland owner must have an incentive to sell TDRs. The county or municipality must downzone farmland in the sending areas when the TDR program begins. Developers must have an incentive to purchase TDRs rather than build under existing zoning regulations. The transfer ratio between the number of TDRs and the number of additional dwellings the developer is allowed to build must be profitable to the developer. Also, there must be an attractive density bonus in the receiving areas for developers who purchase TDRs.
5. *Receiving area strategy:* The county or municipality must adopt planning policies, zoning ordinances, and capital improvements programs that will assure communities in the designated growth areas that a public facility overload will not result from the TDR density bonus. Also, communities must select the housing types: low to moderate income, the mix of single-family, duplex, and multifamily housing.
6. *Political leadership:* Political leaders must show a commitment to the TDR program. The TDR approach takes time to work but performs better if it is mandatory rather than voluntary for landowners in the sending areas and for building higher density housing in the receiving areas.
7. *Public support:* Public involvement and public outreach are essential to generate public support for the TDR program.
8. *A well-trained planning staff:* Transferable development rights programs need people skilled in planning and public relations to create the program, explain it to landowners, developers, politicians, and the public, and make it work.
9. *A TDR bank:* A government agency that purchases and sells TDRs can help keep a program active during slow economic times and provide a floor underneath TDR prices. Developers may find it easier to purchase TDRs from the government agency than from several individual landowners.

[a]Adapted from a presentation by Dennis Canavan, Department of Planning, Montgomery County, Maryland.

understand the costs of continuing to grow as in the past. The problem should be presented as two choices: to continue with conventional growth and development until all buildable land is consumed, or to allow the community to determine where development will occur and where it will not.

One way to help citizens see the light is to develop a build-out map showing all the new homes and businesses that will be in place once all the building potential that currently exists is used up. A series of maps might be done, one for each community, or maybe for just the communities that will be most affected. Build-out maps and fiscal impact studies can illustrate the changes growth will likely bring to the community, first under current zoning and then with a TDR program, so two maps, or sets of maps, are actually created.

The importance of this public relations step should not be underestimated when starting out on a TDR program. People generally love maps, so the build-out map will be an attention getter and a stimulus to getting the work done. It will be proved once again that a picture is worth a thousand words.

The most difficult part of TDR is convincing landowners in the sending area to agree to a downzoning of their property. Most urban fringe areas are "over-zoned" to allow more development than the community wants or can afford. Landowners typically apply pressure on politicians to maintain zoning with a large amount of development potential, such as one dwelling unit per acre or per 2 acres. But most communities have never examined the long-term costs of their current zoning. A 1991 report on TDRs by the Burlington County, New Jersey, planning department for Chesterfield Township found that "TDR can successfully accommodate growth in a manner which is at least as cost effective as that of the existing zoning and will produce an overall lower total cost of providing public services and education."[5]

Landowners in receiving areas must be assured that the increased densities allowed with TDRs will not reduce the enjoyment or value of their properties. Landowners in sending areas must have some assurance that offers to buy TDRs will provide them with reasonable compensation in return for preserving the land for farming and open space. Landowners in sending areas may face a choice: either develop the land according to the existing zoning or sell TDRs. To influence their decision in favor of selling TDRs, the zoning in the sending area should be low density, of one dwelling per 25 to 50 acres.

Developers must be convinced that they can earn a profit by purchasing transferable development rights and building at higher densities in the receiving areas. The cost of the TDRs to developers and the compensation to landowners will be influenced by the transfer ratio: the number of TDRs the developer must purchase in order to build an additional dwelling in the receiving area.

A county or municipal government must be willing to provide staff and resources to educate landowners, developers, politicians, and the public about how TDRs work. Many communities have considered a TDR program and decided it was too complicated without ever fully understanding the process. Staff must also be available to administer and monitor the TDR program.

The History of TDRs: Calvert County, Maryland

Calvert County, Maryland, on the western shore of the Chesapeake Bay, boasts the first local government to create a transfer-of-development-rights program aimed at preserving farmland. Prior to 1977, when Calvert County enacted its TDR program, New York City had used the transfer of development rights to preserve Grand Central Station, and in 1974, Collier County, Florida, helped to transfer 526 development rights to protect 325 acres of saltwater marshes and mangrove swamps. (See table 10.1 for examples of active TDR programs.)

When Calvert County began its TDR program, participation was voluntary, and the county did not designate sending and receiving areas. The county considered both sending and receiving sites on a case-by-case basis. Landowners received one TDR per acre. Developers could add an additional dwelling unit per acre for every five development rights purchased.

In 1993, the county established sending and receiving areas, a county fund to purchase TDRs, and a goal to preserve 36,000 of the remaining 45,000 acres of farm and forest lands. In the 55,000-acre sending zone, a landowner has the option to sell TDRs at the rate of one TDR per acre or to develop the property with houses clustered onto 20 percent of the site. The county zones land in the sending and receiving areas at one dwelling unit per 5 acres and allows increased density in the receiving area when the developer applies TDRs. The Calvert County officials have recognized that the one-house-per-5-acres zone in the sending area could result in too much development and the loss of important farm and forest lands. The county has recently been working to lower the density in the sending area to one dwelling unit per 20 or 25 acres.

In Calvert's 17,000-acre "rural community" receiving area, house lots must be clustered onto 50 percent of the site. Additional dwellings are allowed based on one dwelling unit per five TDRs purchased in the sending area. In 1994, 447 TDRs changed hands on the private market at an average price of $2,305 per TDR.

In 1993, Calvert County set up a Purchase and Retirement Fund of $770,000 to acquire TDRs from landowners who want to sell them when there is no private buyer. Under the Purchase and Retirement Fund program, farmers or owners of farm or forest lands of at least 50 acres can create agricultural preservation districts, making the landowners eligible to sell development rights to the county, to the state farmland preservation program, or to developers.

TABLE 10.1 Examples of Active TDR Programs, July 1996

Location and Date Begun	Acres Preserved	Number of Transfers	Estimated Cost
Calvert County, MD, 1977	6,000	400	$8 million
Manheim Township, PA, 1991	188	2	$0.8 million
Montgomery County, MD, 1982	32,225	5,000	$60 million
Pinelands, NJ, 1981	5,800	250	$2.5 million
Thurston County, WA, 1996	—	—	—

There are now about 16,000 acres enrolled in the preservation districts. Typically, Calvert County purchases only ten to fifteen development rights each year from a preservation district. The purpose of the county fund is to help keep the market for development rights active and to set a floor under TDR prices, but developers are still the main source of money in the program. In 1994, Calvert County purchased 204 TDRs at an average price of $2,350. The county fund allows farmers to sell a few development rights and keep others as future security, including development potential on the farm.

All told, Calvert County has preserved over 6,000 acres through the transfer of development rights. An additional 3,450 acres have been preserved through the state of Maryland's purchase-of-development-rights program. The county hopes eventually to preserve another 22,000 acres through mandatory clustering in the sending and receiving areas.

Montgomery County, Maryland: The Leader in TDRs

Montgomery County, Maryland, is nationally renowned for its transfer-of-development-rights and purchase-of-development-rights programs to preserve farmland. Montgomery's success is particularly noteworthy because it is a major suburb just northwest of Washington, D.C., with a population of 782,000 on an area of 316,800 acres. The county has done a remarkable job of accommodating growth and avoiding suburban sprawl while protecting its farmland resources (See figures 10.2 and 10.3).

Combining its TDR program with the state and county PDR programs, as well as easements donated to the Maryland Environmental Trust, the county has protected over 45,000 acres within the 78,000-acre agricultural reserve (also called the Rural Density Transfer Zone). Through over five thousand TDR transactions, the county has protected 33,000 acres of farmland. Another 12,000 acres are protected through purchase-of-development-rights programs and easement donations. Also, Montgomery County has buffered its agricultural reserve area with over 13,000 acres of county parkland. Meanwhile, since 1980, only about 4,000 acres of farmland have been converted to other uses within the agricultural reserve.

It is important to note that, in Montgomery County, when landowners sell TDRs, they retain the right to build one house per 25 acres. In essence, this is a form of *compensable zoning*. A landowner who sells TDRs receives compensation for the one dwelling unit per 25-acre zoning. In a pure TDR arrangement, no development rights would remain once the TDRs were sold and transferred to a receiving area, and a conservation easement would be placed on the farm in the sending area.

From a land-use planning perspective, Montgomery County successfully combined several techniques: 1) a county comprehensive plan and twenty-six area master plans; 2) agricultural zoning to protect a critical mass of farmland in the agricultural reserve; 3) development permits for approving construction

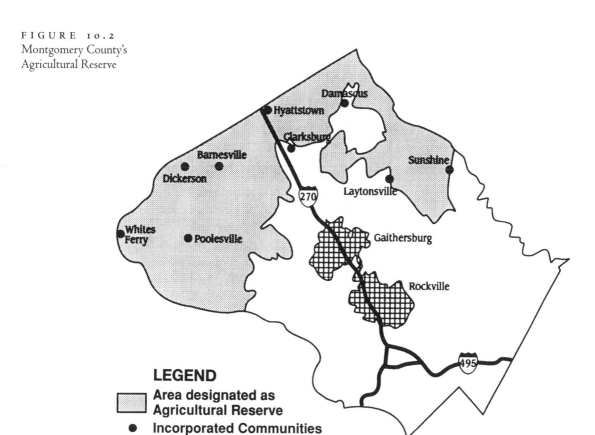

FIGURE 10.2
Montgomery County's Agricultural Reserve

Source: *1980 Montgomery County Master Plan.* Rockville, Md.: Montgomery County Department of Planning.

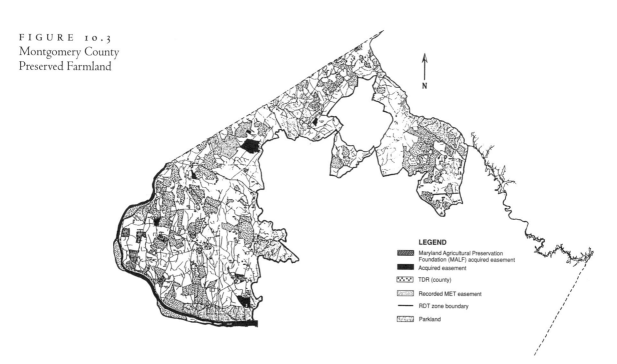

FIGURE 10.3
Montgomery County Preserved Farmland

projects that involve TDRs; 4) capital improvements programs to influence the location and timing of public services, especially roads, sewers, and water lines; 5) a public TDR fund to keep the TDR market active by purchasing TDRs from landowners and selling them to developers; 6) use-value assessment for farmland; and 7) voluntary agricultural districts to encourage farming.

From the developer's perspective, the Montgomery County TDR program works as follows:

1. The developer submits a subdivision preliminary plan to create lots and build houses on a piece of land in a growth area. The developer may build one additional dwelling unit above what the base zone allows for each TDR the developer has purchased, as long as the overall density does not exceed what is allowed in the master plan for that area or will not overburden public services. One TDR represents 5 acres in the agricultural preserve, the TDR sending area. A TDR may cost the developer $15,000 to $30,000, but the profit from an additional dwelling unit usually exceeds this cost. If the developer proposes to use TDRs, the developer must use at least two-thirds of the TDRs that could apply to the development site.
2. Along with the preliminary plan, the developer includes proof of an option to purchase or the actual purchase of enough TDRs. The TDRs purchased refer to the easement that is placed on the land in the sending area, and the developer must show the deed of transfer from the sending land.
3. County officials use the final subdivision plan to record the easement on the sending area land to track how many TDRs are being used and how many remain on a particular farm in the sending area. This record keeping is very important to avoid a duplication of sales of TDRs. After the easement is recorded, the developer's final subdivision plan receives a site plan approval permit. The final plat of the site includes a note of the number of TDRs used and a reference to the easement on the farmland where they came from in the sending area. Now the developer is ready to subdivide the land and build houses in the growth area.

The New Jersey Pinelands TDR Program

The transfer of development rights is perhaps the most challenging farmland preservation technique to design and implement. One farmland protection expert jokingly refers to TDRs as the "nuclear power" of the farmland protection movement: If TDRs worked as well in practice as they do in theory, all of our farmland protection problems would be solved.[6] In reality, the transfer of development rights has not enjoyed as much popularity as the purchase of development rights because of the difficulty in establishing well-defined sending and receiving areas. Landowners may be reluctant to have their property placed in a receiving area because of the increased density of houses that is allowed. Farmers and other large landowners may shy away from a sending area because of uncertainty as to the market value of their TDRs.

This latter problem arose in the New Jersey Pinelands, a seven-county area in southeastern New Jersey, stretching from just east of the New Jersey Turnpike down to Cape May. The Pinelands cover 934,000 acres of pitch pines, cedar swamps, cranberry and blueberry operations, as well as towns and hamlets. Underneath the Pinelands is one of the largest aquifers in the entire Northeast. The ecological importance of the Pinelands was underscored in 1983 when the United Nations designated the Pinelands an International Biosphere Reserve.

In 1978, the U.S. Congress passed a mandate to protect the Pinelands. In 1979, the state of New Jersey established the Pinelands Commission, which drafted a comprehensive plan for the region's fifty-two municipalities (approved by the U.S. Secretary of the Interior), featuring a preservation area and a protection area. The comprehensive plan placed the most environmentally sensitive land in the preservation area and recommended only very limited development. In the protection area, development can occur according to certain performance standards.

In 1981, the Pinelands Commission established a TDR program. Landowners in the preservation area were allocated four development rights (equal to one development credit) for every 39 acres of upland or woodland owned. A developer could purchase four development rights and transfer them to build four additional houses in a designated regional growth area.

Landowners in the protection area were allocated eight development rights per 39 acres of uplands, woodlands, or actively farmed wetlands owned (equal to two development credits). The preservation sending areas have about 24,400 TDRs available for sale, and the receiving areas allow 43,400 more homes to be built than would be allowed under the current zoning. Thus, the market tends to favor the landowners in the sending areas.

The early 1980s saw few transfers of development rights, partly because landowners and developers did not understand the program. To improve the liquidity of the development rights market, the state of New Jersey created the Pinelands Development Credit Bank in 1985. The Credit Bank can purchase TDRs for $2,500 per right ($10,000 per development credit) if no other buyer can be found. The Credit Bank then can sell the TDRs to developers at some future date. Sales of development rights have grown from just 2.25 rights in 1984 to 37 rights in 1992 and 134 rights in 1994. The average cost per right has also increased to an average of $4,000 or $5,000. As of late 1994, 1,800 acres of farmland and 4,000 acres of environmentally sensitive lands in the Pinelands had been preserved. (See figure 10.4 and box 10.2.)

FIGURE 10.4
Application to Sell TDRs to the Pinelands TDR Bank

APPLICATION FOR SALE OF PINELANDS DEVELOPMENT CREDITS TO STATE PINELANDS DEVELOPMENT CREDIT BANK

New Jersey Pinelands
Development Credit Bank
CN-035
Trenton, NJ 08065
(609) 588-3469

For Bank Use Only
App # _____
Date Rec'd. _____
Action _____
Date _____

OWNER INFORMATION

[1] PDC Owner's Name _____ [2] Phone No. _____

[3] Co-Owner's Name _____ [4] Phone No. _____

[5] PDC Owner's Street Address _____

[6] City, State, ZIP Code _____

PINELANDS DEVELOPMENT CREDIT INFORMATION

[7] Has a PDC Certificate been issued for the PDCs which you wish to sell? Yes ___ No ___

[8] If yes, the PDC Certificate number is _____

[9] How many PDCs do you wish to sell to the Bank? _____

Please circle the appropriate answer(s)

[10] The Bank should purchase my PDCs because:

a. Yes   N/A   My property is of significant ecological or apricultural importance.

b. Yes   N/A   My property buffers of complements publicly owned conservation lands.

c. Yes   N/A   The Bank is likely to resell or transfer my PDCs for use in a residential development project which satisfies a compelling public need or which will protect conservation or agricultural lands.

d. Yes   N/A   It will serve as a significant example of the PDC program at work.

e. Yes   N/A   I intend to use the proceeds from the sale to improve, develop, or use my property in a manner consistent with the terms of the deed restriction.

f. Yes   N/A   The purchase otherwise furthers the objectives of the Pinelands Protection Act and Comprehensive Management Plan in an unusual way.

[11] The Bank should purchase my PDCs to help relieve a hardship because:

a. Yes   N/A   My investment in the property represents a substantial portion of my total net worth.

b. Yes   N/A   I have been denied a waiver of strict compliance from the standards of the Pinelands Comprehensive Management Plan.

c. Yes   N/A   I am experiencing an extraordinary and unique financial hardship.

REPRESENTATIVE INFORMATION

[12] Do you authorize a person to act as your reprentative in all matters pertaining to this application? Yes___ No___

[13] Name of representative _____ [14] Phone No. _____

[15] Representative's Street Address _____

[16] City, State, Zip Code _____

[17] Signature of Representative _____

APPLICATION INFORMATION: *Is the following information attached?*

[18] PDC Certificate? Yes ___ No ___ N/A ___

[19] Application for PDC Certificate: Yes ___ No ___ N/A ___

[20] Statement explaining each of the reason(s) you checked to justify Bank purchase: Yes ___ No ___ N/A ___

[21] A net worth statement and a tabulation of my investment in the property: Yes ___ No ___ N/A ___

[22] A copy of the Pinelands Commission's denial of a waiver of strict compliance: Yes ___ No ___ N/A ___

OWNER CERTIFICATION

[23] I certify that the information included in this application is true, and that I will sell to the Pinelands Development Credit Bank the number of Pinelands Development Credits specified in this application if approved for purchase by the Bank.

Signature of Owner (Applicant)   Date   Signature of Co-Owner (Co-Applicant)   Date

Note to Applicants: The State Pinelands Development Credit Bank staff will contact you within two weeks of the receipt of this application.

> **BOX 10.2**
> How to Sell TDRs in the New Jersey Pinelands
>
> 1. A landowner in the sending area asks the Pinelands Commission to determine the number of development rights on a property (Letter of Interpretation).
> 2. The landowner then conducts a title search, obtains subordination agreements from mortgage holders, and records a conservation easement on the property.
> 3. The Pinelands Development Credit Bank issues a Pinelands Development Credit Certificate to the landowner, stating the number of TDRs.
> 4. The Pinelands Development Credit may be sold to the Credit Bank or to a developer. Ideally, the entire process takes about six months.

TDRs in Pennsylvania

In Pennsylvania, development rights can be transferred only within a township, not on a county-wide basis. This has drastically limited the attractiveness of TDRs. Since adopting a TDR program in 1991, Manheim Township of Lancaster County, Pennsylvania, has preserved only two farms totaling 188 acres. This record belies the fact that Manheim Township has 28,000 people, 1,900 acres of land zoned for agriculture (sending areas), and a few thousand acres eligible as receiving areas.

In 1994, Buckingham Township, Pennsylvania, established a TDR program that allows development rights to be transferred within the agricultural sending area with the use of clustering. The agricultural area is zoned at one dwelling unit per 5 acres. The clustering provision allows large density increases of 50 to 80 percent at one dwelling per 2.5 acres or one unit per acre. In addition, the subdivision standards are more flexible, including no minimum lot size. Sending parcels must be more than 25 acres and actively farmed. But Ray Stepnoski, chairman of the township board of supervisors, admitted that transferring the density within the farming area was the down side of his township's TDR program.[7] This use of TDR is not really aimed at preserving farmland but tries to encourage development that fits in with the rural character of the area. So far, three transfers have been completed, with more in the works.

San Mateo County, California

In 1988, San Mateo County, south of San Francisco, adopted a novel TDR effort to preserve agricultural land and promote agriculture as a business. Coastal farmlands were placed in a planned agricultural district (PAD) sending area that limits nonfarm uses at a density of 1 to 40 acres or 1 to 160 acres, depending on the nonfarm use. When a subdivision of farmland occurs in the PAD, a perpetual agricultural easement must be placed on the property.

Farmers in the planned agricultural district receive one extra TDR when farm parcels are consolidated to form a larger, more viable farm, or when they build an agricultural water storage facility.

TDRs may be sent to parcels in the county's rural coastal zone but not onto prime farmland or in scenic corridors. When a TDR is sent, the farm in the sending area is limited by an easement to farm uses and farm labor housing.

To date, only two transfers of bonus TDRs from water storage projects have occurred.

Thurston County, Washington

So far, the transfer of development rights to preserve farmland has mostly been applied in the northeastern states. In 1996, Thurston County, the fastest-growing county in the state of Washington, crafted a transfer-of-development-rights program through downzoning 11,500 acres of farmland for a sending area and then designating receiving areas in the cities of Olympia, Lacey, and Tumwater.

Landowners in the receiving area were downzoned from one dwelling per 5 acres to one dwelling per 20 acres or 40 acres, depending on location. Landowners in the sending areas can petition to have their land removed if surrounding land becomes developed. The sending areas are in ten sections scattered through the southern end of the county and adjacent to zoning of one dwelling per 5 acres.

The three cities have identified over 20,000 acres as receiving areas. In these areas a developer must transfer in one TDR in order to build an additional dwelling per acre. The receiving areas are single-family and multifamily zones. Thurston County hopes to direct about three-quarters of its growth over the next twenty years into the receiving areas.

Workable Innovations in TDR

Few planning concepts could benefit more from innovation than TDR. The transferring of development potential from one area to another is a powerful land management technique. But the obstacle to accomplishing protection of land resources is the political capital and courage necessary to start and maintain the process.

One innovation would be for citizens to become involved in establishing TDR programs, even for modest land preservation objectives. TDR offers an opportunity for citizens to become involved in the development marketplace. With willing buyers and sellers, there is no legal obstacle to prevent a group of citizens from changing the development potential on certain parcels of land. This is certainly an area that is wide open for land trusts to explore.

Meanwhile, the challenge in using TDRs is to rethink how they are carried out. Is there any way to use the idea differently?

A possible area for innovation is to move development potential from farmland to commercial or industrial sites, or to move development potential from one industrial or commercial site to another without having to go through a downzoning process. So far, TDRs used for land preservation have applied almost exclusively to retiring residential development potential in one area and moving that potential to a residential zone.

In San Luis Obispo County, California, the county government hired a private land conservancy in 1993 to help retire unused building rights in a hillside community that looked out over the Pacific Coast. The community had been developed before resource protection laws were established, and local officials wanted to save what open space remained in the environmentally sensitive area.

The effort focused on transferring the building rights away from the coastal community and into less sensitive areas inland. Instead of designating the entire community as a sending area and an entire inland region as a receiving area, it was decided to try to link each sending area with its receiving area.

The idea was to put the protected resource within view of residents in the receiving zone, so that they could see exactly why the increased density in their neighborhood was worth the sacrifice. In other words, officials said: In return for accepting the increased number of homes in your community, you will receive a permanently protected landscape to enjoy on your way home from work. The idea could be called in-area transfers, since the sending and receiving sites are linked more closely, in terms of distance, than in traditional TDRs. San Luis Obispo County called them "married."

While using TDR in this way takes more involvement on the part of an entity acting as the facilitator, this approach could spur true innovation that could make TDRs more widely used. What this approach does is remove the fear of receiving zone opposition and gives encouragement to elected officials that residents in the receiving zone don't have to carry all the burden of TDR without a benefit they can see on a regular basis.

The drawback to the in-area transfer, married-site idea is that resource protection is back to being carried out on a piecemeal basis, a long-standing problem in land preservation. But if you take a county or township and divide it into districts and create community planning councils to explore potential sending and receiving site projects within their districts, the effort could be extended throughout the locality. One benefit to this approach is that citizens will become immediately involved, and there is no better path to innovation. While this approach, basically an abbreviated form of TDR, will likely not be able to save large contiguous masses of farm or forest land, it will empower a community to identify its own resources that are worthy of protection and to build on that newfound power to have some control over its destiny.

The Shortcomings of TDRs

Nationwide, many counties and municipalities have discussed using a TDR program to protect farmland and open space, yet only a handful of places have created workable programs.[8] The following is a brief discussion of problems to be avoided.

Programs not used. Some counties have enacted TDR programs that see little use. This has been the experience in the Maryland counties of Caroline, Queen Anne's, and St. Mary's. In these counties, *TDR is voluntary not mandatory.* Isabella County, Michigan, created a TDR program in 1989 and dropped it a few years later. "We tried it, nobody liked it," says Carolyn Ramsey of the county planning department.[9] She adds that the county failed to set up a good process before trying out the TDR program.

If a local government goes to the trouble of establishing a TDR program, it should put in a good effort to make the program work. The objective is not just to create a program but to create an effective, working program. Also, if the program is to be voluntary, it will probably also be inactive and ineffective.

Lack of qualified planning staff. Some municipalities and small counties approved TDR programs without the necessary personnel, or without hiring qualified consultants to make the program work. These local governments were not willing or perhaps not able to make the investment in staff to produce a successful TDR effort. Moreover, a TDR must be part of a comprehensive planning program, which designates places for development and land to protect along with zoning and policies to locate infrastructure. The planning process for reviewing and approving proposed housing developments must be tied in with the transfer of development rights from the sending areas to the growth areas.

TDR not needed because other existing land protection techniques are successful. Marin County, California, passed an ordinance to allow eleven development rights from the Lafranchi Ranch to be transferred to a single parcel. But no other TDRs followed. Marin's combination of comprehensive planning, agricultural zoning, and purchase of development rights has so far been effective. In other words, a local government may not need a TDR program to protect land and channel development to appropriate locations.

Record-keeping difficulty. In a PDR transaction, all or nearly all of a farm is preserved at once. With TDRs, a landowner can sell off a few TDRs at a time, which can create a record-keeping challenge for public officials who need to know where landowners have sold TDRs and how many TDRs have been sold from a parcel.

Montgomery County adopted an exemption in the sending area for lots created for a landowner's children, at one lot per child, above the one dwelling per 25 acres zoning standard. This family lot exemption weakens the sending area, because houses will be built and sold away from the farm, increasing the number of nonfarmers in the sending area and risking additional conflicts between farmers and nonfarmers. Keeping track of the children's lots also increases the challenge of maintaining accurate landownership records.

Recessions. For a TDR program to work smoothly, there needs to be an active land market with potential development projects. When a recession hits and housing demand falls off, developers buy fewer TDRs and usually at lower prices. Recognizing the fluctuations in the land market, Montgomery County devised a county PDR program to complement the TDR program. This gives landowners a choice in selling their development rights.

A supply-and-demand imbalance. The number of TDRs created in the sending area must be less than or equal to the number of additional dwelling units allowed with TDRs in the receiving areas. For example, Montgomery County initially created 18,000 TDRs in the sending area but only 9,000 TDR residential units in the receiving areas, an unfavorable balance to landowners in the sending areas. As a result, early TDR sales averaged only $600 an acre.

Without some assurance of the value of TDRs, landowners may very well resist the downzoning of their property to a low density, such as one building lot per 25 acres.

Wrong basis for selection. From a land-use planning perspective, a problem with TDRs is that they are based on the number of acres owned and not necessarily on location, soil quality, and access to public services. In Montgomery County, developers first purchased TDRs in the remotest parts of the sending area because those were the cheapest. Manheim Township, Pennsylvania, offers a flat $5,500 per TDR (about $4,200 an acre) from its TDR bank to any landowner in the agricultural sending district. In the interest of farmland preservation, the land along the boundary of the agricultural reserve (sending area) should have the highest priority because that is the land under the greatest development pressure and hence the most valuable for blocking the development of the sending area.

Case-by-case determination. Finally, there is some debate whether developments involving TDRs should be decided by elected officials on a case-by-case basis or administered by the local planning department according to a set of rules over a large area. Many communities use TDRs on a case-by-case basis, possibly

---

BOX 10.3
How to Tell If a TDR Program Is Successful

- Large number of acres preserved in the sending area
- Many contiguous or nearly contiguous acres preserved in sending area
- Small amount of new nonfarm development in sending area
- Takings challenges avoided (i.e., landowners in sending area satisfied with TDR prices)
- Developers able to avoid legal challenges in receiving areas
- Developers satisfied with cost of TDRs and profit from additional development with TDRs
- Low cost to local government to set up and administer TDR program
- Periodic monitoring and adjustment to improve TDR program
- Durability of the TDR program over time

because they have limited farmland and their preservation goals emphasize open space rather than maintaining the agricultural industry. A case-by-case approach probably will have less support from developers. Because time is money, they prefer a strong degree of certainty that their building projects will be approved in a timely fashion.

Using TDRs with Other Farmland Protection Tools

The transfer of development rights, unlike other farmland protection tools, really cannot be used just by itself. In fact, TDR requires several complementary tools in order to be successful. As a first step, the county or municipality must have a comprehensive plan that identifies on a map areas of farmland and open space for protection and places where future growth can be accommodated. Next, zoning ordinances must downzone land in the protection areas (the TDR sending areas) to reduce the possibility of nonfarm development. At the same time, zoning should also encourage development in the growth areas (the TDR receiving areas). Land in the sending areas should qualify for farm use-value property taxation to reflect the limited use of the land. And agricultural districts in the sending areas are useful to strengthen the right to farm and discourage the intrusion of public sewer and water.

The TDR program may require a local government TDR bank and/or a local purchase-of-development-rights program to keep an active market for the sale of development rights.

Keeping farming profitable. If farmland preservation is the goal, a local government will be wise to create incentives for agricultural economic development. Preserving farmland through TDRs will ultimately succeed only if the landowners can farm profitably. One of the problems faced by farmers in Montgomery County, Maryland, is the loss of farm support businesses. This makes farming less convenient and increases the cost of operating a farm. Montgomery County may face the threat of having several of its farms become nothing more than rural estates, which contribute little to the farm economy. Counties and municipalities should be willing to invest in forming farmers' markets and to allow some farm-based businesses in their zoning ordinances to help farmers remain on the farm.

Summary

The transfer of development rights, like the purchase of development rights, avoids much of the debate over private property rights and government compensation for land-use restrictions. While a community might downzone lands to be protected, actual TDR transactions occur between willing buyers and sellers. The market price for TDRs will vary along with economic conditions, and a government TDR bank may help to keep the market active.

Successful TDR programs take time, require some public investment, and work best when there is good comprehensive planning and effective agricultural zoning. TDRs require clearly defined areas for preservation (sending areas) and for growth (receiving areas) and a steady demand for new homes. TDRs are attractive because the cost of preservation falls largely on the private-sector developers, who in turn benefit from being allowed to build at higher densities in the designated receiving areas.

CHAPTER 11

Land Trusts

PRIVATE-SECTOR LAND CONSERVATION

> We would rather have a good reputation as people farmers can work with than accomplish immediately all the protections we would like. . . . A conservation easement is as much a community outreach tool as anything else.[1]
> —Dennis Bidwell
> American Farmland Trust

Protecting farmland and open space through private land trusts has become surprisingly popular over the past twenty years. Many people feel that public land-use planning efforts have not adequately protected valuable farmland and natural areas from becoming gobbled up for residential, commercial, and industrial building sites. Local governments compete for development to expand the property tax base, and yet property taxes continue to climb to pay for new development, placing a heavier financial burden on owners of farmland and open space. At the same time, federal farmland protection efforts have been weak and federal park and wildlife programs weakened.

Many rural landowners simply distrust government; they hold dear their private property rights and bristle at the possibility of tighter land-use regulation. For private landowners who are willing to protect land in a *voluntary* way and outside the realm of government, land trusts are a flexible, creative, and successful means of saving important natural areas and farmland.

There are currently more than 1,100 local land trusts in forty-eight states (not yet in Arkansas or Oklahoma) actively involved in the preservation of resource lands (see figure 11.1). Most land trusts focus on their local community, county, or region. Some land trusts, such as the Vermont Land Trust, have a statewide presence. And perhaps the best-known land trusts, the Nature Conservancy, the Trust for Public Land, the Conservation Fund, and the National Audubon Society, are national organizations with large staffs and multimillion-dollar budgets. All together, land trusts have protected nearly 7 million acres of wildlife habitat, scenic vistas, farmland, forests, and wetlands. Collectively, local, regional, and national land trusts have more than 900,000 members.

A 1994 survey by the Land Trust Alliance, a national umbrella organization, found that over sixty land trusts list preserving farmland as one of their top priorities.[2] Some land trusts specialize in farmland preservation; examples are the Marin Agricultural Land Trust (California), the Lancaster Farmland

Trust (Pennsylvania), the Mesa County Land Conservancy (Colorado), the Massachusetts Farmland and Conservation Trust, the Montana Land Reliance, the Ventura County Agricultural Land Trust and Conservancy (California), and the Wisconsin Farmland Conservancy. The American Farmland Trust, based in Washington, D.C., is the only national land trust devoted to farmland preservation. (See table 11.1.)

The guiding principle of land trusts is stewardship of the land, to protect and pass along special environments in good condition to future generations. By comparison, the free market favors short-term decisions over land resources, based on the perception that land is a commodity to be developed for the highest and best use. The free market is not good at making decisions about resources between generations. For example, prime farmland may be more valuable fifty years hence than it is today, but there is no way for the current market to reflect that.

What Is a Land Trust?

A land trust is a private, nonprofit organization whose primary purpose usually is the direct protection of natural areas and open space. A land trust receives an official tax-exempt status from the Internal Revenue Service under section 501(c)(3) of the Internal Revenue Code. Land trusts are considered educational and "charitable" organizations with bylaws, a board of directors, and a paid or volunteer staff.

A land trust may receive gifts of property and donations of development rights, and a land trust may purchase property and development rights to land. (Land trusts prefer to use the term conservation easements rather than *development rights*.) And a land trust may resell land and *conservation easements*, usually to a local, state, or federal government agency. Because a land trust is a tax-exempt organization, donations of money, property, and conservation easements qualify as federal income tax deductions for the donors.

The mission of each land trust is defined in its bylaws. Some land trusts seek to protect land that has natural, scenic, or recreational value. Other land trusts focus on protecting historic landscapes and buildings. And still others are involved in protecting farm or forest lands. The American Farmland Trust, for

TABLE 11.1
Farmland Preserved by Major Land Trusts

Land Trust	Farmland Acres Preserved
Montana Land Reliance	76,347
American Farmland Trust	40,266
Vermont Land Trust[a]	36,580
Marin (CA) Agricultural Land Trust	25,600
Columbia (NY) Land Trust	6,647
Napa County (CA) Land Trust	6,050
Lancaster (PA) Farmland Trust	4,300

[a]With the Vermont Housing and Conservation Board.

example, has the dual purpose of stopping the loss of productive farmland and promoting farming practices that ensure a healthy environment.

Land trusts are not "no-growth" crusaders. In fact, it is not uncommon for a land trust to become involved in a "limited development" project where some land is protected and some built upon. Some land trusts also offer consulting services, much like a private for-profit firm. Land trusts, like other 501(c)(3) organizations, can engage in legislative lobbying without jeopardizing their tax status (and many do), but they cannot engage in partisan politics or support candidates.

A land trust is not a financial or estate "trust." To avoid this confusion, some land trusts refer to themselves as conservancies, such as the well-known Brandywine Conservancy, which has protected more than 22,000 acres in the Brandywine Valley of Pennsylvania and Delaware. Some land trusts are called foundations, such as the Iowa Natural Heritage Foundation, which has protected over 10,000 acres of forest, prairie, and recreation lands.

The Origin of Land Trusts

The first land trust was started in Massachusetts in 1891 under the name of the Trustees of Reservations. Today, the Trustees has ownership of over 17,000 acres and has protected another 7,000 acres with easements. The Trustees has grown into a professional organization with more than one hundred paid and volunteer staff. It is not surprising that land trusts began in the Northeast and that the majority of land trusts are found there today (see figure 11.1). The Northeast has relatively little publicly owned land and is densely settled, particularly along the Boston-to-Washington corridor. The large, growing, and

FIGURE 11.1
Land Trusts throughout the United States, 1992

affluent population has sprawled across the landscape, making open space and farmland increasingly scarce.

The land west of the Rocky Mountains was largely in government hands in the nineteenth century, and vast amounts of many western states are still under federal control. But the land trust concept spread west in the 1970s and 1980s, and impressive land protection gains have been made. With the help of the state-funded California Coastal Conservancy, land trusts have protected over 18,000 acres of California's scenic coastline. The Jackson Hole Land Trust, founded in 1980, has protected over 5,000 acres of ranch and forest lands near spectacular Grand Teton National Park. The Jackson Hole Land Trust has also pioneered a program to seek "conservation buyers" who are willing to purchase land in the area for limited development or permanent preservation. The Marin Agricultural Land Trust, operating in Marin County, north of San Francisco, has protected over 25,000 acres of farm and ranch land, the second highest number of farmland acres protected among counties nationwide.

One indication of the popularity of protecting land comes from the Colorado Cattlemen's Association, which formed its own land trust in 1995. Director Reeve Brown explains, "The mainstream agricultural community is often at odds with preservationists who say the best way to protect land is to leave it alone. The problem is, that doesn't leave room for agriculture. So despite the merits of land trusts, most ranchers ignored the message because of the messenger."[3]

But now farmers and ranchers whose land is restricted to agricultural uses are no longer oddities.

In 1982, the Land Trust Alliance (originally the Land Trust Exchange) was formed as the national organization to foster the creation and growth of local and regional land trusts. The mission of the Land Trust Alliance is to strengthen the land trust movement through publications, conferences, and legislative action and to function as a national clearing house of information about land trust activities.

Jean Hocker, president of the Land Trust Alliance, attributes the popularity of land trusts to more people coming to understand that open space is more than an attractive landscape; more people are aware of the long-term economic, social, and environmental results of farmland loss in metropolitan regions.

The Land Trust Structure

A land trust should follow some basic rules of organization and management. Those rules will help all those in the land trust to understand their roles and will lead to a coordinated effort to protect land. The following rules of operation are recommended for land trusts:

1. A land trust must have clear goals and purposes. These should be spelled out in the bylaws. Also, staff and the board of directors should regularly review the goals and activities of the land trust.

2. The board of directors must realize that they (not the staff) are legally responsible and accountable for the actions of the organization.
3. Board members must be careful to avoid conflicts of interest. Otherwise, the image of the land trust could be tarnished, hampering the trust's activities.
4. A land trust must understand and fulfill its legal requirements as a nonprofit organization. The land trust must prepare articles of incorporation and bylaws that conform to state law and file them with the state. The land trust must also file an Application for Recognition of Exemption with the IRS in order to receive tax-exempt status as a nonprofit charitable organization. Each year the land trust must file a federal 990 form with the IRS.
5. A land trust must conduct fundraising activities in an ethical and responsible manner.
6. The board of directors must be absolutely certain that the land trust manages its finances and assets in a thoroughly responsible and accountable way. Clear financial records are important, and an annual audit by a certified public accountant is recommended.
7. A land trust with a volunteer staff, supplemented with paid consultants when needed, must have sufficient skills and personnel to carry out its programs.
8. A land trust with paid staff must ensure that staff members are responsible and have the skills and support to do their jobs.

Operating a Land Trust

Perhaps the greatest challenge that land trusts face is raising money. Land trusts rely on membership dues, donations, and whatever foundation or government grants they can bring in. Nearly half of all land trusts operate with a volunteer staff and an annual budget of under $10,000. Most land trusts have preserved fewer than 1,000 acres. But then over one-third of all land trusts have been formed since 1984.

The Marin Agricultural Land Trust (MALT), for example, has attracted money from several sources. Funds from the California Coastal Conservancy helped it get started. Then the voters of California approved Proposition 70 in 1988, which provided money for land conservation projects, including $15 million for the purchase of easements in Marin County. MALT also has received funding from a private foundation. Attempts in 1992 and 1996 to obtain funding through a tax on each land parcel in the county won clear majorities in public referendums but fell short of the necessary two-thirds majority.

Another challenge that land trusts face is educating landowners about the financial benefits of land preservation. A land trust can help landowners obtain assistance in estate planning, tax and conservation law, and environmental and land-use planning.

Land trusts also face the challenge of putting together land protection strategies for a community or region. Because both donations and sales of conservation easements are voluntary, owners of important properties may choose not to participate. This may hinder linking properties to create trails and greenways along rivers and streams. Also, a critical mass of lands should be protected to sustain wildlife habitat or create a core of preserved farmland that will maintain farm support businesses. The voluntary aspect of conservation easements may be especially frustrating in the process of negotiating a mutually agreeable easement price with a landowner.

A land trust must be careful in selecting projects. In accepting or purchasing property or conservation easements, a land trust is assuming a long-term legal obligation to manage land or monitor an easement. A land trust should have the financial and personnel resources to uphold that obligation, or else a property could decline to a point that it no longer meets conservation purposes; similarly, if a land trust does not regularly and properly monitor an easement, the validity of the easement could be challenged in court.

A land trust should set up a long-range management plan and monitoring fund for each property, and many land trusts require that the donor of a property or conservation easement also give money for setting up such a fund. Also, in the easement document, the land trust should name a back-up land trust or government agency that could be assigned the responsibility for monitoring and enforcing the easement if the land trust were to become financially unable or cease to exist.

Before a land trust purchases a property or accepts a conservation easement, the trust staff or an outside consultant should make a thorough inspection of the property and a title search. By purchasing a property or acquiring an easement—an "interest" in the property—a land trust becomes part of the chain of title and hence is potentially liable for accidents or the clean-up of hazardous waste.

The term *due diligence* describes the efforts of the land trust to find out if there are any problems with the property before a purchase or an easement is finalized. Banks commonly require a phase one environmental assessment before making a large farm loan to ascertain if there are environmental problems such as leaking gas tanks, dump sites, or hazardous waste. If the environmental assessment turns up a problem, it must be reported to the appropriate government agency. But even due diligence on the part of the land trust may not be enough. The Land Trust Alliance offers access liability insurance to over 300 land trusts and has endorsed an insurance program designed for land trusts.

Finally, the land trust must make sure that each land protection project is in the public interest. This is important both for public relations and for meeting charitable donation standards. For example, accepting the donation of a conservation easement on a 5-acre piece of farmland will not endear the land trust to the farming community. Similarly, accepting a conservation easement on land zoned for residential use with sewer and water lines available is likely to raise eyebrows, if not the ire of the development community and the disapproval of the IRS. Land trusts are publicly-supported in the sense that landowners who sell or donate an easement or donate land may qualify for income and estate tax benefits. To qualify as a valid easement donation, the land involved must further a public purpose and be consonant with stated public policy, such as an easement on farmland zoned for agriculture.

Land Trust Practices

Land trusts emphasize voluntary as opposed to regulatory approaches to land management and preservation. Land trusts can use several tools to protect land (see box 11.1). But in protecting farmland, land trusts work mostly with conservation easements. By purchasing a conservation easement, by accepting the donation of a conservation easement, or through a bargain sale of part cash payment and part donation, a land trust acquires an interest in a property.[4] This way, the land trust can protect land at a cost far below its fair-market value.

Bob Berner, long-time executive director of the Marin Agricultural Land Trust explains, "The easement is designed specifically to allow for any kind of agricultural operation. By holding the easement we do not become a landlord. We do not determine the kind of agricultural operation."[5]

The conservation easement is a formal document that describes the restrictions on the use of land.[6] The easement is attached to the deed and runs with the land either in perpetuity or for a period of time specified in the easement document. *The landowner still owns the land,* but the easement remains on the deed

PHOTO 11.1
Gene Garber, left, points to the farms that he and his father, Henry, preserved through donating conservation easements to the Lancaster Farmland Trust.

BOX 11.1
Land Trust Land Protection Tools

Registry: The landowner signs a nonbinding agreement with the land trust to manage the land so as to protect the conservation values (productive agricultural soils, scenery, wetlands, and rare and endangered plant species). This agreement is not legally enforceable.

Right of first refusal: The landowner signs an agreement to give the land trust the right to match any valid offer to purchase the property, usually within thirty to ninety days.

Limited development: The landowner or the land trust develops part of a property in order to finance the purchase or preservation of most of the property. Limited development features either large house lots under tight restrictions or clustered housing on the lower-quality land. Limited development is considered the protection tool of last resort.

Lease: The landowner and the land trust sign a formal lease specifying a price and a length of time. The land trust then manages the property according to the terms of the lease.

Option: The land trust pays the landowner a set amount of money in return for an option to purchase the property or an easement on the property for a certain price by a certain date. If the land trust is unable to raise the necessary funds, the option expires and the land trust forfeits the payment for the option.

Life estate: The landowner sells or donates the land to the land trust but retains the right to live on the property and use it throughout his or her lifetime.

Testamentary easement: Upon the landowner's death, through the landowner's will, an easement to the property is conveyed to the land trust. (For an example of a testamentary easement see appendix E.)

Conservation easement: The landowner voluntarily sells and/or donates permanent legal restrictions on the property to limit or prohibit development. The land trust is the "donee" who holds the conservation easement and is responsible for making sure that the current landowner and all subsequent landowners abide by the restrictions of the easement document. The land remains in private ownership.

Undivided interest: The landowner and the land trust become co-owners of the property. This is not very common. The land trust may be taking on liability it doesn't want.

Fee simple ownership: The landowner sells or donates the property to the land trust.

Bargain sale: The landowner sells the property to a land trust at a price below the appraised fair-market value. The landowner may use the difference between the appraised fair-market value and the actual sale price as a charitable deduction for federal income tax purposes.

whether the land is sold or passed on through inheritance. The easement document can be tailored to the specific desires of the landowner who is selling or donating the easement, but generally it will tightly restrict the type and amount of development allowed on the property. Public access to a property under an easement need not be allowed.

Preacquisition: A land trust purchases a property or an easement and then resells the property or easement to a government agency. The land trust should have a written agreement with the government agency before spending money on the preacquisition.

Purchase and resale: A land trust purchases a property, places a conservation easement on it, and then sells the property subject to the conservation easement. A refinement of the purchase and resale is the use of a double-escrow easement in which the land trust buys the land with a downpayment and a note; then the final buyer purchases the property from the land trust (subject to the easement) and assumes the note, payable to the original seller. The title company serves as the escrow holder and transfers title as money changes hands. The advantage of the double-escrow easement is that a land trust can purchase an important property with just a downpayment, rather than the full price. The help of a real estate lawyer is recommended.

Stewardship endowment: When a landowner donates an easement, he or she also donates money to establish a fund to help pay for the monitoring and enforcement of the easement.

Charitable remainder trust: A landowner sets up a charitable remainder trust and through the trust gives property to the land trust. The land trust then sells the property. The landowner receives a tax deduction for donating the property to the land trust, and the proceeds of the sale go to the landowner's trust. The landowner receives the earnings of the trust for his or her lifetime, and upon the landowner's death, the remaining money goes to the land trust.

A remainder trust is best for people with highly appreciated real estate, the sale of which would mean large capital gains taxes; and people with large assets who do not have relatives who are interested in taking over the family farm but want to see the farm stay a farm.

A charitable remainder trust can be combined with the donation of a conservation easement. The landowner could donate an easement on a property to a land trust and receive an income tax deduction. Then the landowner, through a charitable remainder trust, could sell the property to any nonprofit (including the same land trust). In this way, the landowner has preserved the property and set up the remainder trust, taking two income tax deductions and establishing a retirement annuity. The help of an attorney in setting up a remainder trust is essential. (When a nonprofit holds a conservation easement and then receives fee interest, the easement becomes extinguished.)

The easement document is versatile, yet it must be carefully crafted so that the easement will withstand legal challenges. In extreme cases, an easement could be overturned by a court. Assistance from an experienced attorney is advisable, especially in defining permitted uses, subdivisions, and permitted buildings. An indemnification clause to protect against liability for hazardous waste sites is highly recommended. In addition to the easement example in appendix C, sample easement language is contained in *The Conservation Easement Handbook* by Janet Diehl and Thomas S. Barrett (Land Trust Exchange and Trust for Public Land, 1996), available from the Land Trust Alliance.[7]

The purchase or donation of a conservation easement must include a professional appraisal of "easement value," which is the difference between what the property can fetch on the open market without an easement and what the property would be worth with an easement on it.

The Internal Revenue Code permits income tax deductions for easement donations that provide or protect:[8]

- public recreation or education
- important natural habitat
- open space, scenic areas, farmland, and forestland
- historic preservation
- a public benefit (that is, the property must be in keeping with public policy and must not simply be blocking development)

There must also be a real difference between market value and restricted value.

A landowner must donate an easement in perpetuity for the easement to qualify as a deductible gift for income tax purposes.[9,10] The appraisal on which the donation or bargain sale is based must be completed within sixty days of settlement of the easement.

There also may be a reduction in property taxes once a conservation easement has been placed on the property. Property tax assessments are typically based on the highest and best use of a property. If the use of the land has been restricted to agriculture and open space, then the assessed valued of the property should be reduced accordingly.

A third tax benefit, discussed in more detail in chapter 12, is the donation or sale of a conservation easement to lower the value of the property for estate tax purposes.[11] This can be very helpful in passing the farm on to the next generation, with the federal tax rate of 37 percent beginning on estates valued at over $600,000.

The outright purchase of a conservation easement may be expensive, and few land trusts have the funds to do several easement purchases in a year. More often, a land trust will offer a bargain sale of part cash and part charitable donation. A donation or bargain sale makes particular sense for a landowner with a high easement value and a large annual adjusted gross income. The landowner can use the donation portion of the easement value to offset up to 30 percent of adjusted gross income in a year, and the donation can be spread out for up to six years.

A good appraisal is essential to stand up to an audit by the IRS (see box 11.2). The land trust should provide the donor with IRS form 8283 for an easement donation valued above $5,000 (see appendix L). The land trust should include a letter stating that "no goods or services have been provided by the donor." This means that the donor did not entice the land trust into accepting the easement donation.

Once a land trust has acquired an easement, the true land conservation work begins. The land trust assumes a legal responsibility to monitor and enforce the terms of the easement. If an easement is not monitored, violations can defeat the purpose of the easement and a court could overturn it.

Monitoring is time consuming and costly, as the land trust creates and adds to a "baseline documentation" of the property over time. Violation of a provision of an easement could result in the loss of tax benefits received by the landowner, and the land trust holding the easement can require redress of any violation. The land trust may also go to court to compel a landowner to abide by the restrictions of the easement document.

If a land trust acquires a property through purchase or gift, the trust will have to decide whether to manage the property or to sell the property subject to a conservation easement either as an intact property or with limited development. The sale of a protected property to a new owner is quite common.

Steps in Land Protection

No two land protection projects are exactly alike. Each property has unique features, and landowners have different motivations and needs. But there are common steps in how a land trust protects property. Following an orderly process will help make the acquisition of land or conservation easements as smooth and swift as possible.

The first contact with a landowner is very important. Sometimes the landowner will unexpectedly get in touch with the land trust staff. Other times, a member of the land trust board will arrange a meeting between the landowner and the staff. It is usually a good idea to meet at the landowner's property. The landowner will feel more comfortable on his or her own turf. Staff can have a brief tour of the property.

Staff must understand the landowner's motivation in protecting the property along with any financial needs. Notes of meetings and phone calls could prove helpful later when drafting either a sales agreement or a conservation easement. Staff should be prepared to explain the basic tax advantages of conservation easements and how they work with overall estate planning. Staff should not profess to give legal advice but should encourage a landowner to discuss tax and legal questions with an accountant or an attorney. It is helpful to have an attorney, either on the land trust board or as an advisor, who is willing to discuss the legal details of estate planning with landowners.

If it appears that the land trust can play a role, staff should determine the conservation value of the property. This might include: soils, wildlife habitat, wetlands, water resources, historic buildings, topography, farmland, and forestland. Much of this information is available from the county soil survey, county

> **BOX 11.2**
> Conservation Easements and the Appraisal of Property
>
> An appraisal is an educated guess of value. The purpose of an appraisal is to determine how much a property is worth and why. A property appraisal is based on the concept of highest and best use, which according to appraiser Leland T. Bookhout, means "the reasonably probable and legal use of vacant land or improved property, which is physically possible, appropriately supported, financially feasible, and that results in the highest value" (Land Trust Exchange, Fall 1988). The highest and best use should include the current zoning of the property and the likelihood of a zoning change to a more valuable use (such as from agricultural to residential use).
>
> There are four ways to estimate highest and best use: 1) comparable sales—comparing the subject property to sales of at least four similar properties within the general neighborhood or county. The appraiser may make adjustments to the four sales to account for differences in size, location, time sold, soil quality, buildings, etc.; 2) cost approach—estimating the current cost of replacing existing buildings (taking depreciation into account) and adding in land value; 3) income approach—estimating the net cash flow of a property and then using an interest rate to capitalize the income stream over time (Land Value = the sum *n* years Net Income per year divided by the capitalization rate); and 4) subdivision approach—estimating land value when the land can be subdivided and sold. The appraiser determines the potential price of lots minus the costs of creating and selling the lots.
>
> In appraising the highest and best use of a farm property, the appraiser has to choose which one of the four methods to use. The appraiser must also decide whether to include the farm buildings in the appraisal. If the property has significant development potential, the farm buildings (barns, milk house, hog house, chicken house, equipment sheds, grain bins) may detract from the value because they would have to be demolished.
>
> [a]Patrick W. Hancock. "A Question of Value: Appraising for Farmland Preservation," *Farmland Preservation Report*, special report (1992).

planning office, local historical society, or state department of natural resources. Staff should then identify the appropriate land protection tools (see box 11.1).

At each stage, staff should keep the board of directors of the land trust informed. The last thing anyone wants is for staff to make a commitment to a landowner and then have the board decide not to protect the property.

A crucial step is the appraisal of the property's fair-market value or conservation easement value. An appraisal of fair-market value is the basis for negotiating a sale price with the landowner. An appraisal of the conservation easement value will enable the landowner to estimate potential income and estate tax benefits from donating an easement.

It is important to hire a professional real estate appraiser with a good track record. If the landowner will be donating an easement and taking a charitable income tax deduction, *the appraisal could be audited by the IRS*. Also, the landowner should pay for the appraisal, because the landowner will be receiving the tax

Appraising Conservation Easements

The value of a conservation easement is the difference in the value of a property before and after a conservation easement is placed on it. The highest and best use is the before-easement value. An appraiser can estimate the after-easement value using any one of the four appraisal methods.

The shortcoming in using the comparable sales approach for determining the after-easement value is that there may be few or no sales of farms under conservation easements. An appraiser may have to use farms that have no road frontage or are remotely located.

The income approach is often a good reflection of the value of a farm restricted to farm use. The appraiser should be aware, however, that a high market price for farmland can also indicate the scarcity of farmland. According to appraiser Pat Hancock, "The phenomenon of sale prices of easement encumbered properties exceeding value estimates generated by the Income Approach has been demonstrated repeatedly in locations such as New Jersey, Long Island, the Hudson Valley (NY), and Lancaster County, PA."[a]

We recommend the comparable sales approach to estimate after-easement value, or the income approach combined with at least two after-easement sales.

A land trust should discuss the appraisal with the appraiser before presenting it to the landowner. The purpose of reviewing the appraisal is not to make the value as high as possible to please the landowner but to produce a well-supported estimate of value. The IRS may challenge an appraisal, and a land trust will look bad if it develops a record of questionable appraisals. A good appraisal is essential for farmland preservation, whether it is for the purchase of development rights by a government agency or the donation of a conservation easement to a land trust.

BOX 11.3
The Land Protection Process

1. Land trust staff meet with the landowner and visit the property
2. Identify the landowner's needs
3. Research the conservation value of the property
4. Identify the appropriate land protection tools
5. Appraise the fair-market value or the easement value of the property
6. Negotiate the purchase price (bargain sale potential)
7. Negotiate the details of the conservation easement or land purchase
8. Conduct the title search and contact lenders to sign subordination agreements
9. Obtain land trust board approval to accept or purchase the conservation easement or property
10. Develop a management plan for the property
11. Raise the money and purchase the property or easement
12. Monitor the property each year
13. Keep records of visits to the property (baseline documentation)
14. Enforce the terms of the easement, if necessary

benefits based on the appraisal. In order for the landowner to receive an income tax deduction, the appraiser must sign an IRS form 8283, which the landowner submits along with his or her income tax return for the year in which the easement or property donation is made.

Once the financial details have been agreed upon by the landowner and the land trust, staff, with the help of an attorney, should draft the terms of the property sale or sale or donation of a conservation easement. (For a sample conservation easement, see appendix C.)

The terms should spell out what property interest, fee simple or a conservation easement, is being transferred to the land trust. The price or conservation easement sale price of the property, conservation easement sale price, or donation value should be mentioned. The property should be identified by deed book and page and the name and address of the landowner. A legal description of the property, based on a prior or new survey of the property boundaries, should be included as an exhibit to the sale agreement or conservation easement.

The sales agreement or easement document must spell out the restrictions on the use of the property. For example, "no residential, commercial, or industrial structures may be built other than a replacement dwelling."

The next step is to conduct a title search of the property. The title search will show:

- the current legal owners
- previous owners of the property
- any mortgages and the mortgage holders
- any liens, judgments, tax claims, pending legal suits (*lis pendens*)
- a copy of the landowner's deed, which contains a legal description of the boundaries of the property
- whether any mineral rights or water rights have been sold off

It is a good idea to have a computer plot made of the legal description of the boundaries to show how accurate the boundaries are. Most properties that have not been surveyed in over thirty years will require a new survey. If the boundaries have more than one foot of error for every five hundred feet, you should have a new survey done and a new legal description written up. The new survey should be accurate to one foot of error for every ten thousand feet in the boundary. Even though survey work is expensive (usually a few thousand dollars), an accurate boundary plot and legal description will identify without a doubt where a protected property is and will help avoid boundary disputes with neighbors.

If the property is being sold to the land trust, all outstanding liens, judgments, tax claims, and lawsuits against the property must be satisfied prior to the sale. Any mortgages must be removed as part of settling the sale. If a conservation easement is being sold or donated, liens, judgments, tax claims, and lawsuits against the property must be satisfied before the easement can be conveyed to the land trust. Mortgages do not have to be removed, but mortgage holders must sign a subordination agreement. The agreement means that if the mortgage holder were to foreclose on the mortgage, the easement would not be removed. The mortgage holder is subordinating his or her claim on the property behind the land trust's easement.

Occasionally, a title search will show that the mineral rights have been sold off the property. This poses a very real problem, because the holder of the mineral rights has the right to search for minerals on the property. A land trust certainly doesn't want a preserved property being dug up. An attorney may be able to help the land trust find out who owns the mineral rights and whether the holder will sell them or give them back to the current landowner. Some mineral rights will have been acquired many years ago and may be impossible to trace; in that case, a judge may have to declare the mineral rights invalid. (Similarly, in western states, the lack of water rights could jeopardize the use of the property for ranching or farming.)

If fundraising is needed to purchase the property or an easement, the land trust board should approve and assist with the fundraising effort. Special events, solicitations from members or major donors, auctions, or grant proposals can be explored.

If the project is a donation of a property or an easement or if the land trust has the necessary funds on hand for a purchase, the land trust board should formally vote to accept or purchase the conservation easement or property.

All legal documents—the contract of sale, the easement, and any mortgage satisfactions or subordination agreements—should be reviewed by an attorney. It is a good idea to have an attorney representing the land trust present at any land or easement settlements.

For purchasing a property, the following costs can be expected:

- legal fees
- cost of a survey (depending on the accuracy of the boundary)
- cost of an environmental audit
- cost of title insurance
- closing costs (recording the deed, etc.)
- proration of property taxes

The same costs will usually apply for an easement purchase or donation, except for the proration of property taxes.

Staff should draft a management plan for the property and close the deal, either purchasing the property or accepting the easement. In either case, a new deed or deed of easement must be officially recorded with the county recorder of deeds. Any satisfactions of mortgages and subordination agreements must also be officially recorded. Staff should provide the landowner and any mortgage holders with copies of the recorded easement and subordination agreements.

Managing a Property and Monitoring and Enforcement of an Easement

Now the real conservation work begins. If a property has been purchased, it will have to be managed. This may include putting up fencing, building trails, putting up signs, and affording travel to the site. The land trust may also want to consider purchasing liability insurance as protection against any accidents that might occur on the property. If an easement has been acquired, staff should inspect the property at least once a year. A good working relationship with the landowner is key to effective land conservation.

Land trust staff should keep a file of all documents and correspondence about the property, from the initial contact with the landowner to the appraisal of value to the management plan for the property to settling on the property or easement. This record is known as the *baseline documentation* and is key to making an inventory of the conservation value of the property and tracking any changes that are made to the property over time.

The baseline documentation should contain at least the location, acreage, and deed number of the property; photos of the land and buildings, and a map of the property; aerial photos; the recorded easement and subordination agreements (if any); and a copy of the appraisal of the property. New techniques, such as identifying the property on a geographic information system map or through satellite imaging may be helpful.

Staff should visit the property each year to see that the landowner is adhering to the terms of the easement or any management agreements. Friendly relations and open communication between staff and the landowner are essential for long-term land protection. Landowners change over time, and the monitoring process is a useful way to educate new owners about the easement and land management. Another good idea is a land trust newsletter to keep close contact with landowners.

The following easement monitoring procedures are from the Maryland Environmental Trust:

1. Staff contacts the landowner by letter, stating the purpose of the visit, the policy of advance notice, and that it is not necessary to enter the home.
2. Staff makes a follow-up phone call to schedule a visit to the property.
3. Staff reviews the landowner's file, especially photos, the terms of the easement, and the condition of the property at the time the easement was acquired and during subsequent monitoring visits.
4. Staff visits the property and preferably meets with landowner or farm manager.
5. Staff completes a monitoring report form. If the landowner is not present, a follow-up phone call may be needed to complete the report.
6. Staff photographs the property and notes the position and camera direction on a property map. The purpose is to show the condition of the

property at the time of the visit. Changes in structures and land uses are particularly noted. Recently, some land trusts have used video cameras to record visits, especially if a violation of the easement is found.
7. Staff places the monitoring report and photos in the landowner's file and a copy in an annual monitoring file.
8. Staff sends a thank-you letter and a copy of the monitoring report to the landowner. If a violation has been found, staff should attempt to resolve the problem with the landowner before taking the matter to the board of trustees or to court.

No one enjoys playing the role of the enforcer. But the land trust has assumed a legal obligation to ensure that the terms of the easement are being met. Ignoring violations can cause legal problems for the land trust, put easements at risk, set a poor precedent, and give the land trust a bad reputation. If a landowner is reluctant to make changes to redress violations, the land trust has three choices: 1) arbitration by a third party; 2) mediation between the land trust and the landowner; and 3) litigation. It is usually wise to stay out of court, if possible. Both arbitration and mediation show a willingness to work things out. It is important to bargain in good faith. The easement may need to be amended. A good rule of thumb is that amendments to the easement should either be neutral or strengthen the easement.

If it is necessary to go to court, it is very important to have a solid case. Court is expensive, and the reputation of the land trust may be at stake.

Public-Private Partnerships

Public-private partnerships between land trusts and government agencies can be useful and effective in protecting and preserving farmland and open space. Public-private partnerships need to happen! There are strengths and weaknesses to both government and private nonprofit approaches to land protection (see table 11.2). Ideally, the strengths of one organization can cancel out the weaknesses of the other and result in successful land protection.

The U.S. Department of the Interior, through the Fish and Wildlife Service and the western-based Bureau of Land Management, purchases land for public use, as does the U.S. Forest Service in adding to national forests and parks. Often when valuable properties have been threatened with development, government agencies have sought the help of local land trusts to acquire the property for eventual resale to the government.

Since 1965, the federal Land and Water Conservation Fund (LWCF) has fostered cooperation in land protection efforts between land trusts and state and federal government agencies. Funding for the LWCF has dwindled from $800 million a year in the late 1970s to about $300 million a year in the 1990s. The decline in federal funding for land protection is a major reason so many land trusts have been formed.

A State-Level Public-Private Partnership

Perhaps the best example of a state-level public-private partnership is that of the Vermont Housing and Conservation Board, a state agency, and the Vermont Land Trust. An often heard criticism of land protection is that it drives up the cost of housing. But so does a growing and more affluent population.

Agriculture contributes more than $500 million a year to the Vermont economy, and the scenery that farms provide is the foundation for a thriving tourism industry. Since the completion of interstate highways and major ski areas in the 1960s, Vermont has become a popular area for vacation homes owned by residents of the Boston-to-Washington urban corridor. Rural land and housing prices have risen in response, often beyond what local Vermonters can afford.

In 1987, the state of Vermont formed the quasipublic Vermont Conservation and Housing Board (VCHB) to finance both affordable housing and land conservation projects. The VCHB makes grants for low-income housing projects so that preservation is balanced with development for those who would otherwise be excluded from the housing market.

The VCHB has received an annual funding in the state budget from bonds and a share of annual real estate transfer tax revenues. The Vermont Land Trust has served a "packaging role" in presenting preservation projects to the VCHB and arranging financing of the easement purchases.[12] The VCHB approves eligible projects and then funnels state money through the Vermont Land Trust to purchase development rights to farm and forest land. The Vermont Land Trust has helped the VCHB preserve over 24,000 acres of farmland. Since 1980, the Vermont Land Trust, acting on its own and through partnerships with the VCHB and other private land trusts, has protected about 50,000 acres of farmland and more than 50,000 acres of natural areas and lands with scenic, recreational, and historic values. These accomplishments have made the Vermont Land Trust a model of what a land trust can achieve in a relatively short amount of time.

The goal of the VCHB is to preserve a critical mass of farmland to enable farm support businesses to remain profitable. So far, the results have been impressive. The town of Swanton in northwestern Vermont has over 4,000 acres of preserved farmland; and farther south in the Champlain Valley is a contiguous block of nearly 3,200 preserved acres around the town of Orwell. Farmers in these areas can see the future of the land in farming and will be willing to invest in their farms.

Often the VCHB, the Vermont Department of Agriculture, and the Vermont Land Trust jointly hold the easement on a farm. The Vermont Land Trust has formed a subsidiary, Vermont Conservation Lands, Inc., to conduct monitoring and enforcement of preserved properties. The subsidiary separates the Trust's functions and reduces the exposure of the Trust for any liability claims.

The VCHB also makes organization-building grants to land trusts and offers technical assistance, similar to a program of the California Coastal Conservancy. All told, the VCHB has protected over 78,000 acres of land and created or protected 3,700 units of affordable housing.[13]

TABLE 11.2
The Strengths and Weaknesses of Public-Private Partnerships

Strengths	Weaknesses
Government	
Financial resources	Possible voter approval
Regulatory power	Slow response in urgent situations
	Rigid easement
Land Trust	
Ability to respond quickly	Limited funds
Flexible easement document	Provides islands of protection
Private negotiation	
Public-Private Partnership	
Joint funds	Need for agreement on easement language
Advance acquisition	Shared monitoring?
Quick response to urgencies	Different priorities possible
Good public relations	Turf and ego conflicts possible
Client referrals	Overreliance on government funds
Long-term relationship	
Coordinated acquisitions/ common goals and vision	

A Local Public-Private Partnership

Lancaster County, Pennsylvania, and the Lancaster Farmland Trust have developed a public-private partnership to broaden and coordinate farmland preservation efforts. In 1990, the county and the Trust signed a cooperative agreement that has resulted in joint easement purchases on two farms and advance easement acquisition on two other farms. In the advance acquisition arrangements, the Farmland Trust assigned the easement to Lancaster County in exchange for reimbursement of the easement price and related expenses. In one purchase arrangement, the Trust and Lancaster County first signed an option to purchase an easement from the landowner. This "froze" the property until the Trust and the county could come up with the funds to purchase the easement. In addition, easement acquisitions by the two organizations frequently complement each other in creating large contiguous blocks of preserved farmland.

By working together, the Preserve Board and the Farmland Trust have built up strong community support. A 1995 poll by a local newspaper showed that farmland preservation was second only to crime among the concerns of county residents.[14] Farmland preservation scored higher than traffic, education, and taxes.

Several Pennsylvania counties have copied the cooperative agreement, and Chester County and the Brandywine Conservancy have jointly preserved three farms. A copy of the cooperative agreement is included in appendix H, and appendix I shows a sample reimbursable agreement.

One word of caution about public-private partnerships: They should not be based only on a handshake or a conversation. The parties involved should always put in writing what is expected of whom in working out specific land protection deals.

The Quasipublic or Municipal Land Trust

Several communities have recognized the success of land trusts in working with private landowners. Yet the limiting factor for most land trusts is money. By combining the taxing and funding power of government with the grassroots style of the land trust, several communities have created public agencies that operate like land trusts. These agencies have protected thousands of acres of important lands.

Perhaps the most impressive results have been achieved in Boulder, Colorado, where more than 17,000 acres of open space have been protected to form a buffer around the city and highlight the majestic Rocky Mountains. The city of Boulder passed a one-cent local sales tax and sold bonds to raise funds; and the city set up an independent agency, the Open Space Board of Trustees, to purchase and manage the open space.

According to Denis Nock, president of the Boulder Chamber of Commerce, businesses move to Boulder despite the higher land costs and taxes because of the open space and livability.[15]

Limited Development

Limited development on farmland and open space has gained increasing interest among planners, landscape architects, real estate developers, and even preservationists. Limited development generally takes one of two forms. One is the subdivision of part of the farm into a few large residential lots, with the remaining ground kept in farm use (see figure 11.2). The other involves the clustering of several to dozens of residential units on part of the farm and retaining sufficient land for a farm operation or at least preserving open space.

There are two main advantages to the limited development of farmland. First, the integrity of a farm operation can be maintained and even enhanced through the infusion of capital. Second, the rural character of the landscape and environmental amenities can be retained through sensitive design.

One especially noteworthy use of limited development was undertaken by the Brandywine Conservancy in protecting much of the 5,300-acre King Ranch in southeastern Pennsylvania. The Conservancy formed a limited partnership with outside investors and purchased the King Ranch. The partnership then donated 800 acres to the Conservancy, placed a conservation easement on the rest of the property, and sold large house lots restricted by the terms of the conservation easement.

A potential pitfall of blending limited development with a conservation easement is called *private inurement*. For instance, a developer has good intentions in protecting a stream bank corridor that provides wildlife habitat. The developer chooses house sites that are located away from the stream and have a low impact on the environment. The developer then protects a buffer strip of land

FIGURE 11.2
Limited Development by Dutchess County Conservancy, Dutchess County, New York

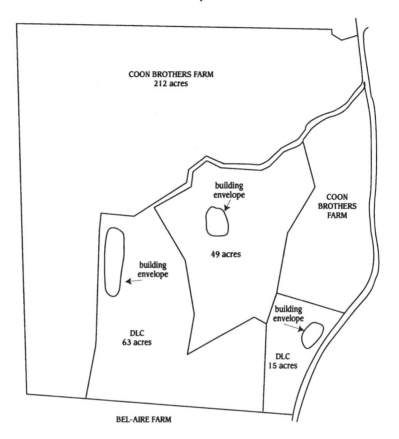

along the stream corridor by donating a conservation easement on the land to a local land trust.

The developer then takes an income tax deduction for the value of the easement on the stream buffer. This income tax deduction helps offset the taxes the developer owes from the sale of the house sites. The price of the house sites may have increased above the typical price in the local market because the sites are next to land that cannot be developed under the easement. If this is the case, the developer has realized an extra cash benefit (an inurement) in addition to the income tax deduction from the easement. The IRS could decide to deny or reduce the value of the conservation easement because it increased the value of the house sites kept out of the easement.[16]

In the land trust community, limited development is considered the protection tool of last resort. There may be financial risks for land trusts and for landowners. Tax deductions for conservation easements may be denied or reduced if the IRS determines that private inurement has occurred in raising the value of building lots kept out of the easement. And there is a potential conflict of interest when a conservation-oriented land trust takes on the role of land developer.

Dennis Bidwell of the American Farmland Trust warns that the public's perception of a land trust can go sour if the trust becomes involved in a limited development. "Some land trusts regret the first deal they did was a limited development deal.... It should be entered into warily after you've done some other deals."[17]

Perhaps the single biggest drawback to limited development projects is that they bring more people out into the countryside. More people bring more pressure on wildlife, water supplies, and open space. Farmers rarely agree on things, but there appears to be a consensus in the farming community that the fewer neighbors, the better. Agricultural and residential uses are simply not very compatible. In short, limited development is a compromise that may or may not work to protect farms and farmland.

Land Trusts in Community Planning

Land trusts can play an important role in comprehensive land-use planning efforts at the local, regional, and state levels. Land trusts can provide leadership and education for citizen involvement in the planning process. Public input on land use and farmland protection is often minimal in the drafting of comprehensive plans. Several land trusts directly or indirectly own resource lands, and intensive development and major public improvements should be kept away. Also, land trusts often have identified vulnerable properties, which should be part of a land-use inventory in the comprehensive plan. Especially at the local and county levels, land trusts can be called upon to assist in the inventory, mapping, and goals and objectives for agricultural and open-space areas.

Land trusts can also acquire property for eventual sale to a public agency for a recreation area or nature preserve; or a trust can purchase development rights to farmland and resell them to a public agency. A trust also can jointly hold development rights to a farm with a government agency.

One criticism of land trust activities is that in preserving land, the value of the land for tax purposes is frozen or reduced. For example, the Maryland Environmental Trust (MET), a state-funded land trust, has acquired easement donations on several thousand acres of farmland. Providing extra incentive to landowners, Maryland provides a 15-year property tax abatement. But given that residential growth does not pay for itself in terms of the tax revenues generated compared to the cost of additional public services, land trusts may actually be providing fiscal benefits. Farmland and open space typically demand few services and generate a net gain in tax revenues over public service costs. A 1995 study in Vermont showed that in towns with a few thousand acres of preserved farmland, the property tax burden for nonfarmers increased by less than $10 a year.[18]

The Future of Land Trusts

The acquisition of properties or conservation easements by land trusts is an important part of a community's effort to manage growth and preserve farmland and open space. Land trusts will continue to acquire conservation easements on farms and on sensitive natural lands that serve as buffers to farmland. Land trusts will also familiarize landowners with conservation easements and other land protection tools that can help with estate planning and farm transfer.

Although most land trusts are rather new, they hold promise for influencing the location, rate, and timing of development by protecting a limited amount of resource lands in areas where zoning, agricultural districts, and property tax breaks may not be sufficient to withstand development pressures. In this way, land trusts can work with local governments in a public-private partnership for managing growth.

Zoning and property tax breaks will continue to be the first line of defense in the effort to protect farmland and open space because these techniques can easily be applied to thousands or millions of acres. Zoning is virtually costless to the public purse, and property tax breaks do not create a major shift in tax burden, at least in the short run. But these public-sector techniques are designed to be flexible and to "manage growth," rather than provide permanent protection of the land base. Land trusts remain attractive because of the long-term protection they can provide for special properties.

As the nation's population rises and the competition for land intensifies, land trusts will continue to grow in number and play an increasingly significant role in land resource management throughout the United States. Land trusts have a proven record of success in channeling the desires of private citizens into effective land protection. All levels of government are struggling to fund a greater variety of public needs. But several states have allocated millions of dollars for land acquisition. Nearly all of these acquisition programs include a role for private nonprofit groups.

Summary

The end of the frontier is a concept that Americans have been reluctant to accept over the past one hundred years. Yet as the nation's population continues to grow, farmland and natural areas will come under more pressure from human activity. The modest and growing success of the land trust movement reflects a widespread desire for the protection of unique lands, plants, and animals well into the next millennium. Land trusts deserve the attention of public officials, private landowners, and citizen groups.

A very real attraction of land trusts is that they may offer more permanent protection of farmland and natural resources than a government agency, public land-use regulations, or fee-simple private ownership. Land trusts can play a complementary role in the comprehensive planning process, especially in determining where development should go and which lands should remain protected from development. Land trusts can be effective partners in protecting and preserving important land resources, and in educating landowners and the public about land stewardship. The federal government and thirteen states have already recognized the important role that land trusts can play and have included funding for land trusts in programs for acquiring and protecting land.

CHAPTER 12

Transferring the Farm and Estate Planning

> How is it possible that a man who owns 240 acres of prime farmland just outside the city, the envy of developers, should worry about passing it along to his children intact?
> —John Hildebrand
> *Mapping the Farm*

> I've gradually paid off more and more of the farm and have devised a long-term estate plan with my parents. My parents' wills reflect their initial warning about being "land rich." I get the remaining farmland and my sister and brother get the cash.
> —Mas Masumoto
> *Epitaph for a Peach*

No one likes to think about death or dispersing one's assets. That is why most people, including farmers and owners of open space, have not done adequate planning for who shall inherit their property or how to structure the transfer of the farm to the next generation. California peach and raisin grower Mas Masumoto notes that even when the parents are no longer operating the family farm, "most of the younger farmers delay confrontation with siblings and allow fate to dictate inheritance."[1]

But with the average age of American farmers at fifty-four and one-fifth of the nation's farmers over sixty-five—including women who have outlived their husbands and hold substantial farm assets—questions about what will happen to the farm are becoming urgent. The answers may take months and even years for a family to work out. But in the end, a well-crafted succession plan or estate plan can remove considerable doubt and conflict about the family farm or ranch. In fact, the future of the farm may very well depend on the family starting an estate plan as soon as possible.

The U.S. Department of Agriculture has estimated that between 1992 and 2002, over 500,000 of the nation's 2 million farmers will retire, and only 250,000 new owners will enter farming.[2] This prediction includes the transfer of millions of acres and hundreds of thousands of farms to new owners. And each transfer is unique.

Nearly all farms and ranches in America are family owned, and there are many challenges in a successful transfer to a family member or sale to someone outside of the family. Rising land values and high federal estate taxes have been the driving forces behind the need to do estate planning. *In many cases, the property has become so valuable that the heirs have to sell it for development in order to pay the inheritance taxes.* Where thirty years ago a simple will may have been enough, owners of farms and ranches are now entering into carefully crafted estate and business transfers.

Why Estate and Transfer Planning Is Needed

A good piece of advice for farmers is to approach estate and transfer planning as part of a total business plan (see box 12.1). Estate planning and transferring the farm involve important and often difficult family decisions.

Parents want to be assured of a comfortable retirement, yet they may want to see the next generation have the farm without assuming a heavy debt load. And the parents want to avoid a hefty tax bite in transferring the farm.

Parents often want to provide an inheritance for all of their children, not just the ones who will get the farm. But equal shares and fair shares for inheritors may be quite different. A very real threat to a farm is a will that simply divides the farm among the heirs, who are then forced to sell part or all of the farm to pay death taxes. Or even if there is enough cash to pay the taxes, the farm may have to be subdivided or sold because one or more heirs want their share in cash, not land.

For one spouse to will the farm tax free to the other spouse is not a long-term solution, either. There is currently no gift or estate tax on transfers from one spouse to another, but when the second spouse dies, the federal estate tax will kick in at 37 percent and up on the value over $600,000 (see table 12.1). Indeed, relying on the transfer of assets to the other spouse can make things worse: assets in the estate of the deceased spouse, which might have been covered by the $600,000 exemption, or taxed at lower estate tax rates, will instead all pile up on top of the second spouse's estate and may be subject to tax at higher rates. Again, the heirs may be forced to sell the farm to pay the death taxes.

TABLE 12.1
Federal Estate Tax Rates and Taxes Due, 1996

Taxable Estate	Federal Estate Tax Rate	Federal Estate Tax
$600,000 or less	0%	$0
$750,000	37%	$55,500
$1,000,000	39%	$153,000
$1,250,000	41%	$255,500
$1,500,000	43%	$363,000
$2,000,000	47%	$588,000
$2,500,000	49%	$833,000
$3,000,000	55%	$1,098,000
>$3,000,000	$1,098,000 + 55% of estate above $3,000,000	

BOX 12.1
Estate Planning: Things to Think About (and Do!)

Where to Start
1. Take an inventory of your assets; find out what they are worth. (A will is not sufficient if you have family assets over $600,000.)
2. Think about long-term goals—retirement, the farm, the next generation.
3. Talk things over with your family and put together an estate plan that meets the family's objectives.
4. Meet with professionals.

Basic Estate Planning Actions to Consider
1. Making a will for each spouse and possibly a living will.
2. Splitting assets between spouses so each can use the $600,000 unified credit.
3. Providing liquid assets for estate tax liability.
4. Establishing a trust, partnership, or corporation.
5. Making yearly gifts of $10,000 to the children ($20,000 if both spouses make gifts).
6. Taking advantage of special-use valuation (Section 2032A) up to $750,000 estate valuation exemption on the farm real estate.
7. Selling or donating a conservation easement on the farm real estate. Real estate is usually valued for estate taxes at its "highest and best" (most profitable) *potential use*, rather than the actual use to which it is currently put.
8. Purchasing life insurance and long-term care insurance. In most cases, life insurance should be counted as part of the taxable estate, although it transfers outside of probate and is income tax free.

A forced sale may compel farming neighbors to buy parts of the original farm just to protect themselves by keeping the land from going into houses. These purchases place an added debt burden on local farmers. On the other hand, an estate sale may result in the farm becoming developed with houses, which makes farming more difficult for the remaining nearby farms.

Farmers may be reluctant to begin estate planning for several reasons. They may not want to endure the costs of an estate plan or share financial information with family members or an outside advisor. Some farmers may feel they are too busy running the farm and can't spare the time.

But not to decide is to decide. John Hildebrand, writing about his in-laws' farm in Minnesota, explains one result of not deciding:

> The only son, who might have taken over the farm at one time, gave up waiting for his father to retire and moved to Wisconsin where he works as an electrical engineer for a super-computer company.[3]

An illness or emergency can strike at any time, and the lack of a farm transition plan or estate plan can cause a family anxiety and feelings of helplessness. As Hildebrand observes:

After years of not thinking about the future, suddenly there's an urgency to act, to do something even if nobody knows exactly what should be done.... Everyone clearly wants to preserve the farm—they don't want to be the ones to undo what three generations before them have accomplished.[4]

The desire to pass the farm on to the children is often the reason that estate planning begins. Each family must think through its wants and needs. Good communication is key. Parents need to talk openly with their children and try to arrive at compatible goals. For example, the parents may be willing to finance a gradual transfer of the farm and even assume chores for an occasional weekend so the younger family can enjoy some time away from the farm. The parents may be willing to expand the farm operation to bring a son or daughter into the farm business. For instance, in 1991, Earl Horning and his son, Jeff, formed a partnership in their Washtenaw County, Michigan, dairy farm. In 1993, the Hornings expanded their herd from 80 to 150 cows and built a double-ten milking parlor and a new loafing barn. The greater herd size enabled the farm to support two families. The transition from the older generation to the younger one is well underway.

A family should seek the advice of an attorney, an accountant, a banker, and a farm financial advisor to explore ways to transfer the farm.

Similarly, an estate plan should be specially crafted for each particular family. Estate tax returns are audited much more frequently by the Internal Revenue Service than individual income tax returns. The family should also keep in mind that once an estate plan is developed, it needs to be updated as circumstances change, such as if one of two heirs decides not to continue farming.

The Basics of Land Transfer and Estate Planning

It is important to have a basic understanding of the challenges and tough decisions that must be made in transferring the farm or open land to the next generation. A good idea is to sponsor a workshop about transfer options and estate planning, featuring presentations by a competent attorney, banker, accountant, and insurance agent.

A good place for landowners to start is to figure out their financial situation:

1. What is the current value of your assets (real estate, equipment, stocks, bonds, CDs, cash, etc.) minus your liabilities (mortgages and other debts)?
2. If you were to die today, what would the federal and state estate taxes be?
3. How would your heirs pay these taxes?
4. What is the value of your assets likely to be if you live another ten or twenty years? Should you consider transfer of the farm now rather than through a will at the time of death?

5. What ways are there to lower the value of your estate so your heirs won't have to sell the land?

The answers to these questions will get landowners thinking about what they want to happen to their land after they are gone. Then they will be prepared to meet with their professional advisors to craft an estate and transfer plan.

Weighing Options and Making Choices

Sooner or later, nearly every farmer—whether operating as a sole proprietor or in a partnership—faces the following choices:

1. Selling the property to the highest bidder for a farm, open space, or for development.
2. Retaining the farm or natural land in the family through:
   - sale to a family member or members or to a young farmer
   - gift to a family member or members
   - combination sale and gift
   - transfer of the property at death by a will
3. Donating or selling a conservation easement to preserve the land for farming or open-space uses.
4. Combining a conservation easement donation with limited development.

We discuss the above choices in general terms, but often the details can be complex, which is one more reason to obtain the guidance of professionals. Each situation is unique, and each option of what to do with the land has different tax consequences.

Selling the Property

If there are no children interested in taking over the farm, or if there are heavy development pressures that impinge on the ability to farm the property, parents may look to sell the farm to the highest bidder. Alternatively, a farm couple may want to help an unrelated young farmer gradually take over their farm on favorable terms. When older farmers consider selling the farm to a family member or an unrelated person, they will typically enter into what is known as a *buy/sell agreement*, which spells out how the transfer will occur.

The two main considerations are the sale price and the income and estate tax consequences. The sale price is a matter of the size of the farm, the location, and the nonfarm uses to which the property could be put according to the local zoning. Tax consequences depend on the landowner's basis and if the sale will be completed in one lump sum or spread out in installments over several years (see boxes 12.2 and 12.3). A farmer may want to consider giving the buyer a mortgage, but that depends on the age of the farmer, his or her financial condition, and the length and size of the mortgage and the interest rate. For example, if the farmer is over sixty-five and in need of cash to pay off debts and set up a retirement fund, then a mortgage of more than ten or fifteen years generally will not be attractive.

Usually, the sale of the farm will qualify for capital gains treatment, which can bring a more favorable tax rate (28 percent) than "ordinary" income. But if the farmer subdivides the farm into several lots and improves the land before sale, at some point the farmer may be viewed by the IRS as a "dealer" in real estate, who does not get the benefit of capital gains rates.

Sale to a Family Member

In selling the farm to a family member, two situations are common. First, the parents have owned the property for a long time, and there is little cost basis left; this means that the parents face hefty federal capital gains taxes if they sell

---

BOX 12.2
Lump-Sum Sale of Farm: Example

	Selling Price	Expense of Sale	Net
250-acre farm × $3,000 per acre	$750,000	$52,500	$697,500
Deduction for fair-market value of farm house if new residence to be purchased or if seller is over 55 years old.	$125,000	$8,750	$116,250
Net selling price of farm	$625,000	$43,750	$581,250
Deduction for basis of farm minus farm house			−$101,250
Gain on sale of farm			$480,000
Federal capital gains tax @ 28%			−$134,400
Net capital gain			$345,600

Net return: $750,000
Expense of Sale − $43,750
Federal Capital Gains Tax −$134,400
$571,850

Note: There may also be state income taxes due on the sale; the level of tax varies from state to state. Depreciation recapture applies generally to equipment and special-purpose buildings, such as poultry houses and hog facilities, to the extent that the depreciation taken was faster than straight line. The sale of standing crops along with the land is taxed as a capital gain. The sale of raised livestock is taxed as a capital gain (though poultry are not considered livestock by the IRS).

the farm in one lump sum. Second, the buyer has limited funds and may not want to buy the property if it will mean carrying a heavy debt load.

A potential solution is an installment sale of the farm. The buyer makes principal and interest payments over a set number of years (usually five to ten years). The seller gets a tax advantage in collecting interest on money that otherwise would have gone up front to pay the gains tax bill. There is no difference between the amount of gains tax paid on a sale for cash and the amount paid on an installment sale, unless gains tax rates change over the installment period (see box 12.3).

A refinement of the installment sale would be to sell the machinery and livestock in installments and then sell the real estate in installments. This

---

BOX 12.3
Installment Sale of Farm: Example

Sale price: $750,000 plus interest payments of $150,000 based on a downpayment of $125,000 and payments of $125,000 a year over five years at 8 percent interest on the balance outstanding.

The farm cost the seller (basis)	=	$101,250
+ the value of the farm house	=	$125,000
+ sales expenses	=	$43,750
		$270,000

Capital gain: $750,000 sale price − cost + value + expenses = $480,000

Ratio of capital gain to the selling price: 0.64

Therefore, each payment represents 64% profit (capital gain) and 36% return of cost and expenses.

Estimated tax due:	Year 1	$22,400
	Year 2	$22,400
	Year 3	$22,400
	Year 4	$22,400
	Year 5	$22,400
	Year 6	$22,400
	TOTAL	$134,400

Capital gains taxes in lump-sum sale: $134,400
Capital gains taxes in installment sale: $134,400

Interest earned (on five years
  of installment payments at
  8% interest): $150,000
Estimated tax on interest: $42,000
Net interest earned: $108,000

arrangement can be an attractive way to transfer the farm gradually to the younger generation. The parents must be assured, however, that the installment payments will be adequate to meet their financial needs.

It is possible to remove the proceeds of the installment sale completely from the parents' estate. The sale can be structured to provide a higher payout while the parents are alive, with the buyer's obligation to make payments automatically ending at their death. This is referred to as a "self-canceling" installment note and can be viewed as a sale of the farm within the family for a life annuity.

Sale to a First-Time Farmer

Some states have an agricultural development authority that can offer special financing for first-time farm buyers through so-called aggie bonds. The farm buyer must have less than $200,000 in net worth. The seller is paid mostly in twenty-year aggie bonds, which produce annual interest exempt from state and federal taxes. The buyer gets an interest rate below the market rate for farm mortgages, and the buyer gets a tax deduction on the mortgage interest paid. The program is a good way to help young farmers get started. If a farm owner does not have children who are interested in the farm and does not want to sell the land for development, sale to a first-time farm buyer may be a way to keep the land in farming. In some states, young farmers looking for land can be matched with retiring farmers looking for buyers through a farm link program, usually operated by the state department of agriculture.

Gift to a Family Member

A gift of land to a family member can be used as part of the donor's $600,000 lifetime unified gift/estate tax credit exemption. A gift of land makes sense only if the parents have considerable assets and could provide for themselves and other heirs without having to sell the farm to the family member.

A one-time gift of land to a family member is very different from the $10,000-a-year gift that can be made by a person to any number of individuals. For example, if each parent gives a $10,000 gift to a child and his or her spouse, a total of $40,000 can be conveyed free of tax *each year*.

Combination Sale and Gift

A combination sale and gift would work as follows. Let's say that the fair-market value of the Bob and Joyce Smith farm is $500,000. The Smiths wish to transfer the farm to their son, Harold, who has been working on the farm since graduating from college ten years ago. The Smiths sell the farm to Harold for $300,000 and use up $200,000 of their lifetime gift tax/estate tax credit exemption.

This method of tranferring the farm is attractive when the parents wish to give favorable treatment to a family member who otherwise could not afford to purchase the farm. The parents should be certain that the sale portion of the transfer will be adequate to meet their financial needs.

Transfer by Will

Transferring the farm to the next generation by a will can have several advantages:

1. The inheritor receives a stepped-up basis to the fair-market value of the farm. That is, let's say that Sam and Sally Jones bought their farm in 1965 for $35,000. On the day of Sam's death, the farm is valued at $450,000. Bill Jones inherits the farm through Sam's will at the $450,000 value. Bill Jones can use this higher value for depreciation purposes as he operates the farm over the years. Also, if Bill were to sell the farm, his basis would be calculated on the $450,000 value minus any depreciation. If he sold the farm for $500,000 shortly after inheriting it for $450,000, he would owe tax on a capital gain of $50,000.
2. Every person is allowed a $600,000 lifetime unified gift/estate tax credit exemption. This means that a husband and wife can pass along up to $1.2 million to heirs without federal estate tax if the estate plan is properly structured. A common mistake is not to use the unified credit for both the husband and the wife, thereby forcing a sale of part or all of the farm to pay estate taxes.
3. Under Section 2032A of the Internal Revenue Service Code, a person who inherits a farm can take the "special use" valuation to exempt up to $750,000 of the value of the farm in an estate under certain conditions:
   - the heirs must elect the special-use valuation within nine months of the owner's death;
   - at least 50 percent of the estate must be made up of farm assets;
   - at least 25 percent of the estate must be farm real estate;
   - the farm must have been actively managed by the immediate family for five of the eight years prior to the owner's death;
   - the farm must remain a working farm and cannot be sold for at least ten years after the inheritor acquires it. If within ten years the farm ceases operation or is sold, the estate taxes that were avoided would have to be paid.[5]

Section 2032A has strict requirements, and we strongly advise consulting an attorney who specializes in estate planning before relying on it.

The $750,000 special valuation can be used in addition to the lifetime gift or estate tax exemption. For example, Joe Wood leaves to his son, Jesse, a farm worth $1,000,000. After Joe's $600,000 unified estate tax/exemption, estate taxes of about $160,000 would be owed on the remaining $400,000. Jesse intends to keep farming, and so he elects to take the $750,000 special valuation. *This way he is able to inherit the farm without paying any federal estate tax.*

If parents properly structure their estate plans, they can take advantage of two $600,000 unified estate tax exemptions and two $750,000 special valuations. Assuming that the farm is worth at least $1.5 million, the parents can transfer $2.7 million to the next generation free of federal estate tax.[6]

4. Even if estate taxes are due, Section 6166 of the IRS Code allows inheritors of a farm to make gradual payment. The catch is that the farm must make up at least 35 percent of the gross estate value. If this test is met, estate tax is deferred for five years, and then the taxes owed can be paid off over the next ten years plus only 4 percent interest.

Section 6166 can help delay or avoid the sale of part of the farm or the entire farm. Inheritors have some time in which to raise cash and arrange their finances to pay the estate taxes.

Life Insurance and Estate Planning

Life insurance can play a helpful role in estate planning by providing cash for heirs to pay estate taxes. There are two main methods for using life insurance in estate planning: 1) an *irrevocable life insurance trust* and 2) a *second-to-die life insurance policy* on two spouses.

The proceeds from a life insurance policy are included in the estate but are deducted under Section 2056 of the IRS Code if the beneficiary of the policy is the spouse of the policyholder. However, if Sally Brown, for example, sets up an irrevocable trust to own the policy on her life, the proceeds are exempted from her estate. The exemption is available only if the insured has no powers of ownership over a policy. Upon her death, the trust can pay interest to her spouse for life and then pass on the principal tax free to her heirs. Or the trust can loan funds to the estate or buy property from the estate for cash, to pay estate taxes.

A life insurance trust is attractive to people with large estates who have illiquid assets, such as a farm, and who don't want heirs to be forced to sell the assets to pay estate taxes. Usually, a life insurance trust will involve a policy worth at least $500,000.

There are several issues to consider about a life insurance trust. First, it is irrevocable. Once the trust is set up and funded, it cannot be changed. Second, trusts are complicated; a good tax advisor will be needed. Third, cash will be needed to pay the life insurance premiums. And fourth, a reliable trustee will be needed to carry out the wishes of the insured after his or her death.

A second-to-die or joint survivorship life insurance policy pays off only after both spouses die. The proceeds from the policy are included in the estate, unless an irrevocable trust is set up to hold the policy. The premiums for the policy are based on the life expectancy of both spouses.

A sizable amount of cash is necessary to fund the policy; for example, a $1 million policy for a healthy couple aged sixty would run almost $25,000 a year for ten years.

Living Trust

A living trust is designed to avoid probate—the court-supervised transfer of ownership that occurs if there is a will or if someone dies without a will—and to make sure that both spouses use their $600,000 exemption. For example, Mr. Jones transfers by deed half of the farm to himself to hold as his own trustee, with full power to amend or revoke the trust at any time. The living trust has no income tax significance; Mr. Jones continues to pay income taxes on the whole farm as before the trust was created. However, because half of the farm is technically owned by the trustee Mr. Jones, rather than by the individual Mr. Jones, when Mr. Jones dies, a successor trustee, named in the trust document, can then take over managing the farm and transfer it to the final heir.

A living trust avoids the expense, delay, and publicity of probate proceedings. It can also help provide continuity in the management of the farm if the person creating the trust suffers disability, insanity, or dementia.

Other Transfer Methods

The downside to transferring property through a will is that inheritors have to wait until the death of the owner to gain control of the property. A gradual transfer of farm assets, first the machinery and livestock and then the real estate, is often a good method. This enables the younger generation to acquire assets without heavy debt, and the older generation retains some control until the younger generation has shown the financial and management ability to take over the entire operation.

A good way to make a gradual transfer is to set up a partnership between the older and younger generations and for the older generation to make $10,000-a-year gifts of farm assets (real estate, equipment, or livestock) to the younger generation.

A similar method is for the parents to form a Subchapter S corporation and transfer the farm assets to the corporation in exchange for shares of stock. This transfer can be done tax free as long as the parents initially retain at least 80 percent of the outstanding shares in the corporation. The parents can then make $10,000-per-child gifts of stock each year. If the parents' assets are large, then upon the death of one parent, $600,000 in stock could be transferred to the heirs free of estate tax, and the balance of the estate could go to the surviving spouse. Upon the death of the second parent, another $600,000 in stock could be transferred free of estate tax.

Most family farm corporations are set up as S corporations, meaning that the shareholders generally pay one income tax, not two—as with the typical stock corporation, where the corporation pays income tax and the shareholders pay tax on dividends. Also, a farm held by a Subchapter S corporation could still qualify for the up to $750,000 estate tax exemption under Section 2032A.

A family corporation or partnership allows ownership to be transferred in small pieces, by gifts of shares or partnership units, without requiring constant filings in the land records. Also, the lack of cash and control can significantly lower the appraised fair-market value for estate tax purposes. Courts and even the IRS have come to accept that owning one-tenth of the stock in a farm corporation is worth less than just the value of the farm divided by ten. In estate planning, the whole and the sum of the parts can be very different.

Estate Planning and the Potential to Merge Private and Public Interests

Land trusts and local and state governments can play valuable roles in helping a family transfer property to the next generation. The sale or donation of a conservation easement on all or part of a property, or limited development together with a conservation easement, can generate tax savings as well as protection for the property. One very real concern that parents may have is that their children not break up or develop the farm they have built up over their lifetimes.

Donation of a Conservation Easement

As mentioned earlier, an easement runs with the land, so that its restrictions apply to future landowners as well as current ones. When a landowner donates or sells a conservation easement, he or she still owns the land and retains the right to sell it and pass it on to heirs. But the land can be used only in ways spelled out in the easement.

A landowner who donates a conservation easement is giving up value. This donation of value will reduce the value of the land in the estate and can lower the estate tax burden. The donation of a perpetual conservation easement may also produce an income tax deduction.

For example, let's say the Fawcett farm is worth $2 million. The Fawcetts then donate a conservation easement on the farm to a land trust. The value of the Fawcett farm after the easement has restricted the use to farming and open space is $1.2 million. The value of the farm for federal estate tax purposes has been reduced by $800,000. The Fawcetts may also use the donation as an income tax deduction subject to a limit of 30 percent of adjusted gross income each year over six years.

The donation of a conservation easement will usually appeal to a high-income landowner for the income tax benefits, or to owners of estates worth over $2.7 million (assuming the use of the unified credit and Section 2032A by both spouses; see page 225).

The value of the conservation easement is the difference between the fair-market value of the property and the value of the property restricted to farming and open-space uses. The easement value must be determined by a qualified ap-

> **BOX 12.4**
> Easement Donation: Example
>
> 200-acre farm:
> $600,000 appraised fair-market value
> $360,000 appraised value restricted to farming or open space
> $240,000 appraised easement value
>
> $200,000 landowner's adjusted gross income (AGI)
> $60,000 maximum deduction (30%) of AGI allowed in one year, but the landowner can spread the donation over up to six years. For example, the landowner deducts $60,000 a year for four years = $240,000 easement value.
>
> The landowner is in the 33 percent income tax bracket, so federal tax savings are worth $80,000. Also, the value of the property on which the easement was donated is reduced for estate tax purposes. Federal estate tax rates begin at 37 percent for an estate valued at more than $600,000. Or if a husband and wife use their unified estate exemptions plus the special valuation, $1.95 million can be exempted from federal estate tax. For example, if the total estate is worth $2.5 million, the $240,000 easement value would save the heirs over $88,000. This would mean total income and estate tax savings of $168,000.

praiser, and the landowner should retain a copy of the appraisal report. The appraisal must include a description of the property, how the fair-market value of the property before and after the easement was determined (fair-market value before − fair-market value after = easement value), a list of the appraiser's qualifications, and how the appraiser was paid by the donor. The appraisal must be completed within sixty days prior to the donation.

The landowner must also file IRS form 8283, "Noncash Charitable Contributions," with his or her income tax return for the year in which the donation occurred (see appendix L).

A conservation easement can also be donated through a will in what is called a *testamentary easement* (see appendix E). A testamentary easement will reduce the value of the land in an estate, but it will not generate an income tax deduction. A testamentary easement is best looked at as a "just in case" technique when a landowner wants to protect the land but may not be able to put a succession or estate plan fully in place before death.

Land Held by a Family Corporation

If a farm or ranch is held by a family corporation (usually a Subchapter S corporation), conservation easements may be helpful in estate and transfer planning, although the donation of a conservation easement is less beneficial for a corporation than for an individual landowner.

The income tax deduction for a corporate easement donation is limited to 10 percent of the corporation taxable income in that year with a five-year carryforward. The extent of the deduction for the shareholders depends on the basis in the stock they own. The donation of a conservation easement may reduce the value of the corporation's assets and lower the estate tax burden on heirs.

Sale of a Conservation Easement

A common problem that farmers face is that the majority of their assets are tied up in land and buildings. The sale of a conservation easement or a bargain sale of an easement (part cash payment and part tax deduction) can enable the landowner to extract cash from the land and buildings without having to sell any property. This leads to greater flexibility in financial and estate planning. The sale of an easement is taxed as a capital gain, and the easement payment can be offset by the basis in the farm land *and* buildings (see boxes 12.5 and 12.6).[7]

For a family looking to transfer the farm, the sale of an easement opens up several possibilities. First, if the parents sell an easement, they then can afford to sell the farm to one of the children at a reduced price. Let's say that Steve and Molly Miller sell a conservation easement on their 150-acre farm for $150,000. Then the Millers sell the farm subject to the easement to their

---

BOX 12.5
Conservation Easement Bargain Sale (part cash, part donation): Example

250-acre farm:
$750,000 appraised fair-market value
$500,000 appraised value restricted to farming or open space
$250,000 appraised easement value
$200,000 bargain sale cash price for easement
$ 50,000 donation

$175,000 basis of the property (cost + improvements − depreciation)
$200,000 bargain sale in cash
−$140,000 deductible basis (80% of $175,000, because $200,000 is 80% of $250,000)

$60,000 taxable capital gain
−$50,000 donation deduction (if AGI is $167,000 or more)
$10,000 taxable capital gain
$ 3,300 tax due @ 33%

Net return is $196,700 ($200,000 bargain sale price − $3,300 capital gains tax due), but no more basis remains in property. Again, the value of the property is reduced for federal estate tax purposes. Also, inheritors get a "stepped-up" basis based on the value of the property at time of death. For example, the property was worth $750,000 before the easement was conveyed through the bargain sale. Five years after the easement was put in place, the owner passes away, leaving just the property to the heirs, and the property is valued by the IRS at $600,000. The basis to the heirs is now $600,000 (not $500,000). The heirs sell the property for $600,000 and pay no capital gains tax.

daughter, Jill, and her husband, Fred, for $300,000. Jill and Fred would have had a hard time buying the farm for $450,000, but they can afford it at the lower price.

The sale of an easement can be combined with other transfer techniques. In the Miller example, the parents could have included a partial gift and conveyed the farm to Jill and Fred for only $250,000. The transfer could have been done on an installment sale over five to ten years. Or the Millers could have formed a partnership with Jill and Fred and made $10,000 annual gifts to both of them over several years. The range of options is another reason that the Millers, their daughter, and their son-in-law should openly discuss their needs and goals and that professional advice is needed.

A bargain sale of an easement may be attractive depending on how much of the bargain sale is cash and how much is income tax deduction. If the cash portion is small, the seller should consider if he or she has sufficient income or estate assets to take full advantage of the income tax donation or estate tax benefits.

Combining Limited Development with a Conservation Easement

In deciding what to do with a farm, the owners may take an "all or nothing" view. That is, they want either to sell the farm as a whole or to pass the entire farm on to the next generation.

In some cases, it may be worthwhile for the landowners to consider the development of a small portion of the farm combined with a conservation easement donation or sale on the majority of the property. The conservation easement reduces the value of the farm "core" for estate tax purposes, may create an income tax deduction, and should protect enough land for a viable farm operation. The land kept out of the easement for limited development, say three house lots, could be sold at some future date to help pay off the estate taxes.

BOX 12.6
Easement Sale:
Example

250-acre farm
    $750,000 appraised fair-market value
    $500,000 appraised value restricted to farming or open space
    $250,000 appraised easement value and cash sale price

    $175,000 basis
    $ 75,000 taxable capital gain

    $ 25,000 tax due @ 33%

The net return on the sale of the easement is $225,000, but no basis remains in the property. The value of the property is reduced for federal estate tax purposes.

> **BOX 12.7**
> Farmland Preservation and Estate Planning: Examples
>
> **Example 1: Like-Kind Exchange with Easement Payment and Sale of Farm to Son**
>
> Joe and Jean Hershey owned a 200-acre dairy farm in Pennsylvania. For several years, the Hersheys' son, John, rented the farm. In 1993, Joe and Jean put the farm into Jean's name only because she was in poor health and the Hersheys' combined assets were over $600,000. Joe and Jean wanted to be able to take full advantage of their $600,000 unified credits ($1.2 million total) in passing on assets to their children.
>
> Later in 1993, Joe and Jean sold a conservation easement on half of the farm to the state of Pennsylvania for $295,000. They used the easement money in a like-kind exchange to acquire rental property to provide a stream of income in their retirement. They paid no capital gains taxes on the easement money. They then sold the preserved half of the farm to John for $250,000, the value of the preserved farmland estimated by the appraiser. Jean paid capital gains tax on the sale of the farm to John, but she was able to use the basis in the farm to offset some of the taxable gain so that the tax due was not large.
>
> *Result:* The Hersheys turned the development rights value in the farm into rental properties to create retirement income and some capital appreciation potential. They also sold half the farm. Their retirement is secure. John Hershey purchased half the farm at an affordable price. The Hersheys may decide to leave the rest of the farm to John and the rental properties to their other children so that all the siblings will feel they were treated fairly in the distribution of their parents' assets.
>
> **Example 2: Spousal Exemption, Easement Sales, and Sale of Farms to Sons**
>
> The Brandt family owned two farms next to each other that were farmed as one. Mr. and Mrs. Brandt applied to sell a conservation easement on their farm, and their sons, Bob and Tom, applied to sell an easement on their farm.
>
> While their applications were being processed, Mr. Brandt passed away. In his

The farm core also could be sold, and, under the terms of the conservation easement, it will remain a farm.

Limited development with the conservation easement can produce a net dollar result that is not much less than the dollars earned from selling the farm to a developer—even though the limited development option carries some extra expenses, such as legal and survey work and possibly the assistance of a land-planning consultant. And with the limited development option, the family retains the majority of the farm for farming or for future sale as a farm.

The limited development option should be chosen only if the development will not impede ability to farm the remaining farm ground. Nonfarm neighbors can be a headache, and the general rule is the fewer, the better. Also, the more house lots that are reserved out of the easement, the less land that remains for

estate, he left his half of the farm to Mrs. Brandt free of federal estate tax (the spousal exemption). Mrs. Brandt then completed the easement sale. The easement funds were free of federal capital gains tax because when Mrs. Brandt inherited half of the farm, it received a new, "stepped-up" basis that was greater than the amount of the easement payment. Mrs. Brandt was able to use the easement payment to set up a retirement fund.

Bob and Tom Brandt then sold an easement on their farm and used some of the proceeds to purchase their mother's farm at a preferential rate of $2,000 an acre. Bob Brandt's two sons are now involved in the farm operation.

Result: The Brandt family used the sale of two easements to establish a retirement fund for Mrs. Brandt and to transfer one farm to the second generation. The third generation of Brandts is now on board as well.

Example 3: Easement Sale and Gifting

Bill and Alice Nelson built up a 200-head dairy operation and did well financially. Shortly before Bill retired, Alice passed away. There are five Nelson children; two of the sons operate the 220-acre dairy.

Upon retiring, Bill sold a conservation easement to preserve the farm for farming. He then used the easement money to begin a gifting program, giving away $10,000 a year tax free to each of his children. When Bill passes away, the conservation easement on the farm will reduce the value of the farm for federal estate tax purposes. Bill's heirs can use the $750,000 estate exemption (Section 2032A) along with Bill's $600,000 unified credit to inherit the farm free of federal estate taxes.

Result: The conservation easement sale enabled assets to be shifted from the older generation to the younger generation. The easement gave Bill Nelson peace of mind because he knew the farm that he loved would remain a farm. The easement will also help to reduce the value of the farm for federal estate tax purposes. This will help Bill's children avoid the common problem of having to sell the farm to pay estate taxes.

---

farming, the lower the value of the conservation easement, and the smaller the reduction in estate value.

Finally, if the landowner claims the conservation easement donation as an income tax deduction, the issue of private inurement may arise. That is, the value of the house lots reserved from the easement may increase because of their proximity to the land protected by the easement. This situation may cause the IRS to challenge the value of the conservation easement donation. In general, the greater the number of house lots reserved, the greater the likelihood for a private inurement situation.[8]

We advise talking with a local or national land trust before embarking on a limited development with a conservation easement sale or donation (see chapter 11). It is important to have a qualified land trust ready and willing to accept the donation of a conservation easement before beginning the limited development. A tax advisor and the land trust can help determine if the easement donation meets the requirements of IRS Code Section 170(h).

Summary

Estate planning and methods to transfer the farm to the next generation are now more important than ever. The increased value of real estate over the past thirty years coupled with high federal estate taxes threatens to force heirs to sell land to pay inheritance taxes. Owners of farms, ranches, and open space should undertake estate and transfer planning as soon as possible to be prepared for the future.

A farm or ranch is a business, and a business succession plan is essential if the business property will be transferred to the next generation.

An important part of the estate or transfer plan is for the family to decide what it wants to do. There are several techniques that landowners can use to minimize estate taxes or transfer the farm to the next generation. The family should consult a good attorney as well as financial professionals to explore options.

CHAPTER 13

Creating a Farmland Protection Package

We have examined public and private farmland protection tools and programs, and we have discussed how different tools can work together for a more powerful and effective approach to keeping land in farming. In this chapter, we summarize the strengths and weaknesses of each tool. We then present examples of strategies for the most common farmland protection situations: the strong farming community, the medium-strength farming community, and the weak farming community. Finally, we warn about some potential pitfalls with farmland protection programs.

The most successful farmland protection programs employ several techniques in a coordinated package and enjoy long-term commitment from landowners, politicians, and the community. Any one technique alone cannot achieve protection for more than the short run. And some techniques, if used alone, can actually encourage development. For example, as described earlier, a property tax break could be used by land speculators to lower the cost of holding farmland while waiting for the value of the land for development to rise.

A strategic package of techniques should be designed to ensure that:

- protection efforts are cost effective: the benefits of protection are achieved at a reasonable cost
- a critical mass of farmland is protected: enough farmland and farmers for efficient farming and to enable farm support businesses to survive
- protection is durable over the long run
- land prices for farm expansion and the entry of new, young farmers are affordable

Each farmland protection technique has advantages and disadvantages (see table 13.1). It is important to understand the potential benefits and drawbacks of these techniques. When a community uses a package of techniques, some of the disadvantages are canceled out. For example, one shortcoming of offering property tax breaks on farmland is that they may go to short-term land speculators; but in states with purchase-of-development-rights programs, the same tax breaks also go to owners of permanently preserved farmland that is limited to farm uses.

Farmland Protection Strategies[1]

Now that you have seen how the individual tools and techniques work, it is time to put them together in a package that reflects a community's goals and

TABLE 13.1
Advantages and Disadvantages of Farmland Protection Tools

Protection Tool	Benefits	Drawbacks
Comprehensive Plan	An organized way to identify good farmland and set growth and protection goals. Serves as the basis for land use regulations.	Not legally binding. May be changed or ignored by planning commission or elected officials as they rule on development proposals.
Differential Assessment of Farmland	Modest incentive to keep land in commercial farming.	Land speculators and hobby farmers may qualify, unless a standard of 25 or more acres or farm income of more than $10,000 a year is used.
Agricultural Districts	Provide exemption from local nuisance ordinances. Often tied to differential assessment. Limits on some other taxes and sewer and water lines. Greater protection from eminent domain.	Strictly voluntary. Landowner may withdraw at any time. Little use near urban areas.
Right-to-Farm Law	Protects farmers from nuisance complaints for standard farming practices.	Does not stop complaints from nonfarm neighbors. May not protect major changes in farm operations or new operations.
Agricultural Zoning	Limits nonfarm development. Can protect large areas of farmland at a low public cost.	Local governments can rezone land out of agriculture or cancel agricultural zoning. Landowners may complain about "equity loss" or the lack of compensation.
Urban Growth Boundaries	Discourage sprawl. Promote more compact development that is cheaper to service.	Agreements on boundaries between cities and counties may be difficult to reach.
Purchase of Development Rights	Provides permanent protection of farmland and pumps cash into the farm and farm economy.	Cost may be high. May be difficult to protect a critical mass of farmland.
Transfer of Development Rights	Developers compensate farmland owners. Creates permanent protection of farmland.	Difficult to establish, especially where development is scattered. Opposition by landowners in receiving areas.
Private Land Trusts and Conservation Easements	Can provide permanent land protection. Can forge public-private partnerships.	Shortage of money. May rely too often on limited development. May create islands of protection, not protect a critical mass or contiguous lands.

objectives of how and where to grow. A "no growth" option is not legal. The package should strike a balance between protecting farmland and accommodating development. This balance depends on local politics, the courts, landowners, the economics of farming, and development pressure.

The three strategies discussed here are designed for the most common farmland protection situations. Above all, a community must decide whether its goal is to protect working farmland or to protect open space. If farmland protection is the goal, the community should employ techniques to encourage the continuation of farming as an industry. Protecting farmland also does protect open space, which is often important for public support. But if farms cannot remain profitable, they will be developed and open space will diminish.

Open space, however, includes a host of nonfarm properties such as small parcels, forests, wetlands, steep slopes, wildlife preserves, parks, and trails. If open-space protection is the true goal, the community should modify the techniques in table 13.1 (especially the type of zoning) in crafting a package for maintaining open spaces.

Strategy One: Maintaining a Strong Farming Community

Area Profile: A strong farming community. Low to moderate development pressure. Land holdings are still mostly in large ownerships of over 100 acres and in contiguous areas of over 1,000 acres. Farm support businesses are adequate. Farmers want to continue farming. This is usually a rural community or an ex-urban community with some long-distance commuting to urban or suburban employment centers.

Recommended Set of Tools:
- comprehensive planning by the township or county
- urban growth boundaries or village growth boundaries
- agricultural zoning of one building lot per 25 acres or per 50 acres; maximum building lot size of 2 acres
- purchase of development rights and/or transfer of development rights
- agricultural districts
- preferential farmland taxation with a stiff rollback penalty for conversion to a nonfarm use
- relief from sewer and water assessments
- right-to-farm law
- agricultural economic development, such as farmers' markets and community-supported agriculture projects

Lancaster County, Pennsylvania

Lancaster County, Pennsylvania, is the only jurisdiction in the nation that has all nine of these farmland protection tools. In 1996, American Farmland Trust recognized Lancaster County's farmland protection efforts with a national achievement award. According to Bob Wagner of the American Farmland Trust,

"Lancaster County is setting the pace for farmland preservation in the United States."[2]

Lancaster covers 603,000 acres of southeast Pennsylvania and contains some of the most productive farmland in the United States. It is the leading agricultural county not only in Pennsylvania but in the entire Northeast, with over $680 million a year in farm goods sold. It is also the nation's number one nonirrigated farming county. Lancaster County farming features the production of dairy products, chickens, and hogs and ranks in the top ten among counties nationwide in the production of these commodities. There are 4,700 farms in the county and about 380,000 acres in farm use. Farm support businesses are strong and serve farmers within a radius of up to 100 miles.

Lancaster County is no stranger to development pressure. The county has a population of 450,000 and is expected to have 545,000 people by the year 2010. It lies sixty miles west of Philadelphia, America's fourth-largest city, and each year the Philadelphia suburbs creep closer. Accommodating development as efficiently as possible without conflicting with the farmland base is a major challenge.

Although county-level planning is essentially advisory, all forty-one townships, nineteen boroughs, and the city of Lancaster have comprehensive plans. The comprehensive plans spell out community goals and objectives for accommodating growth and protecting farmland. The plans show where each community intends to develop land.

To date, a dozen urban and village growth boundaries have been created between boroughs and their surrounding townships. The boundaries indicate where urban-type public services, such as sewer and water, will be provided. Within the boundaries, growth is encouraged, and there should be sufficient buildable land for the next twenty years. The boundaries create a more compact style of development, which is cheaper to service and doesn't waste land. The areas within growth boundaries may eventually serve as receiving areas in transfer-of-development-rights programs. Interestingly, the county's purchase-of-development-rights program has helped to create parts of urban growth boundaries and hence channel development in desired directions, away from productive farmland.

Agricultural zoning is found in thirty-nine of Lancaster County's forty-one townships and covers 320,000 acres or 54 percent of the entire county. Agricultural zoning next to the growth boundaries reinforces the boundaries and makes leapfrog development less likely. Most townships employ a zoning standard of one building lot of up to 2 acres for every 25 acres owned. Agricultural zoning allows some nonfarm uses but of a type and density that does not interfere with neighboring farm operations. Because there have been few rezonings of land out of agriculture, agricultural zoning has dovetailed well with the county's purchase-of-development-rights program and with Manheim Township's local transfer-of-development-rights program. Agricultural zoning helps to buffer "preserved" farms from encroaching development, and it is essential in creating sending areas from which to transfer development rights.

Although agricultural zoning does not offer compensation to landowners for restrictions placed on their land, the purchase or transfer-of-development-rights does provide compensation and has softened opposition to agricultural zoning.

In turn, agricultural zoning has helped to keep the cost of buying development rights within reason. Lancaster County has paid an average of $2,000 an acre for development rights since 1989. This is about one-third the average price paid in neighboring Chester County, which has little agricultural zoning and typically allows one dwelling per 2 acres in rural areas.

The Lancaster County Agricultural Preserve Board and the nonprofit Lancaster Farmland Trust have been aggressive in acquiring development rights to farmland. These organizations have succeeded in preserving over 23,000 acres of farmland, with 19,000 acres held by the county and 4,000 acres held by the Trust. Since 1989, Lancaster County has received over $20 million in funding to buy development rights under a state program, and the county commissioners have added over $7 million. Three blocks of over 1,000 acres of preserved farmland have been formed, and parts of urban growth boundaries have been created by preserving the land for farming. The county has a backlog of 170 landowners who have applied to sell their development rights.

Lancaster County has over 120,000 acres enrolled in voluntary agricultural districts. While agricultural districts are not a substitute for agricultural zoning, they offer attractive benefits at no cost or restriction to the landowner. In Lancaster's case, land enrolled in a district does not qualify for differential assessment. Instead, landowners receive greater protection from eminent-domain actions by government agencies, local governments agree not to enact nuisance ordinances that would restrict normal farming practices, and landowners are eligible to apply to the county government to sell their development rights.

Differential assessment is important for keeping farm property taxes affordable. Pennsylvania has a "clean and green" law that allows counties to assess farmland at its use-value for property tax purposes. In Lancaster County, most of the land that qualifies for use-value assessment is also zoned for agriculture, meaning the tax breaks apply to farmland that will be protected at least for the medium term.

Since 1976, Pennsylvania has provided relief to farmers when water and sewer lines are extended past their property. Usually a landowner must pay a fee based on each foot of line that runs through the property. For a farm owner, this fee can be a whopping burden and has driven out many farmers in the greater Minneapolis–St. Paul area, for example. In Pennsylvania, farmers pay a fee based only on the road frontage of their house.

Pennsylvania has a right-to-farm law, which is further strengthened in agricultural districts to give farmers some legal protection against nuisance complaints by nonfarm neighbors. Though the right-to-farm law is not foolproof, it meshes with agricultural zoning and agricultural districts to indicate that farming is the preferred land use in the area.

PHOTO 13.1
Part of a 1,300-acre contiguous block of preserved farmland in Lancaster County, Pennsylvania.

Agricultural economic development is important to support the profitability of farm operations. Although federal farm programs and interest-rate policies are dominant, state and local governments can promote the marketing of farm produce through farmers' markets, direct sale to consumers, and community-supported agriculture projects. Lancaster County has thriving farm stands and farmers' markets and some community-supported agriculture projects. The Lancaster Chamber of Commerce is the nation's only chapter that has a full-time agricultural services position charged with developing markets for farm products.

The Lancaster County success story depends on leadership from the township officials who adopted and have maintained agricultural zoning, the county commissioners who formed and continue to fund a purchase-of-development-rights program, the state legislators, the citizens who formed the private, non-profit Lancaster Farmland Trust, and the capable farmers who have shown a long-term commitment to agriculture.

Strategy Two: Maintaining Some Farming in an Increasingly Suburban Community

Area Profile: A moderate-strength farming community. Moderate to heavy development pressure. Land holdings are fragmented, with some large ownerships of over 100 acres and a considerable amount of scattered residential and commercial development. Farm support businesses are adequate but may not remain so for long. Some farmers want to continue farming; others don't see much future in it. This is an ex-urban bedroom community within commuting distance to urban and suburban employment centers.

Recommended Set of Tools:

- comprehensive planning by the township or county
- urban growth boundaries or village growth boundaries
- agricultural zoning of one building lot per 20 acres; maximum building lot size of 2 acres
- purchase of development rights and/or transfer of development rights
- agricultural districts
- preferential farmland taxation with a stiff rollback penalty for conversion to a nonfarm use
- relief from sewer and water assessments
- right-to-farm law
- agricultural economic development, such as farmers' markets and community-supported agriculture projects

If a primary goal of the comprehensive plan is to encourage farming and protect agricultural land, keeping land open and scenic is likely to be an important and anticipated benefit. But local residents are going to have to do more than simply rely on zoning. They are going to have to provide financial support for farmland protection and the local farm economy. Agricultural zoning can be used, but some, perhaps most, farmland owners are going to have expectations of selling their land for nonfarm development. Retiring development rights through the purchase of development rights, transfer-of-development-rights, or the donation or bargain sale of conservation easements would provide some compensation to farmland owners to keep their land in farming.

To discourage sprawl, urban and village growth boundaries should be used. Once sewer and water lines are extended next to farmland, it is a signal for the farmer to develop. Right-to-farm laws, differential assessments, and agricultural districts are also important, but these techniques by themselves will not keep land in farming.

Carroll County, Maryland

Carroll County, Maryland, has an agricultural production valued at $67 million a year and uses all of the tools listed above except urban growth boundaries. The county covers 289,000 acres northwest of Baltimore and well within commuting distance of both Baltimore and Washington, D.C. Carroll County farms produce grains, dairy, and poultry.

In 1978, the county set a goal of maintaining at least 100,000 acres of farmland. The county Farm Bureau chapter and extension service agreed on this acreage as providing a critical mass of farms and farmland to sustain farming in the county well into the future. In response, the county commissioners adopted agricultural zoning at one building lot per 20 acres on 188,000 acres, or nearly two-thirds of the county. County commissioners approved the zoning change after studies by the county planning commission and several public meetings at which citizens voiced support for protecting farmland from development.

Carroll County has over 42,000 acres enrolled in agricultural districts

FIGURE 13.1
Farmland Protection
in Carroll County,
Maryland

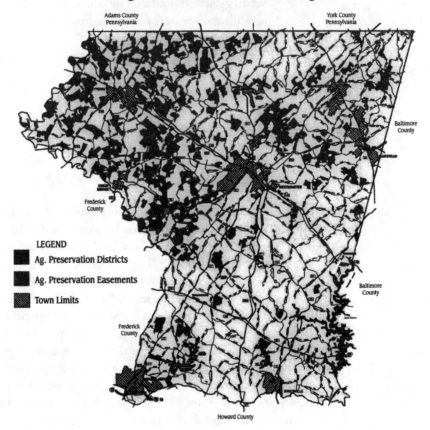

where landowners agree to maintain the land in farm use and not subdivide it for commercial or residential uses for five years (see figure 13.1). The districts make landowners eligible to sell their development rights to the county and state, and Carroll County farmers have sold development rights on more than 24,000 acres. Since 1980, the county commissioners have authorized over $5 million in county funds to match nearly $14 million in state money.

In 1994, Carroll County enacted its own right-to-farm law in addition to the Maryland law. The county requires sellers of real estate to notify potential buyers about nearby farming operations, and the county set up a five-member panel to help settle disputes between nonfarm homeowners and farmers. According to hog farmer John Burton, "There can be a lot of friction between farmers and nonfarming neighbors. It's a lack of communication. The farmer has the responsibility of letting the neighbor know what's going on."[3]

Marin County, California

Marin County, California, is a good example of a steady-to-declining-strength farming area that has fashioned a successful farmland protection program. Although there are fewer than one hundred dairy "ranches" remaining and the county's agricultural output is valued at only $42 million, the grazing land that supports the dairies and the cattle and sheep ranches is well concentrated (see figure 13.2). The ranches are mostly several hundred acres in size and are family owned and operated. The Marin dairy industry supplies much of the greater San Francisco Bay milkshed to the south.

Much of the success or failure in farmland protection depends on the comprehensive planning process in the community. Through its planning process, Marin County identified three general areas: Along the Pacific Coast is the federally controlled Golden Gate National Recreation Area (created in 1972) and Point Reyes National Seashore; to the east is a wedge of grassland and hills where the ranches are located; and farther east is the growth corridor along U.S. Route 101.

In 1972, the county commissioners enacted large-lot zoning of one dwelling per 60 acres in the agricultural area. The county ranchers already had the benefits of state right-to-farm and differential assessment (Williamson Act) legislation. Development would be accommodated in the eastern growth corridor or else forced to locate northward in Sonoma County. And the dividing line between the growth corridor and the agricultural area has in effect served as a growth boundary.

In 1980, the Marin Agricultural Land Trust was formed. Thanks to funding from a private foundation and the California Coastal Conservancy, the Trust began purchasing development rights to the ranch land. As of 1996, the Trust has preserved over 25,000 acres, the most among counties using the PDR technique. Purchasing development rights has helped to stabilize the land base and to strengthen the credibility of the one-to-sixty zoning.

Yet despite receiving millions in state funding through Proposition 70 beginning in 1988, Marin County's PDR effort has slowed in the mid 1990s. In 1992 and 1996, county voters failed to pass a temporary property tax increase of $25 per lot to fund the PDR program. Meanwhile, the county has a backlog of landowners with an estimated $10 to $12 million in development rights to sell. With the passage of a PDR program in California in 1995, Marin is awaiting a new round of state funding.

But the message in the moderate-strength or slowly declining farming community is clear: If something is not done soon, or if protection efforts are not maintained, agriculture will be lost to development. Marin County recognized this situation early on and responded with innovative zoning and purchase-of-development-rights programs. Political leadership from the county commissioners was essential, and though many ranchers were skeptical at first, most came around to support the farmland protection effort. Chief among these ranchers was Ralph Grossi, president of the county Farm Bureau, who helped

PHOTO 13.2
Farmland overlooking Tomales Bay in Marin County, California. (Photo courtesy of Marin Agricultural Land Trust)

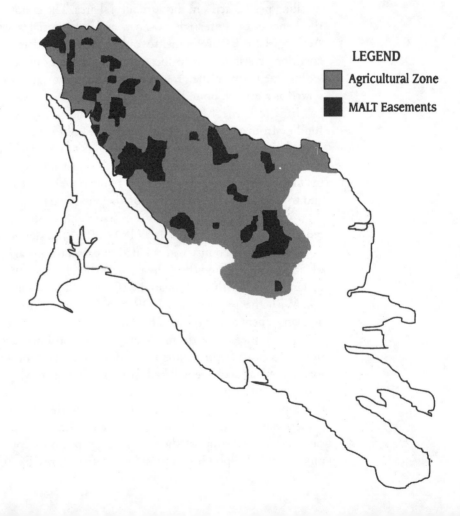

FIGURE 13.2
Farmland Protection in Marin County, California

to forge the Marin Agricultural Land Trust. In 1986, Grossi went on to become the president of American Farmland Trust.

Commenting on his experience in Marin County, Grossi says,

> There is still some perception that Marin is an anomaly. Not so. It's very much a typical case of the problems farmers face everywhere on the urban fringe. And the solutions Marin worked out have meaning elsewhere.[4]

Strategy Three: Maintaining Some Rural Character in a Suburban Community

Area Profile: A weak farming community. Moderate to heavy development pressure. Land holdings are fragmented with few large ownerships of over one hundred acres and widespread residential and commercial development. Farm support businesses are weak; farmers must travel some distance for services. Few farmers want to continue farming. This is a rapidly suburbanizing community within commuting distance of urban or suburban employment centers.

Recommended Set of Tools:
- comprehensive planning by the township or county to accommodate development and maintain open space
- large-lot "rural" zoning of 2- to 10-acre minimum lot size or cluster zoning
- farm property tax breaks only for commercial farming operations that meet a gross revenue standard of $20,000 a year
- right-to-farm law

The goal of the comprehensive plan will probably emphasize maintaining the rural character of the community even though farming is declining. Some agriculture may survive in the form of horse farms, hobby farms, and greenhouse and nursery operations. The appearance of the community is important in providing a good quality of life, as well as supporting residential and commercial real estate values.

Agricultural zoning is probably not realistic or politically possible. Large-lot rural residential zoning or cluster zoning is designed to provide some open space as the remaining farms produce their last crop: houses.

Urban growth boundaries and village growth boundaries may be difficult to put into place because of the already existing sprawl of houses and businesses.

The purchase-of-development-rights is likely to be expensive, well over $5,000 an acre, and at best, a PDR program would create only islands of preserved land. These islands would be surrounded by nonfarm development, which would make farming difficult.

A transfer-of-development-rights will not work unless there is an area of at least 1,000 to 2,000 contiguous acres where the landowners want to remain in farming.

Farmers will be reluctant to enroll their land in agricultural districts unless this is the only way they can receive property tax breaks before they sell their land for development. If the local government has some discretion over the eligibility for farm property tax breaks, those breaks should be limited to commercial farm operations, not hobby farms, which are really rural residences.

As farmers become a smaller minority of the local population, they lose political power. For example, although a right-to-farm law applies statewide, some local governments may attempt to enact nuisance ordinances that restrict farming practices, such as the hours in which farm machinery can operate. Even though these ordinances might violate the state right-to-farm law, farmers may give up and leave the community rather than become ensnared in a costly legal battle.

Montgomery County, Pennsylvania

Montgomery County, Pennsylvania, borders the city of Philadelphia to the east and suburban counties to the north and south. Nearly 700,000 people live in the county, and only the western edge remains viable for agriculture. The county's farming industry produces $34 million a year and only 1 percent of the state's total farm output. Just 226 of the county's 580 farms produced over $10,000 in sales in 1992.

The county has 45,000 acres of farmland, but only 13,400 acres are enrolled in agricultural security areas. There is no agricultural zoning. County Planning Director Arthur Loeben said in 1994, "Ag zoning doesn't save farms. The theory that ag zoning is important is ridiculous."[5] Instead, townships in Montgomery County employ 2-acre zoning in rural areas. The county has encouraged the use of cluster development or open-space zoning in the countryside, but the best this zoning can do is protect some open space rather than working farms.

The county participates in the state purchase-of-development-rights program and has preserved just over 2,500 acres in thirty farms at a cost of over $15 million. A group of fourteen farms has been preserved in a single township in the western area of the county. But two other purchases raised eyebrows: a payment of $2.5 million to preserve a 164-acre farm inside a borough (incorporated village), and $1.5 million to save 96 acres that generated only $5,200 a year in farm production.

Master farmer Larry Weaver, who preserved his two farms in Lancaster County, comments, "What it comes down to is they're trying to preserve open space in those areas. They're paying big bucks to do it. To me, the state has no business preserving farms in places like Montgomery County."[6]

In 1993, Montgomery County voters approved a $100 million bond issue to protect and acquire land resources. Only $4 million of the bond issue was earmarked for farmland preservation.

Farmland Protection Pitfalls to Avoid

There are several potential pitfalls with farmland protection programs. Government programs run the risk of appearing to "throw money at a problem." At some point, the voters will hold politicians accountable and want to know what has been accomplished and at what cost. Also, there is often a large difference between enacting a farmland protection program and actually carrying it out. A major shortcoming with several protection programs is not the techniques chosen but the weak implementation of the techniques.

Timing is crucial. The sooner and more quickly protection programs are put into place, the greater the likelihood of success. For example, the introduction of agricultural zoning in an area already carved into 3- to 5-acre lots makes little sense. Landowners build up expectations about the future use and value of their property. Protection programs usually change those expectations and generate opposition. If protection programs take a long time to implement, some landowners will act to develop their land. For example, it took thirteen years from the time the Oregon Land Use Act was passed to the approval of the last county and city plans in 1986. In the meantime, Oregon experienced the greatest increase of any state in the number of hobby farms (farms of less than 50 acres that produce less than $10,000 a year in farm products).

A potential danger in any farmland protection program is not explaining clearly how various techniques work. Farmers, like anyone, can get confused by the array of incentives and restrictions. And because they hold the land in question, if they don't understand what is being proposed, valuable time, trust, and momentum in putting protection programs into place can be lost.

Another threat common to governments, land trusts, landowners, and the public is a loss of interest and commitment to protection programs. It is often such a struggle to put a protection program together that people become reluctant to make needed changes once the program has been in place for a while. But more dangerous is the fact that elected officials can change with the next vote and undo land-use regulations or cut funding. This loss of support can rob landowners and the public of their enthusiasm for protection and bring on an attitude of "what's the use?"

Another pitfall is a fear of innovating and taking risks. Farmland protection is still a young and evolving practice. In just twenty-five years we have seen the creation of TDRs, PDRs, agricultural districts, and urban growth boundaries, as well as an explosion in land trusts and millions in public funds dedicated for farmland protection programs. What will the next twenty-five years hold?

Let us share some possibilities:

- Land banking could be used by local governments to protect and maintain key farms. In France, each region (called a department) has the power to block the sale of farmland for development or to condemn the land, purchase it, and then sell it to another farmer. A similar program exists in the Canadian province of Prince Edward Island and includes the government buying farmland and leasing it to farmers until they can afford to buy it. A 1992 Connecticut law allows the state government to purchase farmland and then lease the land for farming or resell it with a conservation easement limiting the use to agriculture.

- Building on the 1996 Farm Bill, the federal government could provide $50 million a year to states to purchase development rights to farmland.
- The federal government could make "green payments" to farmers who practice farming that conserves soil and water resources. And these payments could be targeted by location to farms in regions where farming is economically viable.
- State-funded PDR programs could be established throughout the Midwest farm belt. Ohio is already exploring the creation of a PDR program.
- Local tax reform could reduce or eliminate reliance on the property tax. Instead, local income and sales taxes would cover the cost of most local government services and schools.
- The use of impact fees or mitigation could fund farmland protection programs. For each new dwelling, a developer would pay into a local government fund for the purchase of development rights.
- Workable TDR programs could be created in many local jurisdictions.
- The construction of a series of new towns with mixed uses and pedestrian orientation could provide sociable, energy-efficient, and sustainable communities. The federal government could help finance the new towns in joint ventures with private developers. The new towns would have population caps to keep them from getting too big. New towns started in England before World War I and gained popularity between 1945 and 1970. Recently, new towns have again been proposed to absorb England's growing population.
- Reinvestment could finally occur in the inner cities and older suburbs. Tax incentives for refurbishing older buildings and constructing new buildings in cities would go a long way. If urban areas were made into attractive places to live, work, and invest, the invasion of the countryside would slow down dramatically.

Not surprisingly, all of these possibilities involve money, political willpower, and a bit of creative thinking.

Perhaps the greatest potential shortcoming of protecting farmland is that it does not guarantee the financial success of farming in a community or region. Without a doubt, a profitable agriculture industry—from food and fiber producers to the processors, wholesalers, and retailers—is the single best way to ensure that land will continue to be farmed. And if farming is not profitable, it makes little sense to expect people to farm.

The prices farmers receive for their products, the cost of borrowing money, the operating costs, and the ability of new farmers to come into farming are all important issues that are mostly beyond the scope of farmland protection.

State and local governments largely determine farmland policy, while the federal government has authority over farm subsidy, lending, and interest-rate policies. This split between land policy and income policy means a lack of coordination and strategy for overall agricultural policy.

Summary

No single tool can achieve a community's farmland protection goals. The strengths of one tool can cancel out the weaknesses of another, creating a more effective program. A community must tailor a protection program based on the desires of the community and the landowners, the health of the local farming industry, and a strategy of what lands to protect and how.

The purpose of farmland preservation should be not only preserving farmland but ensuring that *commercial* farms can remain in business and contribute to local, regional, national, and international markets. Unfortunately, many communities have not carefully crafted property tax breaks and zoning and have encouraged the creation of hobby farms and low-density rural residential sprawl.

Whatever set of tools a community chooses, the sooner they are put into place, the better. Citizens need to be vigilant about land-use changes and not become discouraged by defeats. Farmland protection doesn't happen overnight. It takes years of careful thought, public investment, and landowner support.

CHAPTER 14

Making the Connection

LAND PROTECTION
AND THE BIG PICTURE

> The problem of excessive population seems to be central to nearly every problem in our State. Too many people means too few jobs and too much competition for them; too many people means too little land for agriculture, and parks, and scenic vistas.[1]
> —George R. Ariyoshi
> Former governor of Hawaii

> We have a moral commitment to leave for future generations a livable environment, even perhaps a better one than the one we inherited.
> —Amatai Etzioni
> *The Spirit of Community*

While land protection programs help to maintain agriculture and open space, such efforts alone are unable to address the larger population forces that drive the consumption of land. It is important for citizens, planning practitioners, and elected officials to understand how population growth, population shifts, and the link between urban decline and suburban expansion can work against community planning and sustaining a good quality of life. As part of the solution, intelligent input and action are needed in the design of new communities, the revival of older ones, and for broader adoption of more environmentally sensitive, more economically sustainable agriculture.

Population Increase and Population Shift

From the beginning of civilization to the 1940s, a total of no more than a few billion people walked on the earth. By the year 2050, just one lifetime from today, the world population is expected to reach *nine billion*.

There are now over 260 million Americans. The U.S. Bureau of the Census estimates that there will be 383 million Americans by the year 2050. Where will these people live and work? The population pressures in certain states will be enormous. Since 1990, Colorado has been growing by 100,000 people a year, a rate twice the national average. The population of Texas is expected to double in the next fifty years. By the year 2005, California is likely to add 2.2 million households, a 20 percent increase over today's 11 million homes.[2] The U.S. Bureau of the Census has predicted that California will have 60 million residents, almost double the current 32 million, in 2050.

Population growth leads to competition for land, which in turn provokes conflicts between competing users of land. About 90 percent of California's cropland depends on irrigation; more residents means fierce competition over water as well as space. On the other side of the country, Florida continues to grow rapidly, adding 800 new residents every day. Agriculture contributes $12 billion a year to the state economy, but the amount of land in farm use is expected to shrink from 10.9 million acres in 1990 to 6.7 million acres by the year 2020.

Population growth in urban-fringe communities is only part of what drives the rapid conversion of landscapes into tract homes and commercial strips. Most of the land conversion in America is caused by shifts in population as people relocate from the city to the suburbs, from one state to another, and from the suburbs to the more rural ex-urbs. Tens of thousands of people are seeking a better quality of life and better home value. Often the quality of life in the former community was degraded by too much development, and the homebuyers are looking for a less crowded community with more open space. Ironically, the home-building industry, which thrives on this type of moving about, repeats the cycle, and another community, another region becomes unbearably congested.

The effect of population shift was studied recently by the *Washington Post*, which found that the closer-in suburbs of Washington, D.C., such as Arlington, Virginia, were losing higher-income residents to new housing developments in counties such as Loudoun County, Virginia, and Calvert County, Maryland, where farmland loss is a hot public issue.[3] Using figures from IRS tax returns to determine who had moved between 1993 and 1994, the *Washington Post* study determined that more than 40 percent of the income pouring into still predominantly rural Loudoun County came from congested Fairfax County, where nearly everyone agrees the quality of life has plummeted from overcrowded highways and public facilities.

It is important to understand that actual population growth in metropolitan areas is much lower than the rate of farmland conversion would seem to indicate. In the Chicago region, for example, a study by the Northern Illinois Planning Commission found that true population growth occurred at just 4 percent between 1970 and 1990, yet land consumed for development increased by an estimated 45 percent. During this period, an estimated 422 square miles of farmland in Chicago's six-county region were lost. Thus, population shift, not population growth alone, was driving the home-building industry.

Urban Decline and Outward Expansion

Sprawl has not only posed a serious threat to land resources, it also has drained financial and human resources from the nation's cities, and in recent decades, from mature suburbs as well. As people continue to relocate from older suburbs to newer ones, they have been the primary customers for the more than 1.4 million new homes that are built in America each year. Hundreds of major corporations have relocated close to these new homes, following their employee base and leaving behind empty shells in the inner city and the older suburbs. In this shell game of relocation, the loss of businesses depletes a city's tax base, making it more difficult to supply quality public schools, streets, and police and fire protection. The old infrastructure decays while brand new (and expensive) schools and roads are built farther and farther from the original urban core.

Most of the nation's population growth between now and the year 2050 is expected to occur in the expanding suburbs. The choices on a national level are clear: a new focus on redeveloping our cities and making suburban growth more compact, or a continued push away from the cities and the gobbling up of farmland and open space. The first choice emphasizes livable communities. The second choice perpetuates the notion of limitless, developable land and measures the nation's wealth by the number of new housing starts.

The nation can't wait for new federal government policies to reverse the sprawl trend. Citizens must take action at the local and state levels to help bring about change. City redevelopment activists and land protection groups need to join forces to work for new private and public investment in existing cities and suburbs so that there is a viable activity for the building industry to pursue, in places other than cornfields.[4]

The idea that urban decline and suburban sprawl are linked has begun to take hold nationally and has presented an opportunity for a variety of organizations to join forces. For example, a coalition of civic action groups in the Chicago region, those working for improvements in inner-city neighborhoods and those working for farmland protection in outlying areas, began working together in 1995. Their objective: to oppose continued infrastructure expansion in Chicago's collar counties and to advocate reinvestment in older communities. Their first target was a proposed third airport for the Chicago region, to be built in rural Will County, that would result in the loss of tens of thousands of acres of farmland. The airport proposal was boosted by the region's business interests as an economic development venture, even though the Federal Aviation Administration showed no urgent need for the facility.

The airport proposal was matched by a push to build an interstate connection to the proposed airport site. Whether the airport and the highway will be built was unresolved as of this writing. Still, the combined development interests of the Illinois State Toll Highway Authority and the airport backers stand as an excellent example of how the political and business worlds can combine to mandate sprawl at the cost of existing cities and suburbs.

David Rusk, the former mayor of Albuquerque, advocates cities annexing suburbs so that the inequalities between city and suburb can be lessened. In his book, *Cities Without Suburbs,* Rusk shows how several cities in the western United States have successfully used powers of annexation to expand their city limits. In the process, these cities have maintained far better social and economic performance than the older cities in the East, where little increase in physical size has occurred in the past fifty years and where suburbs are "cannibalizing" the inner cities.

Minnesota state legislator Myron Orfield has been a pioneer in showing how fiscal disparities between urban and suburban communities drive sprawl. For decades, counties surrounding major cities have competed against each other to entice corporate and business relocations, usually away from the urban core. Elected officials feel it's necessary to boost the tax base with commercial development to make up for the residential development that has burdened urban-fringe jurisdictions.

According to Orfield, tax-base sharing has a "powerful impact on land use" by making exclusive "fiscal zoning" less attractive to local governments.[5] He has long advocated revenue sharing both as a means to reduce fiscal disparities between high-growth and stagnant or declining communities, and to help curb sprawl. Under tax-base sharing, localities will feel less compelled to offer incentives, such as big tax breaks, to entice companies to locate within their borders.

Minnesota is a good place to start in the study of how to alleviate the disparities created by the outward expansion of suburbs. In 1971 the Minnesota legislature created the Fiscal Disparities Program to try to reduce inequalities in property tax bases between communities with markedly different growth rates. The program affects the seven counties that make up the greater Minneapolis–St. Paul area.

Orfield authored legislation to expand the program in 1995. It passed the legislature but was vetoed by the governor. While the current program affects only commercial and industrial taxes, Orfield's bill would have established tax-base sharing on residential property values above $200,000. His legislation had the strong support of diverse community and environmental interest groups as well as a solid coalition of legislators from both urban and suburban jurisdictions.

Under the current Minnesota program, communities contribute 40 percent of all new commercial and industrial development revenues into a pool; the monies are then redistributed based on population and per capita real estate property values measured against the area average. Communities with below-average values receive a greater share of pooled funds.

Most of the 187 metro-area communities that participate are net gainers in the equation; 138 received more tax base from the shared pool in 1995 than they contributed. The other 49 were net losers. Among the 58 communities of over 9,000 population, 35 were net gainers and 23 were net losers.

Supporters of the program have been concerned about the revenue pool because pool funds have been declining since the early 1990s. The reason for the decline is believed to be a large number of court-ordered reductions in assessed property values, mainly on office buildings in downtown Minneapolis and on interstate corridors.

The Orfield bill would have created "a fiscal enhancement" of the Fiscal Disparities Program, with revenues in the pool growing gradually over the years to add 40 percent. Fiscal experts said disparities would be reduced from a ratio of 12:1 to 7:1.

Tax-based sharing demonstrates the important political reality that America's suburbs are not all the same, says Orfield. Each suburb has different social needs, a different business climate, and different political situations. So when it comes to promoting regional solutions to the critical problems facing America's cities, citizens should not allow politicians to form suburban alliances against cities. It is time to look at the bigger picture, and to see that a strong inner core of urban cultural, economic, and civic activity is necessary for our society to sustain itself. We must rebuild our cities in order to save our countryside. We all must work together for regional solutions.

National fiscal experts have called the Minnesota program successful and a positive factor that creates a practical financing option for state and local governments. According to Standard and Poor's, tax-base sharing has regional advantages, in that it reduces competition for ratables among neighboring municipalities and therefore serves to limit the tax incentives localities use to attract industry.[6]

In metro areas across the United States, dozens of small, inadequate town and county governments have been left to struggle with development problems that are too big for them to handle. Tax-base sharing could help to foster regional infrastructure and land-use planning, as well as allow local governments to share state-mandated facility spending.

Sustainable Development and Growth

While farmland protection advocates cannot expect to stop growth from occurring, even in places where agriculture reigns, they can promote energy-efficient, environmentally sound development and land-use patterns. Sustainable development has become a buzz phrase for the kind of growth that is appropriate for a world where resources are limited. Sustainable development is not harmful to the environment, pays its own way in the services it uses, and will last for a long time. Much of the growth and development that occurs in America has been sustained thus far only because of cheap and plentiful energy supplies. With the United States importing over $50 billion worth of oil each year and racking up huge trade deficits, we must ask how long can this continue?

Most new homes being built in America will require those who reside in them to own at least one car, and often two, since a two-income household is usually necessary to own a home in the suburbs. Dependency on cheap oil, then, has been a significant factor in the outward expansion of suburbs into the countryside.

An innovative development design trend, called neotraditional, actually looks backward to traditional towns, where buildings were built on a human scale, where stores and shops were within walking distance of homes, where streets were narrow, forcing automobiles to move along slowly, where corner markets were acceptable to residents, and where even apartments above the stores were seen as affordable housing for young people just starting out or for the elderly. These needs were understood and accepted as a way of community life. This old-style small-town America was left behind when automobiles made it possible to move outward, where anyone who was gainfully employed could afford a home on an entire acre or half-acre all to themselves, and live on a street that had only local residential traffic. All commercial establishments were segregated from the homes, as if this was a better arrangement. Cars were required in order to buy anything.

Today, having a grocery store within walking distance has become a quaint anachronism that only people living in upper-income city neighborhoods enjoy and appreciate. But many other people would prefer the old style of community and town life if it was available. Thousands of people are moving back to old towns, or to neotraditional towns such as Seaside, Florida, and Kentlands, Maryland, that have been built using old ideas of how town life should function. These new "old" towns can only be built where zoning regulations allow them. Believe it or not, in most places in the United States, it is illegal to build an old-fashioned town. While we don't advocate that new towns be built in the middle of a farming region, of course, these are exactly the type of sustainable designs America needs within its suburbanizing regions where services such as water and sewer exist to support them.

One older design idea that still has appeal is the system of Garden Cities envisioned by the Englishman Ebeneezer Howard. A Garden City could be built integrating homes, stores, businesses, and green space in a circular pattern around a city core. A Garden City would have no more than thirty thousand people. A set of Garden Cities could be built in the countryside with farmland and open space between the cities.

Several Garden Cities were built in England, and while they are still attractive places to live, they have become mainly commuter suburbs around greater London. In the United States, Radburn, New Jersey, designed in 1929, was the model Garden City and remains a comfortable place to live. However, America's experiment with Garden Cities was cut short by the Great Depression. Today, the concept of Garden Cities holds promise for alleviating the pressure on land resources. A Garden City would take up far less space to house and employ thirty thousand people than the current pattern of sprawl. With strict zoning in place that would limit outward expansion, there would be less conflict between farmers and nonfarm neighbors, and less impact on natural areas.

Population Growth, Agriculture, and a Global Perspective

Although population shifts are a major culprit in the pressures placed on land resources in our metropolitan regions, population growth in America and in the rest of the world could have devastating effects on agricultural lands. In 1994, the Union of Concerned Scientists warned that unless population growth slows and agricultural practices improve, the world will face a food shortage by 2050.[7] According to the Census Bureau's 1995 *Population Profile of the United States*, the nation's annual births exceed its annual deaths by about 1.7 million. In addition, an estimated 750,000 legal immigrants enter the United States each year. This combined population increase is the same as adding the number of people in San Francisco, Dallas, *and* Washington, D.C., *each year.*

The U.S. Department of Agriculture's veteran demographer, Calvin Beale, notes that "areas that have subtropical winters—the Florida peninsula, the lower Rio Grande Valley, Southern California—all of those areas are having very rapid growth, almost runaway population growth."[8] And these are exactly the places we depend on for important fruits and vegetables.

In addition to our own growth burdens, the United States is part of a global economy. American farmers export about $50 billion worth of food and fiber products each year, while American consumers buy about $25 billion worth of foreign-produced agricultural goods. Americans consume about $145 billion in food and fiber products, so our dependence on foreign growers is relatively small, roughly 17 percent of all food and fiber products.

But the rest of the world has much less capacity to feed the United States than the United States has to feed the rest of the world. Americans make up only 6 percent of the world's population. Each year, there are 90 million new mouths to feed in developing countries. A very real demand on America's farmland and food supplies will come from increasing populations in China and other developing nations as they adopt American dietary habits and eat more grain-fed meat. Meanwhile, world fish stocks are in serious trouble, with declining yields in many traditionally productive waters. With the current world population of 5.8 billion likely to reach 9 billion by the year 2050, American farmers will need a reliable source of high-quality land to meet both domestic and foreign food demands. Yet the federal government has taken almost no direct action to protect even our prime agricultural lands.

Sustainable Agriculture

Protecting farmland from development makes economic sense only if the land will be used productively, responsibly, and profitably over the long run. One reason that farmland protection, in some regions, has not received the support given to the protection of open space and natural areas is a public perception that many farming practices are harmful to the environment.

This perception has some basis in fact, yet most farmers see themselves as stewards of the land and recognize that soil and water conservation are essential to high levels of crop production over time.

Agriculture is the largest user of water in the United States. Farming and ranching use 85 percent of America's annual consumption of groundwater to grow crops and feed livestock. In the Great Plains and the Southwest, access to water makes the difference between productive cropland or grazing land and wasteland. And given the high rate of population growth in the Southwest, the competition over water will become more intense in the future.

But agriculture also has great potential for water conservation. For example, the replacement of flood irrigation with surge irrigation, which alternates water between crop rows, or drip irrigation, long used successfully in Israel, could cut water use by more than half.

Some farming practices have caused water pollution. Soil erosion from overgrazing livestock and plowing steep slopes has led to a build-up of silt in rivers, streams, and lakes. Manure runoff from feedlots and fertilizer use on fields has also polluted both surface water and groundwater. About 150 million pounds of pesticides are sprayed on crops each year, and some pesticides have caused drinking water problems in several states.

The establishment of huge animal confinement units, particularly the notable increase in hog and chicken facilities, is a trend in agriculture that is creating severe waste disposal problems. Operations with many thousands of hogs or several hundreds of thousands of chickens create manure disposal problems such as runoff into rivers and streams and overloading land with manure, resulting in the pollution of groundwater and wells. On some farms, manure lagoons have ruptured, sending millions of gallons of liquid manure into nearby rivers.

Factory farming may appear to be more efficient financially, but the burden on the environment can be enormous. This type of agriculture is facing greater public scrutiny.[9] Environmental soundness and sustainable farming practices should be requirements in farmland preservation programs.

Farmers will continue to face the challenge of following environmentally sound and financially rewarding practices. There is plenty of room for consumers to play a role in making this beneficial agriculture a reality, even if it means paying higher prices for food.

Protecting America's Farms and Farmland: A Summary

The effort to protect America's farms and farmland has matured from its infancy in the 1970s and 1980s. Many techniques and programs have been tried, evaluated, and refined. In many places, farmland protection is no longer an experiment: it is an accepted and proven method to safeguard both the agricultural industry and the overall quality of life.

Americans have lately been expressing a strong desire for a greater sense of community, of belonging to a place. Too many cities and towns look like Anywhere, USA, with the same style of houses and commercial strip development. Hopefully, by protecting farmland and open space, a community can maintain a character—an appearance and a feeling—that makes it distinctive. And this protection can be a very real source of community pride.

People will seek to move to states and localities that have protected their land resources. Those places active in promoting economic development will be more attractive if they have shown a caring attitude toward their environment and considered how new growth fits in.

In those communities where sprawl is a growing problem, farmland protection must be implemented quickly, or the effort will be too late. Farmers are far less skeptical about farmland protection and more open to protection techniques than they were only five to ten years ago. This is especially the case with voluntary programs such as agricultural districts and the purchase of development rights.

But it is increasingly clear that communities cannot allow landowners simply to keep their options open. The conversion of large tracts of farmland into scores of houses places both a financial and an environmental burden on a community. Growth will need to occur more carefully, guided by considerations of property tax impacts and burdens on roads, schools, and sewer and water capacity. In this way, farmland protection and growth management are being seen as one and the same. But growth management is not without costs. If farmers are expected to keep their land undeveloped, they must have incentives to do so. Regulations must be compatible with the need to earn a profit from the business of farming.

Uniting Citizens, Farmers, and Politicians for Farmland Protection

Farmland protection is a local and regional issue that could have national and even global importance in the long run. By allowing so much farmland to be developed without protecting the best remaining land, Americans are taking a risk with their ability to manage growth and, perhaps in time, with the nation's food supply. How large is the risk? No one can say for sure. But as more farmland is lost and America's (and the world's) population increases, the risk increases. Farmland protection efforts are an attempt to manage that risk.

The first step in becoming active in farmland protection is to become informed about the many issues that affect farmland conversion, the techniques that exist to protect farmland, and farming practices and business operations.

The second step is to keep an open mind, to listen to all sides of an issue, and to keep your eyes open to what is happening to the land resources in your community.

The third step is to organize and participate in local and regional planning and development decisions that affect the future of farming and open space. Although land planning is a political process, supporters of farmland protection cover the political spectrum. This is not a liberal versus conservative issue. Many proponents of farmland protection vote for liberal causes, and many are politically conservative.

The fourth step is to educate the general public, landowners, public officials, and lawmakers about the importance of protecting farmland and open space, and about how to achieve protection through public and private efforts.

The fifth step is to support your local farmers through farmers' markets, zoning ordinances that allow additional on-farm businesses, and helping find amicable solutions to disputes between farmers and their nonfarm neighbors. Farmers must feel wanted and valued in the community.

Appendices

SAMPLE FORMS AND AGREEMENTS

APPENDIX A

Model Agricultural Zoning Ordinance

An agricultural zoning ordinance is part of a county or township zoning ordinance that also includes residential, industrial, commercial, and public zoning districts. The model agricultural zoning ordinance presented here is adapted from the Lancaster County, Pennsylvania, model ordinance and is designed to create an A-Agricultural Zoning district within an existing county or township zoning ordinance.

1. The following defined terms are hereby added to their Section of the county or township zoning ordinance.

    Church—A separate detached building devoted primarily to religious worship.

    Farm or farm parcel—A tract or parcel of land containing at least 25 acres, devoted primarily to agricultural uses, together with a dwelling or other accessory uses.

    Farm-related business—A business operated on a farm parcel, related to or supportive of agricultural activities, such as blacksmithing, farm implement repair, or roadside sale of agricultural products.

    Intensive agricultural use—Intensive Agricultural Uses include, but are not necessarily limited to: a) slaughter areas; b) areas for the storage or processing of manure, garbage, or spent mushroom compost; and c) structures housing more than fifty animal units. (Under the Pennsylvania Nutrient Management Act of 1993, a farm with more than two animal units per acre, with an animal unit defined as one thousand pounds, must have a management plan.)

    Nonfarm lot—The following constitute nonfarm lots within an agricultural zoning district: a) a lot or parcel containing less than 25 acres and containing one or more dwelling units—agricultural activities may be carried on as part of the use of a nonfarm residential lot; b) a lot within an agricultural zoning district devoted to uses other than agricultural or residential.

    Parent tract—Each tract of land located within an A-Agricultural District on the effective date of this ordinance, and held in single and separate ownership is a parent tract.

2. The following new Article IV (A-Agricultural District) is hereby added to the County or Township Zoning Ordinance:

Article IV

A—Agricultural District

The following provisions shall apply to all land within an A-Agricultural District:

Section 401. *Purposes.* The purposes of the agricultural district are:
  a) To protect and promote the continuation of farming in areas with prime soils (SCS Class I and II) and soils of statewide importance (SCS Class III) where farming is a viable component of the local economy, and to promote the continuation of farming in areas where it is already established;
  b) To permit, with limited exceptions, only agricultural land uses and activities;
  c) To separate agricultural land uses and activities from incompatible residential, commercial, and industrial development, and public facilities. The protection of land for agricultural purposes is a legitimate zoning objective under the State Planning Statutes, which the regulations set forth in this Section seek to achieve.
  d) To put into action the local comprehensive plan, which contains the goal of protecting agricultural lands and promoting agriculture as a component of the local economy.

Section 402. *Use Regulations.*

Section 402.1 Uses by Right

  a) All forms of agriculture (except new Intensive Agricultural Activities, see Section 402.2 below), horticulture, and animal husbandry, including necessary farm structures;
  b) Forestry uses, including sawmills;
  c) Farm dwellings;
  d) Production nurseries and production greenhouses;
  e) Wildlife refuges and fish hatcheries;
  f) Private elementary schools, which may include classes only through grade nine;
  g) The following uses accessory to a principal use:
   i) roadside stands for the sale of agricultural products, at least half of which are grown on the premises. Off-road parking shall be provided for all employees and customers, and the stand shall be set back at least twenty (20) feet from all property lines and road rights of way. The stand shall not be more than three hundred (300) square feet in size.
   ii) garages (see Section _____);
   iii) signs pursuant to the local sign ordinance, Section _____;
   iv) beekeeping;
   v) manure storage facility (see Section _____);

vi) noncommercial recreation (family pool, tennis court, etc.);
vii) noncommercial antenna or communications facility (see Section ____);
viii) worship services in the home which require no exterior modification of the dwelling.

402.2. *Uses Permitted by Special Exception (Requiring Approval of the Township Zoning Hearing Board)*

a) Temporary farm housing, provided that this use:
  i) takes place on a farm of at least twenty-five (25) acres;
  ii) utilizes mobile home or manufactured housing;
  iii) is used only to house farm laborers;
  iv) is removed when farm laborers no longer occupy the housing;
b) New intensive agricultural uses, where adjacent to a residential zone, shall not be located closer than two hundred (200) feet from any residentially zoned property line or residence on an adjacent property;
c) Home occupations (see Section ____);
d) Family care for fewer than six (6) children or adults (see Section ____);
e) Bed and breakfast inns (see Section ____);
f) Granny flats and ECHO housing (see Section ____);
g) Animal hospitals, veterinary facilities, and kennels (see Section ____);
h) Cemeteries and necessary incidental structures of no more than one (1) acre;
i) Riding schools and/or horse boarding stables (see Section ____);
j) The conversion of a single-family dwelling to a two- or three-family dwelling;
k) Water treatment and transmission facilities and wastewater collection facilities;
l) Public utilities;
m) Farm-related businesses that are conducted outside the home, subject to the following regulations:
  i) no more than two (2) acres of land shall be devoted to such use, including areas used for structures, parking, storage, display, setbacks, landscaping, et cetera. Any lane serving the farm-related business and a home and/or farm contained on the same lot shall not be included as lot area devoted to the farm-related business. No additional lane or curb cut to access the farm-related business shall be allowed; and
  ii) no more than fifty percent (50%) of the area devoted to a farm-related business shall be covered by buildings, parking lots, or any other impervious surface; and
  iii) the owner or occupant of the farm must be engaged in the farm-related business; and

       iv) no more than two (2) full-time and two (2) part-time persons, other than individuals who reside on the farm, may be employed in the farm-related business; and
       v) the use must be conducted within a completely enclosed building typical of farm buildings; and
       vi) any out-building used for the farm-related business shall be located behind the principal farm residence on the site, or shall be located at least two hundred (200) feet from the closest street right-of-way; and
       vii) any outdoor storage of supplies, materials, or products shall be located behind the building in which the farm-related business is conducted.

n) Churches, subject to the following regulations:
       i) a minimum of two (2) and a maximum of four (4) acres of land shall be devoted to such uses including areas used for structures, parking, storage, display, setbacks, landscaping, et cetera; and
       ii) no more than forty percent (40%) of the area devoted to a church shall be covered by buildings, parking lots, or any other impervious surface;
       iii) for other criteria, including accessory uses, see Section_____.

o) Dwellings located on nonfarm lots, provided that:
       i) each application includes a scaled drawing indicating the location of the proposed dwelling to the surrounding farms;
       ii) the dwelling is sited on that portion of the lot which separates it as much as possible from adjacent farming, including minimizing the length of property lines shared by the residential lot and adjoining farms;
       iii) the dwelling and its lot are located on the least productive farmland wherever practical;
       iv) the dwelling is sited on the smallest practical area to satisfy the requirements of this Ordinance and on-site sewage disposal regulations.

p) Communications antennas, towers, and equipment, provided that:
       i) antennas greater than forty-five (45) feet in height and tower-mounted antennas shall require a statement from a registered professional engineer regarding the structural integrity of the antenna and/or tower;
       ii) antennas that are capable of transmitting signals shall not create electrical, electromagnetic, microwave, or other interference off-site;
       iii) communications structures shall have a setback from all property lines equivalent to the height of the structure, but in no event shall an antenna or tower exceed a height of two hundred (200) feet from grade;
       iv) the applicant must demonstrate that the proposed location on

agricultural land is necessary for the efficient operation of the communication system, and that alternative locations outside of the agricultural district are not available;

v) the applicant must submit notice of approval for the proposed installation from the Federal Aviation Administration and the Federal Communications Commission.

402.3 Incompatible Uses

Uses not specifically permitted under Subsections 402.1 or 402.2 above are not permitted in the agricultural zone. In general, uses and activities that induce nonfarm development, generate large amounts of traffic, require substantial parking, or could pose a threat to agricultural water supplies are inconsistent with the purposes of the A-Agricultural District.

Without limiting the foregoing, the following specific uses are deemed by the Township Board of Supervisors / County Commissioners to have the effect of altering the essential character of the A-Agricultural Zoning District and causing substantial and permanent impairment to the prevailing agricultural uses within this zoning district and hence, are detrimental to the public welfare, and such uses would therefore be in contravention of the requirements set forth in the State Planning Statutes for consideration of variances: landfills, quarries, golf courses, sewage treatment plants, airports, and country clubs.

Section 403. Area Requirements and Limitations on Subdivision of Parent Tracts.

Section 403.1 Minimum Area

a) Except when conducted as an accessory to a residential use of a non-farm lot, agricultural uses shall require a minimum area of twenty-five (25) acres, and no farm parcel shall be subdivided from a parent tract unless it shall meet the minimum area requirement for agricultural uses.

b) A nonfarm lot subdivided from a parent tract shall have a minimum of one (1) acre and a maximum of two (2) acres.

c) All other uses permitted by right, special exception, or conditional use shall have a minimum lot area of one (1) acre.

403.2 Limitations on Subdivision of Parent Tracts

a) In order to protect agricultural uses within the A-Agricultural District, it is the intent of this provision that the creation of non-farm lots and the subdivision of farm parcels from parent tracts shall be limited, in order to provide for the retention of tracts of sufficient size to be used reasonably for agricultural purposes.

b) Each parent tract containing fifty (50) or more acres shall be permitted limited rights of subdivision. Each parent tract of 50 or more acres shall be permitted to subdivide a combination of one or more farm parcels and/or nonfarm lots up to, but not in excess of, a total of one such nonfarm lot or one such farm parcel for each fifty (50) acres of area within the parent tract. For example, a parent tract having one hundred twenty-five (125) acres is permitted two subdivisions into a total of three lots or parcels: the remainder of the parent tract and i) two farm parcels, ii) one farm parcel and one nonfarm lot, or iii) two nonfarm lots.

c) A single-family detached dwelling may be erected on any single undeveloped lot of record (parent tract) as of the effective date of this Ordinance, notwithstanding the limitations imposed by Section 403.2(b). Such lot must be a parent tract in single ownership and not contiguous with other tracts in the same ownership. The parent tract must meet applicable requirements for minimum lot size, and any buildings erected on the lot must meet yard setback, lot coverage, and height regulations.

d) The provisions of this section shall apply to all parent tracts as of the effective date of this Ordinance. Regardless of size, no parcel or lot subsequently subdivided from its parent tract shall qualify for additional single-family detached dwellings or lots pursuant to this section. All subsequent owners of parcels of land subdivided from a parent tract shall be bound by the actions of the previous owners of the parent tract. Any subdivision or land development plan hereafter filed for a parent tract in the A-Agricultural Zoning District shall specify which lot or lots shall carry with them the right to erect or place thereon any unused quota of single-family detached dwellings or farm parcels as determined and limited by the provisions of this section.

e) In the event a tract of land not originally classified as part of the A-Agricultural Zoning District on the effective date of this Ordinance is hereafter classified as part of the A-Agricultural Zoning District, the size and ownership of such tract of land and its classification as a parent tract shall be determined as of the effective date of the change in the zoning classification to A-Agricultural.

Section 404. Yard, Coverage, and Height Requirements.

404.1 All lots or parcels shall have a minimum width of one hundred (100) feet at the building setback line and sixty (60) feet at the street right-of-way line.

404.2 All structures located on nonfarm lots shall have a minimum front and rear yard of fifty (50) feet, respectively, and a minimum side yard of twenty-five (25) feet on each side.

404.3 All structures located on farm parcels shall have front, rear, and side yard setbacks of at least fifty (50) feet. New intensive agricultural uses shall be set back an additional distance as required by Section 402.2(b) of this Ordinance.

404.4 The total impervious coverage, including both buildings and other impervious surfaces, of a nonfarm lot shall not be more than twenty percent (20%); the total lot coverage of a farm parcel shall not be more than ten percent (10%).

404.5 The maximum height of a residential building shall be thirty-five (35) feet. The maximum height of all other buildings shall be seventy-five (75) feet, excluding silos and windmills, which shall, however be set back a distance at least equal to their height from all property lines.

Section 405. Vegetation Setback Requirement.

Section 405.1 On any separate nonfarm parcel, no shrub or tree shall be planted within twenty (20) and thirty (30) feet, respectively, of any land used for agricultural purposes.

Section 406. Required Conservation Plan.

Any agricultural, horticultural, animal husbandry, or forest use that involves earth-moving activities or commercial harvesting of trees shall require the obtainment of an approved conservation plan by the _____ County Conservation District, pursuant to Chapter 102, Erosion Control, of Title 25, Rules and Regulations, State Department of Environmental Resources. All on-site activities shall be in compliance with the approved conservation plan.

Section 407. Agricultural Nuisance Disclaimer.

Lands within the A-Agricultural District are used for commercial agricultural production. Owners, residents, and other users of this property may be subjected to inconvenience, discomfort, and the possibility of injury to property and health or even death arising from normal and accepted agricultural practices and operations, including, but not limited to, noise, odors, dust, the operation of machinery of any kind, including aircraft, the storage and disposal of manure, the application of fertilizers, herbicides, and pesticides. Owners, residents, and users of this property should be prepared to accept these conditions and are hereby put on official notice that Section 4 of Pennsylvania Act 133 of 1982, the "Right-to-Farm-Law," may bar them from obtaining a legal judgment against such normal agricultural operations.

Section 408. Required Nutrient Management Plans.

All agricultural uses shall comply with the State of _____ Nutrient Management Act of 19__ and subsequent amendments.

APPENDIX
B

Agricultural Use Notice / Nuisance Disclaimer

Agriculture as practiced today is an industrial process that involves the use of chemical fertilizers, herbicides, pesticides, and heavy machinery. Although farms are attractive to look at, there may be some inconveniences and even hazards in living next to one. All states have "right-to-farm" laws, which generally offer farmers some protection against lawsuits brought by neighbors in complaint about normal farming practices (called nuisance suits). These laws, however, have not been widely tested in the courts.

A better way to reduce conflicts between farmers and nonfarm neighbors is by including in the town's zoning ordinance an Agricultural Use Notice. Prospective buyers and current landowners of any property located in or adjacent to an agricultural zone must be forewarned thereby that farming operations generate noise, dust, odors, and sprays that often spill over onto neighboring lands. The following Agricultural Use Notice is strongly recommended:

> All lands within the Agricultural Zone are located in an area where land is used for commercial agricultural production. Owners, residents, and other users of this property or neighboring property may be subjected to inconvenience, discomfort, and the possibility of injury to property and health arising from normal and accepted agricultural practices and operations, including but not limited to noise, odors, dust, the operation of machinery of any kind, including aircraft, the storage and disposal of manure, the application of fertilizers, soil amendments, herbicides, and pesticides. Owners, occupants, and users of this property should be prepared to accept such inconveniences, discomfort, and possibility of injury from normal agricultural operations, and are hereby put on official notice that the state Right-to-Farm Law may bar them from obtaining a legal judgment against such normal agricultural operations. (Warwick Township, Lancaster County, Pennsylvania)

Resource Management Easements

Another technique, employed in Bonneville and Fremont counties in Idaho, is to require resource management easements for new residential development in or adjacent to an agricultural zone. The easement recognizes that the proposed residence is located in or next to an agricultural area and may be subjected to noise, odors, dust, and other impacts of normal farming operations. *The easement waives the homeowner's legal right to object to lawful farming operations on adjacent lands.* The easement is recorded as part of the landowner's deed before a building permit is issued and before any construction begins.

It is important to keep in mind that right-to-farm laws, zoning disclaimers, and resource easements protect only normal and legal farming operations. Farming practices that violate state or federal laws, such as water pollution from feedlot runoff, are grounds for lawsuits by nonfarm neighbors.

Town Resolution

Finally, the town government can pass a resolution not to enact nuisance ordinances that would restrict normal farming practices.

APPENDIX C

Model Agricultural Conservation Easement

The model agricultural conservation easement presented below is for a lump-sum payment.

To use a bargain sale (part-payment, part-charitable donation), add the following language in paragraph 3 after "WHEREAS, the value of this grant of easement is defrayed by consideration to the GRANTOR of _____Thousand Dollars ($_____)":

> and said consideration is below the appraised fair-market value of the agricultural conservation easement. The GRANTOR intends that the difference between the consideration and the fair-market value of the easement be a charitable gift to GRANTEE; and

For an installment sale, you can list the schedule of interest and principal payments on a settlement sheet, in a separate contract of sale, or in the Grant of Easement. For a like-kind exchange, an exchange agreement must be drafted by an attorney in addition to the model grant of easement.

GRANT OF EASEMENT
County of _____ Farmland Preservation Board

This grant of easement in the nature of a restriction on the use of land for the purpose of preserving productive agricultural land and open space is made by and between _____ ("GRANTOR") of _____ County, State, and the _____ County Farmland Preservation Board ("GRANTEE"), an agency of _____ County with its offices at _____.

    WHEREAS, GRANTOR is the owner in fee of a farm located in _____ County, State, more fully described in a deed recorded in the Office of the Recorder of Deeds of _____ County, State in Deed Book ___, Page ___, and attached hereto as Exhibit A (the "Property"). The Property consists of _____ acres on which is located one (1) single-family dwelling unit; and

    WHEREAS, the value of this grant of easement is defrayed by consideration to the GRANTOR of _____Thousand Dollars ($_____); and

    WHEREAS, the Legislature of the State of _____ (hereinafter "Legislature") authorizes the State of _____ and counties thereof, as well as nonprofit conservancies, to preserve, acquire, or hold lands for open-space uses, which specifically include farmland; and that actions pursuant to these purposes are for the public health, safety, and general welfare of the citizens of the State of _____ and for the promotion of sound land development by preserving suitable open spaces; and

WHEREAS, the Legislature has declared that public open-space benefits result from the protection and conservation of farmland, including the protection of scenic areas for public visual enjoyment from public rights-of-way; that the conservation and protection of agricultural lands as valued natural and ecological resources provide needed open spaces for clean air as well as for aesthetic purposes; and that public benefit will result from the conservation, protection, development, and improvement of agricultural lands for the production of food and other agricultural products; and

WHEREAS, GRANTEE has declared that the preservation of prime agricultural land is vital to the public interest of _____ County, the region, and the nation through its economic, environmental, cultural, and productive benefits; and

WHEREAS, GRANTOR desires and intends that the agricultural and open-space character of the Property be preserved, protected, and maintained; and

WHEREAS, GRANTOR, as owner in fee of the Property, intends to identify and preserve the agricultural and open-space values of the Property; and

WHEREAS, GRANTOR desires and intends to transfer those rights and responsibilities of protection and preservation to the GRANTEE in perpetuity; and

WHEREAS, GRANTEE is a qualified conservation organization under State of _____ acts and the Internal Revenue Code, whose primary purposes are the preservation and protection of land in its agricultural and open-space condition; and

WHEREAS, GRANTEE agrees by acquiring this grant of easement to honor and defend the intentions of GRANTOR stated herein and to preserve and protect in perpetuity the agricultural and open-space values of the Property for the benefit of this generation and the generations to come; and

WHEREAS, the specific agricultural and open-space values of the Property are documented in an inventory of relevant features of the Property, dated _____, 19__, on file at the offices of the GRANTEE and incorporated by this reference ("Baseline Documentation"), which consists of reports, maps, photographs, and other documentation that the parties agree provide, collectively, an accurate representation of the Property at the time of this grant of easement and which is intended to serve as an objective information baseline for monitoring compliance with the terms of this grant of easement;

NOW, THEREFORE, in consideration of the foregoing and intending to be legally bound, the undersigned GRANTOR grants and conveys to GRANTEE an easement on the Property for which the purpose is to assure that the Property will be retained forever in its agricultural and open-space condition and to prevent any use that will impair the agricultural and open-space values of the Property. To carry out this purpose, the following deed restrictions are recorded.

COVENANTS, TERMS, CONDITIONS, AND RESTRICTIONS

With the preceding background paragraphs incorporated by reference and intending to be legally bound, GRANTOR declares, makes known, and covenants for himself, his heirs, successors, and assigns, that the land described in the deed book and page mentioned above shall be restricted to agricultural and directly associated uses as hereafter defined. However, more restrictive applicable State and local laws shall prevail in the determination of permitted uses of land subject to these restrictions.

1. Agricultural uses of land defined for the purposes of this grant of easement, as: the use of land for the production of plants and animals useful to man, including, but not limited to, forage, grain, and field crops; pasturage, dairy and dairy products; poultry and poultry products; other livestock and fowl products, including the breeding and grazing of any or all such animals; bees and apiary products; fruits and vegetables of all kinds; nursery, floral, and greenhouse products; silviculture; aquaculture; and the primary processing and storage of the agricultural production of the Property and other similar and compatible uses.
2. Directly associated uses are defined as customary, supportive, and agriculturally compatible uses of farm properties in _____ County, State, and are limited to the following:
    a. the direct sale to the public of agricultural products, of which at least half of the proceeds are accounted for by products produced on the farm;
    b. any and all structures contributing to the production, primary processing, direct marketing, and storage of agricultural products at least half of which are produced on the farm, so long as the total surface coverage of the farm by impervious surfaces for existing pavement, buildings, and all other permitted structures does not exceed two hundred thousand (200,000) square feet of the Property;
    c. structures associated with agricultural research;
    d. the provision, production, and sale of, by persons in residence, or agricultural goods, services, supplies, and repairs—including the conduct of on-farm businesses, traditional trades, home businesses, and the production and sale of home occupation goods, arts, and crafts—so long as these uses remain incidental to the agricultural and open-space character of the farm and are limited to occupying residential and/or principally agricultural structures of the Property;
    e. structures associated with the production of energy for use principally on the farm, including wind, solar, hydroelectric, methane,

wood, alcohol fuel, and fossil fuel systems and structures and facilities for the storage and treatment of animal waste;
   f. structures and facilities associated with irrigation, farm pond impoundment, and soil and water conservation;
   g. the accommodation of tourists and visitors within existing residential and/or agricultural structures (see exhibit B) of the farm Property so long as this use is incidental to the agricultural and open space character of the Property;
   h. religious uses including the conduct of religious ceremony on the Property;
   i. other similar uses considered upon written request to the GRANTEE.
3. Residences permitted on the Property subject to these restrictions are: 1) the preexisting single-family dwelling, or, in the event of its destruction, its replacement with a single-family dwelling; and 2) one (1) additional single-family dwelling permitted for use by someone directly involved in the farm operation or for use as a principal residence for the owner of the Property. Other residential uses are prohibited. Residential subdivisions are prohibited, except for one lot of no more than two (2) acres with the existing dwelling or additional permitted dwelling.
4. Agricultural subdivisions are subject to the prior written approval of the GRANTEE. Agricultural subdivisions shall be compatible with the "Subdivision Guidelines for Land Subject to an Agricultural Easement," published by the _____ County Farmland Preservation Board, 199__, as revised. Such Guidelines are made a part hereof in the Baseline Documentation, which is on file at the offices of the GRANTEE and is incorporated by this reference.
5. Any conveyance of land from the Property shall include a clause in the Deed of Conveyance indicating the disposition and allocation of permitted residences between the subdivided portion and the balance of the Property. Upon a conveyance of all or a portion of the Property, any new Deed shall recite verbatim the terms of the Easement.
6. Institutional, industrial, and commercial uses other other than those associated uses described in restrictions 1 and 2 are prohibited.
7. Commercial recreational development and use, involving structures or extensive commitment of land resources (i.e., golf courses, racetracks, and similar uses), shall be prohibited.
8. The commercial extraction of minerals by surface mining and the extraction and removal of topsoil from the Property are prohibited. The extraction of subsurface or deep-mined minerals, including natural gas and oil, and the noncommercial extraction of minerals, including limestone, shale, and other minerals for on-farm use, shall be permitted, but may occupy, at any time, no more than one percent (1%) of the total surface acreage. GRANTOR shall promptly repair any damage to the Property caused by the extraction of subsurface or deep-mined minerals and replace the

surface of the ground to the state that existed immediately prior to the mining so as not to affect the agricultural viability and uses of the Property.

9. Use of the Property for dumping, storage, processing or landfill of non-agricultural solid or hazardous wastes generated off-site is prohibited, including, without limitation, municipal sewage sludge and/or bio solids application.

10. Signs, billboards, and outdoor advertising structures may not be displayed on the Property; however, signs, the combined area of which may not exceed twenty-five (25) square feet, may be displayed only to state the name of the Property, to announce the existence of this conservation easement, and to commemorate the importance of the Property, the name and address of the occupant, to advertise an on-site activity permitted herein, and to advertise the Property for sale or rent.

11. All agricultural production on the Property shall be conducted in accordance with a soil and water conservation plan approved by the _____ County Conservation District or the GRANTEE. Such plan shall be reviewed and updated every ten years and upon any change in the basic type of agricultural production being conducted on the Property. Agricultural lands shall be managed in accordance with sound soil and water conservation practices in a manner that will not destroy or substantially and irretrievably diminish the productive capability of the Property.

12. GRANTOR reserves to him- or herself, and to his or her personal representatives, heirs, successors, and assigns, all rights accruing from his or her ownership of the Property, including the right to engage in or permit or invite others to engage in all uses of the Property that are not expressly prohibited herein and are not inconsistent with the purpose of this Easement.

GENERAL PROVISIONS

1. No right of public access is provided for, nor will result from the recordation of these restrictions.

2. The GRANTEE, its successors or assigns, jointly or severally, shall have the right to enforce these restrictions by injunction and other appropriate proceedings, including, but not limited to, the right to require the GRANTOR to restore the Property to the condition existing at the time of this Grant in order to correct any violation(s) of this Grant of Easement. Representatives of the GRANTEE, their successors or assigns, may at reasonable times and after appropriate notice to the GRANTOR and any persons residing on the Property, enter the Property from time to time for the purposes of inspection and enforcement of the terms of the Easement.

3. Any cost incurred by GRANTEE in enforcing the terms of this easement against GRANTOR, including court costs and attorneys' fees, and any cost of restoration necessitated by GRANTOR's violation of the terms of this Grant shall be borne by the GRANTOR.

4. The restrictions contained herein shall apply to the land as an open-space easement in gross in perpetuity. The covenants, terms, conditions, and restrictions of this easement shall be binding upon and inure to the benefit of the parties hereto and their respective personal representatives, heirs, successors, and assigns and shall continue as a servitude running in perpetuity with the Property. A party's rights and obligations under this Easement terminate upon transfer of the party's interest in the Easement or Property, except that liability for acts or omissions occurring prior to transfer shall survive transfer.
5. If circumstances arise in the future such as to render the purposes of this easement impossible to accomplish, this easement can only be terminated or extinguished, whether in whole or in part, by judicial proceedings in a court of competent jurisdiction.
6. If the easement is taken, in whole or in part, by the exercise of the power of eminent domain, GRANTEE shall be entitled to compensation in accordance with applicable law.
7. GRANTOR agrees to incorporate the terms of this easement in any deed or other legal instrument by which he or she divests him or herself of any interest in all or a portion of the Property, including, without limitation, a leasehold interest. GRANTOR further agrees to give written notice to GRANTEE of the transfer of any interest at least ten (10) days prior to the date of such transfer. The failure of GRANTOR to perform any act required by this paragraph shall not impair the validity of this Easement or limit its enforceability in any way.
8. GRANTOR shall hold harmless, indemnify, and defend GRANTEE and its members, directors, officers, employees, agents, and contractors and their heirs, personal representatives, successors, and assigns (collectively "Indemnified Parties") from and against all liabilities, penalties, costs, losses, damages, expenses, causes of action, claims, demands, or judgments, including without limitation, reasonable attorneys' fees arising from or in any way connected with: (1) the result of a violation or alleged violation of any state or federal environmental statute or regulation including, but not limited to, the Act of October 18, 1988 (P.L. 756, No. 108), known as the Hazardous Waste Sites Cleanup Act, and statutes and regulations concerning the storage or disposal of hazardous or toxic chemicals or materials; (2) injury to or the death of any person, or physical damage to any property, resulting from any act, omission, condition, or other matter related to or incurring on or about the premises, regardless of costs, unless due solely to the gross negligence of any of the Indemnified Parties; and (3) existence and administration of this easement.
9. GRANTOR retains all responsibilities and shall bear all costs and liabilities of any kind related to the ownership, operation, upkeep, and maintenance of the premises, including the maintenance of adequate comprehensive general liability insurance coverage and payment, as and when due, of all real estate taxes.

10. GRANTEE shall record this instrument in a timely fashion in the official records of _____ County, State, and may re-record it at any time as may be required to preserve its rights in this easement.
11. This Grant of Easement in the nature of a restriction is intended to be an easement in gross so as to qualify for a Qualified Conservation Contribution under the applicable provisions of the Internal Revenue Code.
12. GRANTEE agrees that they will hold this easement exclusively for conservation purposes and that they will not assign their rights and obligations under this easement except to an entity (a) qualified, at the time of the subsequent assignment, as an eligible donee under applicable state and federal statutes and regulations to hold and administer this easement, and (b) that has the commitment, resources, and ability to monitor and enforce this easement so that the purposes of this easement shall be preserved and continued. GRANTEE further agrees to obtain the new entity's written commitment to monitor and enforce this Grant of Easement.
13. If any provisions of this easement, or the application thereof to any person or circumstances, is found to be invalid, the remainder of the provisions of this easement, or the application of such provision to persons or circumstances other than those as to which it is found to be invalid, as the case may be, shall not be affected thereby. TO HAVE AND TO HOLD unto GRANTEE, its successors, and assigns forever.

IN WITNESS WHEREOF GRANTOR AND GRANTEE have set their hands on the day and year written.

Signed _____   _____
             GRANTOR   GRANTEE

Date _____   Date _____

Notary Public _____

Attach Exhibit A—Description of the Property

Attach Exhibit B—Sketch of Property Layout

APPENDIX D

Sample Governor's Executive Order

The following sample governor's executive order on farmland preservation policy was adapted from the executive order signed in May of 1994 by Pennsylvania Governor Robert Casey. The governor's executive order puts in place at the state level the same coordination and review by state agencies as those that federal agencies are supposed to follow under the Farmland Protection Policy Act of 1981 to ensure that government projects do not cause the unnecessary conversion of prime farmland.

STATE OF _____
OFFICE OF THE GOVERNOR

WHEREAS, the State of _____'s agricultural land has sustained farm families, farm operations, and rural communities for generations; and

WHEREAS, increased land development and farm costs have caused the State of _____ to lose more than __ percent of its farmland since 1950; and

WHEREAS, the citizens of _____ have invested tax revenues to protect agricultural land through property tax reductions to farmers who put their land to agricultural uses; and

WHEREAS, as part of the State of _____'s continuing efforts to conserve its farmland, assist farm operations, and preserve the quality of life in rural communities, it is in the best interest of all citizens of _____ that the State refine its Agricultural Land Preservation Policy; and

WHEREAS, all state agencies under the Governor's jurisdiction should work together to preserve agricultural lands with a common definition of primary agricultural land and a common vision.

NOW, THEREFORE, be it resolved that I, _____, Governor of the State of _____, by virtue of the authority vested in me by the Constitution of the State of _____ and other laws, do hereby order and direct that all agencies under my jurisdiction seek to mitigate and protect against the conversion of primary agricultural land and adopt policies herewith.

1. It shall be the policy of the State of _____ , through the administration of all agency programs and regulations, to protect the State's prime agricultural land from irreversible conversion to uses that result in its loss as an environmental and essential food production resource.
2. State funds and state-administered federal funds shall not be used to

encourage the conversion of prime agricultural land to other uses when feasible alternatives are available.

3. The prime agricultural land to be protected under this Executive Order shall include lands: 1) in active agricultural use (not including the growing of timber); 2) devoted to active farm use the preceding three years; and 3) in at least one of the categories of agricultural land described below. State agencies shall provide protection to prime agricultural land under this Executive Order based on the following levels of priority:

   a. **Preserved farmland (highest priority).** Farmland that is restricted to agricultural use by an agricultural conservation easement that has been recorded in the appropriate county land records office.

   b. **Farmland planned for agricultural use and subject to effective agricultural zoning (second priority).** Farmland designated for agricultural use in a comprehensive plan and a zoning ordinance adopted pursuant to the State Planning and Zoning Enabling Act, as amended.

   c. **Farmland enrolled in agricultural districts (third priority).** Farmland approved for agricultural districts by local governments after public review and comment.

   d. **Land Capability Classes I, II, and III Farmland and Unique Farmland (fourth priority).** Land Capability Classes I, II, and III Farmland are mapped by the U.S. Department of Agriculture Natural Resources Conservation Service and published as county soil surveys. Unique Farmland is defined as land other than prime farmland that is used for the production of specific high-value food and fiber crops.

4. All agencies under the governor's jurisdiction shall prepare and submit to the governor, with a copy to the Department of Agriculture, no later than six months from the effective date of this order, a document entitled "Guidance for Implementation of the Agricultural Land Preservation Policy." The Department of Agriculture will review the guidance document for conformity with the purposes and goals of this order and advise the Governor's Office as to the consistency of agency actions with the policy established by this order. The guidance document shall include:

   a. A listing of agency actions including land acquisitions, planning, construction, permit review, and financial assistance that may directly or indirectly impact prime agricultural lands.

   b. A statement of agency guidelines and procedures that have been or will be instituted to eliminate or minimize impacts detrimental to the continued use of prime agricultural lands.

   c. A description of any changes in statutes or regulations needed to implement the intent of this order.

5. The following state agencies shall participate in an interagency committee, chaired by the Department of Agriculture, to solve mutual problems in meeting the objectives of this order:

- Governor's Policy Office
- Governor's Budget Office
- Department of Agriculture
- Department of Commerce
- Department of Community Affairs
- Department of Corrections
- Department of Environmental Resources
- Department of General Services
- Department of Transportation

6. The State of _____ Department of Agriculture shall be the lead agency for implementing this policy. All agencies under the governor's jurisdiction shall fully support this agricultural land preservation policy and shall cooperate with the secretary of agriculture by providing assistance and information, as necessary, to carry out the functions and responsibilities hereunder.

7. This order shall take effect immediately.

APPENDIX E

Sample Testamentary Easement

The following language was suggested by Attorney Stephen J. Small for a testamentary easement on real estate. The easement may either be included in a new will or added as a codicil to an existing will. The terms of the easement take effect upon the death of the donor.

Before the donor passes away, the Grantee (either a qualified land trust or a government agency) of the easement should write a letter thanking the donor and notifying the donor that the land trust or government agency will accept the easement.

There should be language in the codicil allowing the executor of the donor's estate to make adjustments to the easement language but nothing that would weaken the easement.

Finally, the value of the easement is determined on the donor's date of death.

CODICIL

I, _____, make this codicil to my last will dated _____, hereby ratifying, confirming, and republishing my said last will in all respects except as modified by this codicil and by codicils dated _____.

1. I give, grant, and devise a conservation easement upon the real property I own in __(County and State)__, in form and content as set forth in the Deed of Conservation easement immediately following with __(Name of Trust or Government Agency)__, (referred to herein and in the Deed of Conservation Easement as "Grantee"), provided that at the time of my death Grantee constitutes a "qualified organization" as defined in Section 170(h)(3) of the Internal Revenue Code of 1986, as amended, and the regulations thereunder (the "Code"), or comparable provision of successor federal revenue laws. If Grantee does not constitute an organization that conforms to the requirements of the preceding sentence at the time of my death or does not receive the foregoing devise for any other reason, then I give, grant, and devise said conservation easement to such one or more qualified organizations as my personal representatives shall select.

   I intend that the foregoing devise be deductible as a charitable gift under Section 2055(a) of the Code as in effect at the time of my death. I authorize and direct my personal representatives to execute, seal, acknowledge, deliver, and record such Deed of Conservation Easement and such confirmatory instruments and to take such other actions as my personal representatives may deem appropriate to effectuate my intentions in this regard, any and all of which instruments and actions shall be effective and binding as of, and relate back to, the time of my death. I further authorize and direct my personal representatives as follows:

(a) To make such modifications, if any, in the terms and conditions of the conservation easement as may be necessary to conform to the requirements of Sections 170(h) and 2055 of the Code or to carry out my intentions, but no such modification shall permit additional residences to be constructed on said land other than residences permitted by the Deed of Conservation Easement set forth below.

(b) To secure the acceptance and approval of the grant, if necessary or desirable, by Grantee and any appropriate governmental authority.

(c) To cause surveys or plans to be made of said real estate or any portion thereof if and to the extent my personal representatives determine the same to be useful or advisable to clarify or otherwise effectuate this grant in any respect, and to pay the costs thereof, and of any service they deem necessary to implement the easement, as an expense of administration. The foregoing shall include specifically (but without limitation) the authority to determine and lay out by survey the locations of and the definitive boundary lines between the various lots and/or Limited Building Areas contemplated by or referred to in the Deed of Conservation Easement set forth below.

2. The interpretation and construction of the provisions of this codicil shall be governed by the laws of the State of _____.

APPENDIX F

Agreement Creating an Urban Growth Boundary

Agreement between the City of Halsey, Oregon, and Linn County, Oregon, for the joint management of the Urban Growth Boundaries, the Plans for the Urban Growth Area, and the Area of Mutual Interest.

WHEREAS, the City of Halsey, Oregon, and Linn County, Oregon, are authorized under the provisions of ORS 190.003 to 190.030 to enter into intergovernmental agreements for the performance of any or all functions that a party to the agreement has authority to perform; and

WHEREAS, ORS 197.175, 197.190, 197.250, 197.275, and 197.285 and OAR 660-03-010 require counties and cities to prepare and adopt comprehensive plans consistent with statewide planning goals, and to enact ordinances or regulations to implement the comprehensive plans; and

WHEREAS, Statewide Planning Goal Number 14 requires that establishment and change of Urban Growth Boundaries shall be a cooperative process between the city and the county that surrounds it; and

WHEREAS, the City of Halsey and Linn County recognize a common concern regarding the accommodation of population growth and utilization of lands adjacent to the City; and

WHEREAS, the City of Halsey and Linn County have adopted coordinated and consistent comprehensive plans that establish urban growth boundaries, a plan for the Urban Growth Area, and policies related to urban development and the provision of urban services within the Urban Growth Area; and

WHEREAS, the City of Halsey and Linn County recognize that as their comprehensive plans and implementing ordinances are amended from time to time they shall remain consistent and coordinated with each other; and

WHEREAS, the City of Halsey and Linn County recognize that it is necessary to cooperate with each other to implement the urbanization policies of their comprehensive plans.

NOW, THEREFORE, THE PARTIES DO MUTUALLY AGREE AS FOLLOWS:

1. *The Intent of Agreement.*
   a. The City of Halsey, Oregon, and Linn County hereby agree to establish a joint management procedure for the implementation of the Halsey Urban Growth Boundary and plan for the Halsey Urban Growth Area, both of which form an integral part of the Halsey Comprehensive Plan. The Halsey Urban Growth Boundary is at-

tached to this agreement as Exhibit A. The area situated inside the Halsey Urban Growth Boundary and outside the Halsey city limits shall be referred to as the Urban Growth Area.
  b. The procedures for implementation of the Urban Growth Boundary and the plan for the Urban Growth Area shall be specified in this agreement.
  c. The City and County further agree to utilize the provisions of this agreement, the Halsey Comprehensive Land-Use Plan, and the Linn County Land-Use Plan, as amended, as the basis for review and action on Comprehensive Plan Amendments, development proposals and implementing regulations that pertain to the Urban Growth Area.
2. *Comprehensive Plan Amendments.*
  a. An amendment to the following comprehensive plan provisions shall be enacted only after agreement by both parties in accordance with plan amendment procedures as established by both jurisdictions and as specified in this section.
      1) An amendment to the City of Halsey Comprehensive Plan text and map as they pertain to the Urban Growth Area, the Urban Growth Boundary, and urbanization policies.
      2) An amendment to the Linn County Comprehensive Plan text and map as they pertain to the Urban Growth Boundary and urbanization policies.
  b. An amendment request may be initiated through either the City or the County. Whichever jurisdiction initiates the plan amendment request shall forward the request to the other jurisdiction within ten (10) days after the request has been initiated. The responding jurisdiction shall be given thirty (13) days to complete its review and make an initial decision on the request. Additional time may be provided at the request of the responding jurisdiction and with the concurrence of the initiating jurisdiction.
  c. If the initial decisions of the City and County are in agreement, they shall be final.
  d If the initial decisions of the City and County differ, a joint meeting or meetings of the City Council and the Board of County Commissioners, or their designers shall be held to resolve the difference. A maximum of 45 days from the date of the initial action by the initiating jurisdiction shall be used to resolve the differences.
  e. If after the forty-five-day (45-day) period is over and concurrence cannot be achieved, the amendment request shall be considered denied. Either party may seek review of the denial from the Land-Use Board of Appeals.
  f. If both parties agree to approve the amendment, the City and County shall then formally amend their comprehensive plans to reflect the agreed upon change.
3. *Review Process for Land-Use Activities.* The City and County shall use the

following process for review and action on development proposals and implementing programs and projects in the Urban Growth Area:
  a. The City shall make recommendations on development proposals, and implementing programs and projects in the Urban Growth Area, including the following:
     1) Amendments to the text or map of the zoning ordinance.
     2) Amendments to the subdivision ordinance.
     3) Conditional use permits.
     4) Planned unit developments.
     5) Land divisions.
     6) Plans, or amendments to plans, for economic or industrial development.
     7) Functional plans, or amendments to plans, for utilities, drainage, recreation, transportation, or other similar activity.
     8) Recommendations for the designation of health hazard areas.
     9) Requests for amendment or establishment of special districts.
  b. The County shall make recommendations on development proposals, and implementing programs in the Urban Growth Area, but which are a responsibility of the City, including the following:
     1) Transportation facility improvements or extensions.
     2) Public water supply, sanitary sewer, or drainage system improvements or extensions.
     3) Public facility or utility improvements or extensions.
     4) Requests for annexations.
  c. The jurisdiction, City or County, that has authority for making a decision regarding a specific development proposal, implementing ordinance, or program, shall formally request the other jurisdiction to review and recommend action for consistency with its comprehensive plan. This request shall allow the reviewing jurisdiction thirty (30) days within which to respond. If the positions of the two jurisdictions differ, every effort will be made to arrive at an agreement.
4. *Area of Mutual Interest.*
  a. The City and County agree to establish an Area of Mutual Interest outside of the Halsey Urban Growth Boundary. A map of the Halsey Area of Mutual Interest is attached to the agreement as Exhibit B. The County shall give the City a minimum of twenty (20) days to review and submit recommendations to the County with regard to the following activities that will apply to the Area of Mutual Interest:
     1) Provisions of the County comprehensive plan or amendments to the plan.
     2) Amendments to the text of the County zoning ordinance and to the zoning map.
     3) Conditional use permits.

4) Planned unit developments.
5) Subdivisions.
6) Major public works projects, including transportation projects.
7) Formation of, or changes of, the boundary or function of special service districts.
8) Other plans or proposals similar to the above.

5. *Special Provisions.*
   a. *Annexations.*
      1) The City of Halsey shall consider annexation of land only under the following circumstances:
         a) The land is contiguous with the city limits.
         b) The development of the property is compatible with the rational and logical extension of utilities and roads to the surrounding area.
         c) The City is capable of providing and maintaining its full range of urban services to the property without negatively impacting existing systems and the city's ability to adequately serve all areas within the existing city limits.
      2) Annexation proposals to the City that are for areas outside the Urban Growth Boundary shall be considered as a request for an amendment to the Urban Growth Boundary and shall be subject to the approval of the City and County as an amendment to the comprehensive plans of each jurisdiction and shall be subject to the provisions of each comprehensive plan.
   b. *Urban Services.*
      1) Extensions of City water and/or sewer services shall be permitted when they are consistent with the policies and proposals of the comprehensive plan.
      2) City services such as sewer, water, and street maintenance shall be provided only to those subdivisions or other development projects that annex to the City.
   c. *Extension of Services Beyond the Urban Growth Boundary.* Provision of City sewer and/or water service capable of supporting development at urban densities shall occur beyond the Urban Growth Boundary only after a determination by affected agencies that a "danger to public health" exists, as defined by ORS 413.705 (5). The service thus authorized shall serve only the area in which the danger exists.

6. *General Provisions.*
   a. *Severability.* The provisions of this agreement are severable. If any sentence, clause, or phrase of this agreement is adjudged by a court or board of competent jurisdiction to be invalid, the decision shall not affect the validity of the remaining portions of this agreement.
   b. *Review and Amendment.* This agreement may be amended at any time by mutual consent of both parties, after public hearing and referral to the City and County planning commissions for a recommendation. Any modifications in this agreement shall be consistent with the

comprehensive plans of the City of Halsey and Linn County.
c. *Termination.* This agreement may be terminated by either party under the following procedure:
   1) A public hearing shall be called by the party considering termination. The party shall give the other party notice of hearing at least forty (40) days prior to the scheduled hearing date. The forty-day period shall be used by both parties to seek resolution of differences.
   2) Public notice of the hearing shall be in accordance with applicable statewide and local goals and statutes.
   3) An established date for termination of the agreement shall be at least one hundred and eighty (180) days after the public hearing in order to provide ample time for reconsideration and resolution of differences.

Attached Exhibit A—Map of the City of Halsey Urban Growth Boundary

Attached Exhibit B—Map of the City of Halsey and County of Linn

Area of Mutual Interest

APPENDIX G

Sample Easement Sale Application Ranking System and Application

County of _____

Farmland Preservation Board

RANKING SYSTEM FOR CONSERVATION EASEMENT SALE APPLICATIONS

Introduction

The ranking system is used to rate and set in priority applications for conservation easement sale. The main purpose of the easement program is to preserve high-quality farms in large blocks. Preference is given to farms under moderate development pressure.

Criteria

Quality of the Farm. 50% of the overall points. Five factors are related to the productivity of the farm and stewardship and historic features. Each factor is assigned a weight (from 1 to 10) and a range of possible point values (from 1 to 10).

The sum of all the factors yields a score for the Quality of the Farm category. That score is adjusted to reflect the Quality category points on a scale from 0 to 50 possible points.

Likelihood of Conversion to Nonfarm Use. 50% of the overall points. Five factors are related to the development pressure on the farm. Each factor is assigned a weight (from 1 to 10) and a range of possible points (from 1 to 10). The weight times the points determines the value for a factor.

To find the total points for a farm, add the points for the Quality of the Farm to the points for the Likelihood of Conversion. Quality of the Farm + Likelihood of Conversion = TOTAL SCORE.

QUALITY OF THE FARM

FACTORS	Weight	Point Value	Score
1. *Size of Farm:*			
100 acres or more	6	10	60
75 to 99.9 acres	6	7	42
40 to 74.9 acres	6	4	24
Less than 40 acres			0
2. *Soils:*			
75% or more Class I, II	10	10	100
50–74% Class I, II	10	8	80
50% or more Class I–III	10	5	50
less than 50% Class I–III			0
3. *Gross Annual Farm Product Sales:*			
$65,000 or more	5	10	50
$25,000–$64,999	5	7	35
less than $25,000 a year	0		
4. *Stewardship:*			
NRCS soil conservation plan on farm	3	10	30
no plan on farm	0		
5. *Historic and Environmental Value:*			
exceptional	1	10	10
significant	1	6	6
some	1	3	3

TOTAL Maximum points for Quality of the Farm: 250 points multiplied by the adjustment factor (1/5) – 50 points maximum

LIKELIHOOD OF CONVERSION TO NONFARM USE

FACTORS	Weight	Point Value	Score
1. *Nonfarm Development in the Area:*			
10 or more nonfarm lots adjacent	10	7	70
20 or more nonfarm lots within 1/2 mile	10	10	100
scattered nonfarm lots within 1 mile	10	4	40
no significant nonfarm development in area			0
2. *Zoning:*			
residential, commercial, or industrial zoning within 1/4 mile	5	6	30
between 1/4 and 1/2 mile away	5	10	50
agricultural or rural zoning within 1/2-mile radius	5	4	20
agricultural zoning covering more than 1/2-mile radius			0
3. *Distance to Sewer Service* (existing or planned within 5 years):			
existing capacity within 1/4 mile	5	6	30
existing capacity within 1/2 mile	5	10	50

existing capacity within 1 mile	5	4	20
no capacity within 1 mile			0

4. *Road Frontage:*

over 1/4 mile of buildable frontage	5	10	50
400–1,320 feet of buildable frontage	5	5	25
less than 400 feet of buildable frontage	0		

5. *Distance to a Farm with an Easement or Easement Sale Application:*

adjacent	10	10	100
within 1/2 mile	10	7	70
more than 1/2 mile			0

TOTAL Maximum points for Likelihood of Conversion: 350 points multiplied by the adjustment factor $(1/7) = 50$ points maximum

Example: Farm A is a 250-acre dairy farm with 50% Class II soils and gross annual sales of $200,000. The farm has a soil conservation plan and provides beautiful scenic views. It is not near a farm under easement and is beyond 1 mile of sewer service. There are some scattered nonfarm houses in the area. The farm has 4,000 feet of road frontage.

Quality of the Farm Factors	Score	*Likelihood of Conversion Factors*	Score
1. Size	60	1. Nonfarm development	40
2. Soils	80	2. Zoning	0
3. Sales	50	3. Distance to Sewer	0
4. Stewardship 30		4. Road Frontage	50
5. Environment 10		5. Distance to Easement	0
TOTAL	230	TOTAL	90
times 1/5 =	46	times 1/7	12.86
TOTAL SCORE	46	+12.86 =	<u>58.86</u>

Office Use: File #_____

County of _____

CONSERVATION EASEMENT SALE APPLICATION

SEND TO: Farmland Preservation Board
222 West Street
Springfield, (State)
Telephone (007) 876–5390

I/We, _____, landowner(s), hereby offer to sell a perpetual agricultural conservation easement on farm property located at _____ in _____ County, (State). The property is identified in the land records of _____ County, Deed Book, ____, page ____. The Conservation Easement Sale to the _____ County Farmland Preservation Board is offered in consideration of

1) an amount to be based on an appraisal and acceptable to buyer and seller or
2) $_____ (circle choice).

Signatures of Landowner(s): _____

Address: _____
_____

Telephone: _____   Date: _____

Social Security Number(s):
_____   _____

Total Acreage of Farm: _____
Acres Proposed for Easement Sale: _____
CROPS GROWN ON LAND PROPOSED FOR EASEMENT SALE FOR THE PAST SEASON

19__	CROP	ACRES GROWN	YIELD PER ACRE
1.			
2.			
3.			
4.			

NUMBER AND KIND OF LIVESTOCK ON THE FARM:
_____

GROSS FARM INCOME IN LAST YEAR: _____
LIST ANY MORTGAGE OR LIEN HOLDER: _____
THE DATE OF THE USDA SOIL CONSERVATION PLAN, IF ANY:
_____

IF FARM IS NOT OWNER OCCUPIED, TENANT'S NAME:
_____

NAME, ADDRESS, AND TELEPHONE NUMBER OF PERSON TO BE CONTACTED TO VIEW THE FARM (if different from landowner):
_____

- -

FOR OFFICE USE ONLY:
Date Received: _____
USGS Topographical Map Showing the Farm: _____
Property Tax Map with Tax Parcel Number of Farm: _____
County Soil Survey Map Number: _____

APPENDIX H

Cooperative Agreement for Public-Private Partnership in Farmland Preservation

COUNTY OF LANCASTER, LANCASTER COUNTY AGRICULTURAL PRESERVE BOARD, AND LANCASTER FARMLAND TRUST

Farmland Preservation Coordination and Cooperation Points of Agreement:

1. *General*

    The parties agree that in Lancaster County, Pennsylvania, both public and private programs are needed to preserve and protect our most productive agricultural lands. Such compatible and complementary efforts enhance our ability to ensure that Lancaster County's exceptionally fertile soil, magnificent landscape, and farming traditions are passed on for generations.

2. *Board of County Commissioners, Lancaster County Agricultural Preserve Board, and Lancaster Farmland Trust, Board of Trustees*

    The parties agree to hold, at least once annually, a meeting of representatives of each of the Boards to review progress and to enhance coordination of farmland preservation activities. An annual January meeting is suggested.

3. *Program Management/Administration*

    The staffs of the Lancaster County Agricultural Preserve Board and the Lancaster Farmland Trust shall remain in close and frequent communication regarding land preservation projects and prospects in order to coordinate the respective organizations' land preservation investments and interests, to avoid the perception of and/or any actual market competition for land preservation interests, to maximize the success of combined efforts of public agencies and private organizations to preserve and protect farmland resources, and to assure the complementary and compatible nature of public and private farmland preservation activities.

4. *Landowners and Preservation Options*

    The parties agree that, in both written material and in person, landowners should be informed of the full array of appropriate conservation tools and options and various public agencies and private organizations that may provide farmland preservation services and investments. The parties agree to refer landowners to one another's programs, when appropriate, to ensure that any farmland preservation transaction makes use of the best possible conservation tools.

5. *Municipal Communication*

    The parties agree to communicate on a periodic basis with officials of the municipalities in which farmland preservation interests are sought or secured concerning the compatibility of the respective organizations' land conservation activities with adopted municipal plans and policies and to report on the nature and location of land conservation interests held. When a prospective land conservation interest is not compatible with adopted or proposed municipal plans and policies, but may be justified by compatibility with other adopted public plans and policies, the parties agree to communicate with elected municipal officials in advance of securing such a conservation interest. Further, it is agreed that the organizations may provide technical and material assistance to municipalities and landowners toward the voluntary establishment of Pennsylvania Agricultural Security Areas (agricultural districts) to the extent that resources permit.

6. *Cooperative Land Conservation Projects*

    The parties agree to consider, when appropriate, and on a case-by-case basis, the possibilities of implementing cooperative farmland preservation projects. Loans, grants, jointly held conservation interests, and reimbursible acquisitions between public agencies and private nonprofit conservation organizations are possible. Each such cooperative endeavor involving actual land conservation interests will be based on a separate and subsequent agreement specifying terms and interests of the respective organizations.

This cooperative agreement shall take effect on the 29th day of August, 1990. In order to further our jointly held goal: the protection of Lancaster County's rich farming heritage.

Signed: Lancaster County Board of County Commissioners
Lancaster County Agricultural Preserve Board
Lancaster Farmland Trust

APPENDIX I

Sample Reimbursable Conservation Easement Acquisition Agreement

This Agreement entered into this ___ day of _____, 19__, between the County of Lancaster, its agency the Lancaster County Agricultural Preserve Board, the Lancaster Farmland Trust, and John and Jane Smith witnesseth that:

WHEREAS, Pennsylvania regulations for the purchase of conservation easements authorize County Boards to request a qualified nonprofit land conservation organization to acquire conservation easements; and

WHEREAS, the County of Lancaster's Agricultural Conservation Easement Program was approved by the State Agricultural Land Preservation Board and the Lancaster Farmland Trust is an approved 501(c)(3) nonprofit land conservation organization; and

WHEREAS, the John and Jane Smith farm property, hereafter "Farm," in _____ Township is pending for sale and time is of the essence to protect the Farm from potential development; and

WHEREAS, the Farm consists of 124.5 acres and predominantly Class II soils, which are regarded as prime, and the preservation of the Farm will contribute significantly to the agricultural productivity of _____ Township and Lancaster County; and

WHEREAS, the parties hereto desire to cooperate to acquire expeditiously a conservation easement on the Farm;

NOW THEREFORE, in consideration of the foregoing and intending to be legally bound, the parties hereto agree as follows:

1. The Board of Commissioners for the County of Lancaster and its agency the Lancaster County Agricultural Preserve Board request the Lancaster Farmland Trust to purchase a conservation easement on the John and Jane Smith farm in _____ Township for a cash price to be determined by appraisal, but in any event not to exceed One Hundred Forty Thousand Dollars ($140,000.00).

2. The Lancaster Farmland Trust will not acquire the easement unless the Farm is located in an Agricultural Security Area (agricultural district) that has been acted upon officially by the _____ Township Board of Supervisors by ___(date)___.

3. If the easement closes, the County of Lancaster will reimburse the Lancaster Farmland Trust for the purchase price of the easement equal to the appraised value of the easement, but not to exceed One Hundred Forty Thousand Dollars ($140,000.00), and in addition, financing costs,

appraisal costs, necessary legal costs, and recording fees. The Lancaster Farmland Trust has been advised by the attorney for the Lancaster County Agricultural Preserve Board that the boundary of the Farm has been plotted and is of sufficient accuracy that a new survey will not be necessary. If the Smiths decide not to accept the easement offer, the Smiths will pay the cost of the appraisal.

4. The Lancaster Farmland Trust will acquire a conservation easement that meets Pennsylvania Agricultural Conservation Easement standards, including the acquisition of a qualified appraisal and a title search. This Agreement is expressly made contingent upon the Smiths holding good and marketable fee simple title to the Farm and the Farm being free and clear of liens and encumbrances.
5. The Commonwealth of Pennsylvania is intended to be the final holder of the conservation easement on the Farm. However, the decision by the Commonwealth to purchase the easement is not a contingency of this Agreement. The County of Lancaster will purchase the easement from the Lancaster Farmland Trust for the consideration contained in paragraph 3 on or before the time set for reimbursement of the purchase price contained in paragraph 6.
6. The County of Lancaster will reimburse the Lancaster Farmland Trust on or before _____(date)_____ .
7. Time is of the essence in this Agreement.
8. The County of Lancaster and its agency the Agricultural Preserve Board have reviewed the Smith farm and determined that the Farm meets the Lancaster County Agricultural Conservation Easement Program eligibility requirements for purposes of acquiring a conservation easement. The County of Lancaster's and the Agricultural Preserve Board's obligations under this Agreement are not contingent upon the occurrence of any event or the satisfaction of any condition except as stated herein.
9. The Smiths warrant that there are no outstanding agreements of sale or options on the Farm.

signed and notarized:
—John and Jane Smith
—Board of Commissioners of Lancaster County
—Chairman, Lancaster County Agricultural Preserve Board
—President, Lancaster Farmland Trust

APPENDIX J

Sample Assignment of Conservation Easement

This Assignment is made this _____(Date)_____ , 199__ , by and between the Lancaster Farmland Trust, a qualified private nonprofit organization created and organized under the laws of the Commonwealth of Pennsylvania, with its mailing address at P.O. Box 1562, Lancaster, Pennsylvania ("Assignor"), and the Lancaster County Agricultural Preserve Board, its successor, nominee, or assign, an agency of the County of Lancaster, a third-class county, created and organized under the laws of the Commonwealth of Pennsylvania, with its offices at 50 North Duke Street, P.O. Box 83480, Lancaster, Pennsylvania ("Assignee").

Background
By Grant of Easement dated _____(Date)_____ , 199__ , the terms of which are fully incorporated herein by reference, John and Jane Smith conveyed an agricultural conservation easement, as defined in the Agricultural Area Security Law (P.L. 128, No. 43), ("Easement"), to Assignor, which instrument is recorded at Book ____ , Page____ , in the Lancaster County Office of the Recorder of Deeds.

Under the terms of the Easement, the Smiths consented to the subsequent assignment of the Easement to Assignee, provided that Assignee assumes all rights, obligations, and responsibilities of Assignor under the Easement, as if Assignee had been originally named as Grantee under the Easement.

In consideration of and incorporating the foregoing, and of the sum of $_____ , receipt of which is hereby acknowledged, and intending to be legally bound, the parties agree as follows:

1. Assignor assigns all its rights, title, and interest in and to the Easement to Assignee.
2. Assignee accepts the assignment and agrees to assume and perform all the duties, obligations, and responsibilities to be performed by Assignor under the terms of the Easement and to observe the terms of the Easement as if it had been named Grantee therein.
3. Assignee agrees to indemnify and hold Assignor harmless for any cost, expense, or liability, including reasonable attorney fees, for performance or nonperformance of the duties and obligations assumed by it.
4. Assignor warrants that it is the sole owner of the interest assigned hereby.

5. Assignor warrants, to the best of Assignor's knowledge, that the Easement has not been altered, modified, or amended in any way.
6. Assignor warrants, to the best of the Assignor's knowledge, that it is not in default under any terms, conditions, or covenants of the Easement.
7. Assignee shall have full power to reassign the interest herein conveyed to the Commonwealth of Pennsylvania without further consent from or notice to Assignor.
8. This Assignment, together with the agreements and warranties contained in it, shall inure to the benefit of Assignee and shall be binding upon the parties hereto, their successors, and assigns.

Signed and notarized:
—John and Jane Smith
—President, Lancaster Farmland Trust
—Director, Lancaster County Agricultural Preserve Board

APPENDIX K

State Agricultural District Laws

STATE	STATUTE
California	CA Gov. Code 51230 to 51298
Illinois	IL Ann. Stat. 1001 to 1020.3
Iowa	IA Code Ann. 176B.1 to 176B.13
Kentucky	KY Ch. 262.850
Maryland	MD Ann. Code 2-509 to 2-515
Minnesota	MN Stat. Ann. 472H.01 to 473H.18
New Jersey	NJ Stat. Ann. 4:1B-1 to 4:1B-15
New York	NY Code Art. 25AA 300 to 309
North Carolina	NC Gen. Stat. 106-735 to 106-743
Ohio	OH Rev. Code Ann. 929.01 to 929.05
Pennsylvania	PA Code 901-915
Virginia	VA Code 15.1-1507 to 15.1-1513

APPENDIX L

IRS Form 8283 for Easement and Property Donations and Bargain Sales

APPENDIX L. IRS FORM 8283 FOR EASEMENT

Form **8283** (Rev October 1995) Department of the Treasury Internal Revenue Service	**Noncash Charitable Contributions** ► Attach to your tax return if you claimed a total deduction of over $500 for all contributed property. ► See separate instructions.	OMB No. 1545-0908 **55**
Name(s) Shown on Your Income Tax Return		Identifying Number

Note: *Figure the amount of your contribution deduction before completing this form. See your tax return instructions.*

Section A — List in this section **only** items (or groups of similar items) for which you claimed a deduction of $5,000 or less. Also, list certain publicly traded securities even if the deduction is over $5,000. (see instructions).

Part I Information on Donated Property — If you need more space, attach a statement.

1	(a) Name and address of the donee organization	(b) Description of donated property
A		
B		
C		
D		
E		

Note: *If the amount you claimed as a deduction for an item is $500 or less, you do not have to complete columns (d), (e), and (f).*

	(c) Date of the contribution	(d) Date acquired by donor (mo, yr)	(e) How acquired by donor	(f) Donor's cost or adjusted basis	(g) Fair market value	(h) Method used to determine the fair market value
A						
B						
C						
D						
E						

Part II Other Information — Complete line 2 if you gave less than an entire interest in property listed in Part I. Complete line 3 if restrictions were attached to a contribution listed in Part I.

2 If, during the year, you contributed less than the entire interest in the property, complete lines a - e.

   a Enter the letter from Part I that identifies the property ► _____. If Part II applies to more than one property, attach a separate statement.

   b Total amount claimed as a deduction for the property listed in Part I: (1) For this tax year ► _____
     (2) For any prior tax years ► _____

   c Name and address of each organization to which any such contribution was made in a prior year (complete only if different than the donee organization above).
     Name of Charitable Organization (donee)

     Address (number, street, and room or suite no.)

     City or Town   State   ZIP Code

   d For tangible property, enter the place where the property is located or kept. .... ►

   e Name of any person, other than donee organization having actual possession of the property.... ►

3 If conditions were attached to any contribution listed in Part I, answer questions a - c and attach the required statement (see instructions):   Yes No

   a Is there a restriction, either temporary or permanent, on the donee's right to use or dispose of the donated property?............

   b Did you give to anyone (other than the donee organization or another organization participating with the donee organization in cooperative fundraising) the right to the income from the donated property or to the possession of the property, including the right to vote donated securities, to acquire the property by purchase or otherwise, or to designate the person having such income, possession, or right to acquire?..

   c Is there a restriction limiting the donated property for a particular use?...

BAA For Paperwork Reduction Act Notice, see separate instructions.   Form **8283** (Rev 10-95)

Form **8283** (Rev 10-95)   Page 2

Name(s) Shown on Your Income Tax Return   Identifying Number

Section B — Appraisal Summary — List in this section only items (or groups of similar items) for which you claimed a deduction of more than $5,000 per item or group. **Exception.** Report contributions of certain publicly traded securities only in Section A. If you donated art, you may have to attach the complete appraisal. See the **Note** in Part I below.

Part I   Information on Donated Property — To be completed by the taxpayer and/or appraiser.

4 Check type of property:

- [ ] Art* (contribution of $20,000 or more)
- [ ] Art* (contribution of less than $20,000)
- [ ] Real Estate
- [ ] Coin Collections
- [ ] Gems/Jewelry
- [ ] Books
- [ ] Stamp Collections
- [ ] Other

* Art includes paintings, sculptures, watercolors, prints, drawings, ceramics, antique furniture, decorative arts, textiles, carpets, silver, rare manuscripts, historical memorabilia, and other similar objects.

Note: *If your total art contribution deduction was $20,000 or more, you must attach a complete copy of the signed appraisal.*

5	(a) Description of donated property (if you need more space, attach a separate statement)	(b) If tangible property was donated, give a brief summary of the overall physical condition at the time of the gift	(c) Appraised fair market value
A			
B			
C			
D			

	(d) Date acquired by donor (mo, yr)	(e) How acquired by donor	(f) Donor's cost or adjusted basis	(g) For bargain sales, enter amount received	See Instructions	
					(h) Amount claimed as a deduction	(i) Average trading price of securities
A						
B						
C						
D						

Part II   Taxpayer (Donor) Statement — List each item included in Part I above that is separately identified in the appraisal as having a value of $500 or less. See instructions.

I declare that the following item(s) included in Part I above has to the best of my knowledge & belief an appraised value of not more than $500 (per item). Enter identifying letter from Part I and describe the specific item. . . . . ▶ _____

Signature of Taxpayer (donor) ▶   Date ▶

Part III   Declaration of Appraiser

I declare that I am not the donor, the donee, a party to the transaction in which the donor acquired the property, employed by, or related to any of the foregoing persons, or married to any person who is related to any of the foregoing persons. And, if regularly used by the donor, donee, or party to the transaction, I performed the majority of my appraisals during my tax year for other persons.

Also, I declare that I hold myself out to the public as an appraiser or perform appraisals on a regular basis; and that because of my qualifications as described in the appraisal, I am qualified to make appraisals of the type of property being valued. I certify that the appraisal fees were not based on a percentage of the appraised property value. Furthermore, I understand that a false or fraudulent overstatement of the property value as described in the qualified appraisal or this appraisal summary may subject me to the penalty under section 6701(a) (aiding and abetting the understatement of tax liability). I affirm that I have not been barred from presenting evidence or testimony by the Director of Practice.

Sign Here   Signature ▶   Title ▶   Date of Appraisal ▶

Business Address (including room or suite no.)   Identifying Number

City or Town   State   ZIP Code

Part IV   Donee Acknowledgement — To be completed by the charitable organization.

This charitable organization acknowledges that it is a qualified organization under section 170(c) and that it received the donated property as described in Section B, Part I above on ▶ _____ (Date)

Furthermore, this organization affirms that in the event it sells, exchanges, or otherwise disposes of the property described in Section B, Part I (or any portion thereof) within 2 years after the date of receipt, it will file **Form 8282**, Donee Information Return, with the IRS and give the donor a copy of that form. This acknowledgement does not represent agreement with the claimed fair market value.

Name of Charitable Organization (donee)   Employer Identification Number

Address (number, street and room or suite no.)   City or Town   State   ZIP Code

Authorized Signature   Title   Date

Notes

Chapter I
1. Quoted in *Farmland Preservation Report* 3, no. 2 (November 1992): 3.
2. American Farmland Trust, *Alternatives for Future Urban Growth in California's Central Valley: The Bottom Line for Agriculture and Taxpayers* (Washington, D.C.: American Farmland Trust, 1995).
3. Bank of America, Greenbelt Alliance, California Agency of Natural Resources, and the Low Income Housing Fund, *Beyond Sprawl* (1995). Available from Bank of America, Environmental Policies and Programs, #5800, P.O. Box 37000, San Francisco, CA 94137.
4. Quoted in Dan Looker, *Farmers for the Future* (Ames: Iowa State University Press, 1996), 141.
5. See Peirce Lewis, "The Urban Invasion of Rural America: The Emergence of the Galactic City," in E. Castle, ed., *The Changing American Countryside* (Lawrence: University of Kansas Press, 1995), 39–52.
6. Quoted in *Farmland Preservation Report* 3, no. 6 (May 1993): 2, 7.
7. Figures from U.S. Department of Agriculture, *1992 Summary Report, Natural Resources Inventory* (Washington, D.C.: U.S. Department of Agriculture, 1994).
8. Peter Wolf, *Land in America: Its Value, Use, and Control* (New York: Pantheon Books, 1981), 531.
9. American Farmland Trust, *Farming on the Edge* (Washington, D.C.: American Farmland Trust, 1993).
10. Aggregating farmland loss on a national basis can be misleading. The loss of an acre of farmland in rural Massachusetts is not the same as the loss of a farmland acre in Fresno County, California, the nation's leading agricultural county.
11. In 1995, Utah Governor Michael Leavitt held a three-day growth summit to find solutions to the burgeoning growth in his state. New Jersey voters approved a $50 million bond measure for purchasing development rights to farmland. Colorado voters in Aspen, Boulder, and three other towns put slow-growth initiatives on the ballot. Voters in Portland, Oregon's, metropolitan service district approved a $135 million bond issue to buy park land, natural areas, and open space. Montgomery County, Maryland, adopted a 100 percent abatement on county property taxes on lands under a perpetual conservation easement. The Delaware legislature authorized $12 million to implement the state's purchase-of-development-rights program to preserve farmland for farming. Voters in Bainbridge Island, Washington, passed a $2.57 million bond to purchase a 327-acre lakeshore area. Scottsdale, Arizona, voters approved a one-fifth of 1 percent (.2 percent) sales tax for the next thirty years to raise $240 million to create a 2,860-acre desert preserve. The Michigan Farm Bureau called farmland preservation its number one issue. New York appropriated $300,000 for county farmland protection planning grants. Virginia Beach, Virginia, began the first purchase-of-development-rights program to preserve farmland in Virginia. A unique coalition of the California Farm Bureau Federation, the Sierra Club, and Tuolomne County won a judgment in the county's superior court against developers who proposed to build condominiums

on land protected under the state's contract property tax relief program for farmland (known as the Williamson Act).

Chapter 2
1. Quoted in *Farmland Preservation Report* 3, no. 12 (September 1993): 4.
2. Quoted in *Farmland Preservation Report* 6, no. 2 (November 1995): 3.
3. Phillip H. Allman, *The Economic Structure of San Joaquin County and the Impact of Agricultural Land Displacement on County Income* (Stockton, Cal.: San Joaquin County, 1993).
4. Henry L. Diamond and Patrick F. Noonan, eds., *Land Use in America* (Washington, D.C.: Island Press, 1996), 239.

Chapter 3
1. Diamond and Noonan, eds., *Land Use in America*, 240.
2. Perhaps the most striking example of development rights being a creation of government comes from Britain, where in 1947 all private development rights in land were nationalized. A landowner retains the right to use the land in its current use. But for any development, other than farm buildings, a landowner must ask for and receive permission from the local government. A landowner in Britain has no specific right to develop his land, and the government need not grant permission to develop land.
3. Quoted in *Farmland Preservation Report* 3, no. 8 (May 1993): 7.
4. For an introduction to the planning process, including the comprehensive plan, zoning, subdivision regulations, and capital improvements plan, see T. Daniels, J. Keller, and M. Lapping, *The Small Town Planning Handbook*, 2nd edition (Chicago: Planners Press, 1995).
5. Nearly all larger communities and counties have professional planners on staff to assist the planning commission and elected officials. Some Maryland counties do not appoint a planning commission, but rather rely on a professional planning staff.
6. Shrewsbury Township, York County, PA. *Comprehensive Plan*, 1980, p.16.
7. In Ohio, township and county zoning ordinances must be approved by a voter referendum.
8. California Farm Bureau Federation News Release, February 24, 1994.
9. Quoted in Land Trust Alliance, *Rally '91 Workbook* (Washington, D.C.: Land Trust Alliance, 1991), 327.
10. See Jay E. Closser, "Assessing Land under Conservation Restrictions," *Assessment Journal* (July-August 1994): 20–24.
11. Communities throughout the United States have attempted to slow the rate of residential growth to keep a handle on local finances and to protect open space. For example, limits on the number of building permits issued in a year have been enacted in Petaluma, California, and Litchfield, New Hampshire. Ramapo, New York, will not build infrastructure until there is sufficient demand to pay for it. Many communities have from time to time imposed sewer moratoria, halting new hook-ups to sewer lines until new sewer capacity can be built.
12. Quoted in *Farmland Preservation Report* 3, no. 6 (March 1993): 4. From, Ad Hoc Associates, *Land Conservation and Local Property Taxes* (Montpelier, VT: Vermont Natural Resources Council and Vermont League of Cities and Towns, 1990).

Chapter 4
1. Quoted in *Farmland Preservation Report* 6, no. 2 (November-December 1995): 4.
2. Robert West Howard, *The Vanishing Land* (New York: Ballantine Books, 1985), 281.
3. Quoted in *Pennsylvania Farmer* 227, no. 1 (January 1996): 22.
4. Dixon Esseks, "The Politics of Farmland Prservation," in Don F. Hadwiger and William P. Browne, eds., *The New Politics of Food* (Lexington, Mass.: Lexington Books, 1978), 210.

5. *Agins v. City of Tiburon* 447 US 255 (1980).
6. Quoted in *Farmland Preservation Report* 4, no. 6 (May 1994): 3.

Chapter 5

1. Quoted in *Farmland Preservation Report* 2, no. 7 (April 1992): 2.
2. Ibid.
3. Quoted in *Farmland Preservation Report* 1, no. 7 (April 1991): 3.
4. U.S. Department of Agriculture, *Recommendations on Prime Lands* (Washington, D.C.: U.S. Department of Agriculture, 1975).
5. In Britain, by comparison, there has been a national policy to protect as much farmland as possible since the Town and Country Planning Act of 1947. Local governments have carried out this policy in three ways. First, farmers do not pay property taxes on their land or on most farm buildings. Second, local governments often cause developers to look for land other than farmland to develop. And third, local government comprehensive plans generally designate farmland for farm uses only.
6. Specifically, farmland is defined as prime, unique, statewide, and locally important. Prime farmland is Class I and Class II land, defined by the Natural Resources Conservation Service. Prime farmland is the easiest land to farm in terms of gentle slope, soil conservation practices needed, and fertilizer and energy inputs. Unique farmland produces special high-valued crops, such as orchard land, which may have a steep slope (15–25 percent), classified as Class IV farmland by the NRCS. Farmland of statewide and local importance is usually Class III land with a slope of 8–15 percent.
7. *Farmland Preservation Report* 2, no. 8 (May 1992): 2.
8. Keith Wiebe, Abebayehu Tegene, and Betsey Kuhn, *Partial Interests in Land: Policy Tools for Resource Use and Conservation* (Washington, D.C.:U.S. Department of Agriculture, Economic Research Service Report Number 744, 1996), 21.
9. Wiebe, Tegene, and Kuhn, p. 23.

Chapter 6

1. In 1987, the Indiana Court of Appeals upheld the state's right-to-farm law in *Shatto v. McNulty* 509 NE 2nd 897 (IN 1987). McNulty had been raising hogs since 1956. His farm was located in a farming area and zoned for agriculture. Shatto purchased 15 acres in 1970 and built a house directly across the road from the hog pens. Shatto then filed suit against McNulty, claiming that the hog operation was a nuisance. The court noted that the McNulty hog enterprise preexisted the Shatto house, and there was no evidence that McNulty had operated the farm improperly.

    In 1982, a Michigan appeals court supported the state's right-to-farm law in *Rowe v. Walker*, No. 81-228769, slip op., Mich. App., 1982. Walker operated a grain-drying facility that produced enough noise to bother neighbor Rowe. Rowe claimed the grain-drying facility was not a farm enterprise and hence did not qualify for protection under the right-to-farm law. The court found that the grain dryer was located in an agricultural area, preceded Rowe's move to the neighborhood, and had been operated properly. Above all, the court pointed to the language of the right-to-farm law, which applied to "land, buildings, and machinery used in commercial production of farm products."

    In 1982, a Connecticut superior court cited the state's right-to-farm law in upholding a dairy farmer's practice of spreading manure (*DeCapu v. Cella*, No. 19-85-59, slip op., New Haven Super. Ct., 1982). A neighbor, DeCapu, complained that the smell of manure and the resulting flies were offensive. The court determined that the dairy farmer, Cella, was spreading manure only in the summer and therefore it was not continuous or substantial. Also, the court found that the spreading of manure

on the land preceded the building of the housing development in which DeCapu lived. Finally, Cella was operating according to proper farming practices.
2. See D.A. Bradbury, "Agriculture Law: Suburban Sprawl and the Right to Farm," *Washburn Law Review* 22, no. 2 (1983): 448–468.
3. *Burlington Free Press*, Burlington, Vermont, June 3, 1994, 1.
4. Most state laws require local governments to conduct a new assessment of property value every five years or so. In a rural farming community, the highest and best use is not likely to change much over five years. In growing communities, however, the value of highest and best use of land (for building sites) can skyrocket. Hence, use-value taxation of farmland provides some protection against a soaring tax bill based on highest and best use. Even so, the tax rate may have to rise to pay for additional public services demanded by the growing population.
5. J. MacKenzie and G. Cole, "Use-Value Assessment as a Farmland Preservation Policy," in W. Lockeretz, ed., *Sustaining Agriculture Near Cities* (Ankeny, Iowa: Soil and Water Conservation Society, 1987), 251–262.
6. Quoted in *Farmland Preservation Report* 2, no. 7 (April 1992): 7.
7. Property tax abatement is not a new concept. For decades, English farmers have paid no property taxes on their farmland and only some taxes on their farm buildings.

Chapter 7

1. Quoted in Robert E. Coughlin and Stanford Lembeck, *Agricultural Protection Zoning: A Key Element in an Effective Agricultural Protection Program for Pennsylvania Municipalities* (Harrisburg: Center for Rural Pennsylvania, 1995), 35.
2. Each state has a zoning enabling act, which gives a county or municipality the power to use zoning. The act may or may not expressly authorize the use of agricultural zoning. In Pennsylvania, the Municipalities Planning Code says that zoning ordinances can be designed to "preserve prime agriculture and farmland considering topography, soil type, and classification, and present use"—Pennsylvania Statutes Annotated, Section 11604(3). Connecticut law does not specifically allow agricultural zoning.
3. In *Lucas v. South Carolina Coastal Council*, 112 S. Ct. 2886 (1992), the U.S. Supreme Court stated that "if a regulation denies a landowner all economic use of his or her land, then the regulating government must pay compensation."
4. Legal cases in California, Illinois, Maine, Maryland, New Jersey, Oregon, and Pennsylvania have upheld the validity of agricultural zoning.

   In *Gisler v. County of Madera*, 38 Cal. App. 3d 303, 112 Cal. Rptr. 919 (1974), an appeals court found that Madera County's 18-acre minimum parcel size in the agricultural zone did not result in a taking of private property.

   In *Wilson & Voss v. McHenry County*, 416 N.E. 2d 426 (1981), an Illinois appeals court upheld the county's 160-acre minimum lot size requirement as being in the public interest.

   In *Gardner v. New Jersey Pinelands Commission*, 125 NJ. 193, 593 A.2d 251 (1991), a New Jersey court ruled that agricultural zoning did not constitute a taking of private property, nor did it violate the right to equal protection.

   In *Giguere v. Inhabitants of the City of Auburn*, 398 A 2d. 514 (1978), the Supreme Court of Maine upheld the right of a local government to use agricultural zoning.

   A Maryland circuit court concluded that agricultural zoning in Montgomery County did not result in a taking and was permissible under Maryland law (*Dufour v. Montgomery County*, Law no. 56964, Montgomery County Circuit Court, 1983).

   An Oregon appeals court determined that the downzoning of land from low-density residential to farm and forest uses was not a taking of private property because the land can still be "beneficially" used for farming (*Joyce v. City of Portland*, 24 Or. App. 689, 546 P.2d 1000, 1976).

The Pennsylvania Supreme Court upheld a sliding scale agricultural zone in *Boundary Drive Associates v. Shrewsbury Township Board of Supervisors*, 507 Pa. 481, 491 A.2d 86 (PA 1985). The Pennsylvania Commonwealth Court (an appeals court) found that fixed-area-ratio zoning in *Codorus Township v. Rogers*, 492 A.2d 73 (Pa. Commw. 1985), served a public purpose and did not amount to a taking of private property without just compensation.

A Wisconsin appeals court found that agricultural zoning did not mean a taking in *Petersen v. Dane County*, 402 N.W.2d (Wis. App. 1987).

5. Some states exempt agricultural-related construction from local zoning ordinances. North Carolina, for example, does not require even a 10,000-head hog operation to meet local zoning standards. While we believe that agricultural uses should be allowed in agricultural zones, large and intensive livestock operations should be required to show how proper manure disposal and water quality will be maintained.
6. Exclusive Farm Zones have been used on a large scale in Carbon County, Wyoming; Utah County, Utah; Boone County, Illinois; Crook County, Oregon; Santa Cruz County, California; and the city of Auburn, Maine.
7. Sharon L. Emelock, "Wisconsin: Managing Growth and Limiting Taxes," in Hal Hiemstra and Nancy Bushwick, eds., *Plowing the Urban Fringe* (Fort Lauderdale, Fla.: Florida Atlantic University, 1989), 9–21.
8. In New England, Connecticut and Rhode Island do not specifically allow agricultural zoning.
9. Charles Johnson and Linda H. Smith, "When Urban Sprawl Chokes Agriculture," *Farm Journal* (March 1996): 27–28.
10 Quoted in *Farmland Preservation Report* 4, no. 1 (October 1993): 7.
11. Quoted in *Farmland Preservation Report* 4, no. 1 (October 1993): 6.
12. Quoted in James E. Peters, "Saving Farmland: How Well Have We Done?" *Planning* 56, no. 9 (1990): 16.
13. Quoted in *Farmland Preservation Report* 5, no. 10 (September 1995): 2.
14. Quoted in *Farmland Preservation Report* 5, no. 10 (September 1995): 4.
15. Quoted in *Farmland Preservation Report* 5, no. 10 (September 1995): 5.
16. See ORS Chapter 215. Goal 3 of the Oregon Land Use Act states:

> To preserve and maintain agricultural lands.
>
> Agricultural lands shall be preserved and maintained for farm use, consistent with existing and future needs for agricultural products, forest and open space. These lands shall be inventoried and preserved by adopting exclusive farm use zones pursuant to ORS Chapter 215. Such minimum lot sizes as are utilized for any farm use zones shall be appropriate for the continuation of the existing commercial agricultural enterprise within the area.

The legislative "Agricultural Land Use Policy" (ORS 215.243) states that since exclusive farm use zoning substantially limits alternative uses of agricultural land, incentives and privileges are justified in order to hold such land in exclusive farm use zones, including:

- deferred property taxation
- right-to-farm protection
- farm use-value taxation for inheritance tax purposes
- assurance that only compatible nonfarm uses will be allowed within the exclusive farm use zone.

17. U.S. Department of Commerce, Bureau of the Census, *Census of Agriculture, 1992* (Washington, D.C.: U.S. Government Printing Office, 1994).
18. See T. Daniels and A. Nelson, "Is Oregon's Farmland Preservation Program Working?" *Journal of the American Planning Association* 52, no. 1 (1986): 22–32.
19. Oregon Department of Land Conservation and Development, *Development Activity in Exclusive Farm Use Zones* (Salem, Ore.: Department of Land Conservation and Development, 1991).

Chapter 8

1. Urban Land Institute, *The Costs of Alternative Development Patterns: A Review of the Literature* (Washington, D.C.: Urban Land Institute, 1992).
2. Quoted in *Farmland Preservation Report* 2, no. 8 (May 1992): 4.
3. V. Gail Easley, *Staying Inside the Lines: Urban Growth Boundaries,* Planners Advisory Service Report Number 440 (Chicago, Ill.: American Planning Association, 1992), 5.
4. Oregon Department of Land Conservation and Development, *Oregon's Statewide Planning Goals* (Salem, Ore.: Oregon Department of Land Conservation and Development, 1988), 13.
5. Gerrit J. Knaap and Arthur C. Nelson, *The Regulated Landscape: Lessons on State Land Use Planning from Oregon* (Cambridge, Mass.: Lincoln Institute, 1992), 137.
6. Arthur C. Nelson, "Policies to Preserve Prime Farmland in the U.S.A.," *Journal of Rural Studies* 6, no. 2 (1990): 151–166.
7. Quoted in the *Philadelphia Inquirer,* Dec. 1, 1991, 6-B.
8. California Government Code Section 56377.
9. Yolo County, California, Local Agency Formation Commission Agricultural Conservation Policy, March 1994.
10. New Jersey Office of State Planning, *The Final Report of the Economic Impact Assessment of the New Jersey Interim State Plan* (Trenton: Office of State Planning, 1992).

Chapter 9

1. Quoted in *Lancaster Sunday News,* Sept. 17, 1989, 9.
2. Purchase of development rights is referred to as the purchase of agricultural conservation easements (PACE) by the American Farmland Trust. We use the terms *development rights* and *conservation easement* to mean the same thing. Technically, a government or land trust purchases a "negative easement in gross" on the property. Whereas a right-of-way easement, such as for a power line, grants access across a property to a non-owner, a conservation easement restricts nearly all development on the property.
3. Only Alabama, Oklahoma, West Virginia, and Wyoming do not have laws allowing governments to acquire development rights.
4. Quoted in Vermont Land Trust, *1996 Summer Quarterly Report,* 4.
5. In general, we believe that if the cost of the development rights exceeds 60 percent of a property's fair-market value, then PDR is probably not the right technique. A government should consider buying the entire property fee simple or else buying development rights to other, more affordable properties and more acreage.
6. For example, voters in Buckingham Township, Pennsylvania approved a $4 million PDR program in November of 1995. The compelling reason for the program was a need to control school populations to avoid having to build expensive new schools. According to Bucks County PDR administrator Rich Harvey, "The underlying theme is it costs more to build a school than preserve the land." Quoted in *Farmland Preservation Report* 6, no. 4 (February 1996): 2.
7. Personal communication, Albert J.A. Young, Attorney, Bel Air, MD, April 1994.

8. Quoted in *Farmland Preservation Report* 3, no. 9 (May 1992): 3.
9. The state of Maryland exempts the sale of a preserved farm from the .5 percent real estate transfer tax.
10. If a county or state uses the securitized installment purchase agreement, it is wise to avoid a written policy on discounting offers to landowners. One of the attractions of the discount is presenting a landowner with a potential income tax deduction. If there is a written policy to offer landowners only 60 percent of the appraised easement value, the Internal Revenue Service could conceivably disallow the deduction on the remaining 40 percent. The IRS could reason that the county or state has an established policy and the landowner has no choice but to abide by the county policy, and therefore there is no "donative intent" in the landowner's donation of 40 percent of the easement value.

    A similar problem may exist in counties or states that have a written policy for a cap on the amount per acre for easement payments when a landowner attempts to use the appraised amount above the cap as an income tax deduction. For example, if the appraised value of the easement is $3,000 per acre but the county has a $2,000-an-acre cap on easement payments, the landowner may try to use the $1,000-per-acre difference as an income tax deduction. Landowners should be warned that any claim of a charitable deduction for income tax purposes is a matter between the landowner and the IRS.
11. Internal Revenue Service Letter Rulings 9215049 and 9232030. A private letter ruling is a response from the Internal Revenue Service to a specific tax question posed by an attorney on behalf of a client. The private letter ruling cannot be cited as precedent by other taxpayers, yet the ruling indicates how the service would likely view similar tax situations.
12. The following language could be used in a voter referendum on the sale of bonds to finance a purchase-of-development-rights program: "Do you favor the incurring of indebtedness by the State of _____, the County of _____, of $X,000,000 for the purchase of agricultural conservation easements for the preservation of agricultural land in perpetuity?"
13. *Farmland Preservation Report* 6, no. 9 (July-August 1996): 5.
14. *Dolan v. City of Tigard.*

Chapter 10

1. John J. Costonis, *Space Adrift: Landmark Preservation and the Marketplace* (Washington, D.C.: National Trust for Historic Preservation, 1974).
2. See Amanda Jones Gottsegen, *Planning for Transfer of Development Rights: A Handbook for New Jersey Municipalities* (Mount Holly, N.J.: Burlington County Board of Chosen Freeholders, 1992). This is, to our knowledge, the most complete guide available on designing and implementing a TDR program. It is available from: Burlington County Land Use Office, 49 Rancocas Road, Mount Holly, NJ 08060, or call (609) 265–5787.
3. County and municipal governments may claim that they have the authority to create a TDR program as an extension of their state's planning and zoning enabling legislation and the Tenth Amendment to the Constitution, which allows local governments to use "police power" to protect the public health, safety, and welfare.

    A TDR ordinance is similar to a zoning ordinance, because of the sending and receiving areas involved. And, like zoning, a TDR ordinance must be consistent with the community's comprehensive plan. In fact, a TDR program is perhaps best understood as a hybrid of zoning and the purchase of development rights.
4. Gottsegen, *Planning for Transfer of Development Rights*, 31.

5. Amanda Jones Gottsegen and Charles Gallagher, *The Fiscal Impacts of Growth under Three Development Patterns: Transfer of Development Rights, Existing Zoning Ordinance, and Agricultural Zoning: Case Study: Chesterfield Township, New Jersey* (Mount Holly, New Jersey: County of Burlington Office of Land Use Planning, 1991).
6. Attributed to Bob Wagner of the American Farmland Trust.
7. *Farmland Preservation Report* 6, no. 4 (February 1996): 2.
8. Places that have seriously examined a TDR program include: San Luis Obispo County and Santa Barbara County, California; Dade County and Palm Beach County, Florida; Scott County, Kentucky; Baltimore County, Harford County, and Talbot County, Maryland; Chesterfield Township in Burlington County, New Jersey; Wake County, North Carolina; East Hempfield Township in Lancaster County, East Nantmeal Township in Chester County, and Oley Township in Berks County, Pennsylvania. In 1990, the state of Kentucky passed legislation enabling its counties to create TDR programs.
9. Quoted in *Farmland Preservation Report* 6, no. 2 (November-December 1995): 7.

Chapter 11

1. Quoted in *Farmland Preservation Report* 6, no. 2 (November 1995): 2.
2. Reported in *Farmland Preservation Report* 1, no. 10 (July 1991).
3. Quoted in the *Christian Science Monitor*, Jan. 4, 1996, p. 10.
4. Conservation easements were first used in and near Boston in the 1880s to protect parks designed by Frederick Law Olmsted. In the 1930s, the National Park Service acquired extensive easements along the Blue Ridge Parkway in Virginia. But easements did not really become popular until the 1970s, when several states passed legislation allowing the use of easements. Today, nearly all states allow landowners to sell or donate easements.
5. Quoted in *California Farmer*, June 1, 1985, 37.
6. There are "positive" and "negative" easements. A positive easement allows someone to cross a property that he or she does not own. For example, a common positive easement is a right-of-way for a utility power line to cross private property. Another positive easement is an access right-of-way, such as a driveway, to provide access to a parcel of land that has no access to a public road. A conservation easement is a "negative easement in gross" because it restricts the use of a parcel of land.
7. Occasionally, a land trust may need to amend the terms of an easement. The following questions are important for the land trust to discuss in amending the easement: What precedents might you set? What do you have to gain or lose? and What is the landowner's interest?

    As much as possible, the land trust should stick to the letter of the easement. If the land trust decides to amend the easement, there are five general types of changes: 1) to clarify the language of the easement; 2) to make corrections in the easement (e.g., to cite the correct acreage or include a correct legal description); 3) to place additional conditions in the easement to make it more strict; 4) to make needed changes in the management of the property; and 5) to extinguish the easement if the impacts of land uses on surrounding lands render the conservation purpose of the property impractical.

    A land trust should have an amendment policy, such as: 1) the change will improve the easement or will be neutral; 2) the amendment should be made in the form of a written agreement between the land trust and the landowner; 3) the amendment does not affect the value of an easement donation or the land trust; 4) the change is consistent with the easement; and 5) the amendment will be included in the easement and recorded with the recorder of deeds.

8. See Internal Revenue Code Section 170(h), especially Section 170(h)(4). For farmland easements, see IRS Letter Rulings 8422064, 8544036, 862307, 8711054. See also Treasury Regulations Sections 1.170A-14 and Stephen J. Small, *The Federal Tax Law of Conservation Easements* (Washington, D.C.: Land Trust Alliance, 1995).
9. If the land is held by a corporation or trust, an easement donation may be used on up to only 10 percent of that year's income. But the easement will reduce the value of the stock of the corporation for estate tax purposes.
10. If a landowner preserves land through donating a conservation easement to a land trust and then gives the preserved land to the trust, the deduction will be limited to the restricted value of the land. The better way to accomplish the landowner's goal is for the landowner to give the land to the land trust and have the land trust put on the conservation easement. That way the landowner can receive an income tax deduction based on the fair-market value of the land when the landowner gave it to the land trust (Internal Revenue Service Revenue Ruling 85–99).
11. A donation of cash may be used as an income tax deduction on up to 50 percent of adjusted gross income. A donation of appreciated stock may be used as an income tax deduction on up to only 30 percent of adjusted gross income. See also Internal Revenue Service code section 2055.
12. Eve Endicott, ed., *Land Conservation through Public/Private Partnerships* (Washington, D.C.: Island Press, 1995), 196.
13. Vermont Housing and Conservation Board, *1995 Annual Report* (Montpelier: Vermont Housing and Conservation Board, 1996).
14. *Lancaster New Era*, March 21, 1995, p. 1.
15. Elizabeth Brabec, "On the Value of Open Spaces," *Scenic America Technical Bulletin* 1, no. 2. (Washington, D.C.: Scenic America, 1992).
16. See Internal Revenue Regulations Sections 1.170A-14(h)(3) and (4).
17. Quoted in *Farmland Preservation Report* 6, no. 2 (November 1995): 3.
18. Reported in the Vermont Land Trust newsletter, Montpelier, VT, Winter 1996.

Chapter 12

1. David Mas Masumoto, *Epitaph for a Peach* (New York: HarperCollins, 1996), 105.
2. U.S. Department of Agriculture, *The New Generation of American Farmers: Farm Entry and Exit Prospects for the 1990's* (Washington, D.C.: U.S. Department of Agriculture, 1995).
3. John Hildebrand, *Mapping the Farm* (New York: Alfred A. Knopf, 1995), 4–5.
4. Hildebrand, pp. 235–236.
5. A potential trap is that if the inheritor uses Section 2032A and then sells or donates a conservation easement on the farm within ten years of inheriting the farm, it could trigger estate tax recapture as if it were the sale of the farm. See IRS Letter Rulings 87-31001 and 89-40011. Lately, however, the IRS position has become less clear. In 1992 and 1996, two New Jersey farms under Section 2032A sold conservation easements, and the IRS did not require estate tax recapture. Unfortunately, the IRS did not render an official ruling in either case. But in a recent Maryland case, landowners who sold an easement while under the ten-year shadow of 2032A were ruled to be subject to estate tax recapture. See *Farmland Preservation Report* 6, no. 8 (June 1996): 4.
6. The 2032A special valuation need not apply to the entire farm property but only to as much of the farm as is worth $750,000. Farmland not subject to Section 2032A could be sold to help pay estate taxes, or a conservation easement could be sold or donated within ten years without triggering estate tax recapture.
7. IRS Revenue Ruling 77–414 allows a landowner to use the basis in the land and

buildings to offset the taxable capital gain from the sale of a conservation easement (development rights).
8. According to Section 1.170A-14(h)(3) of the Income Tax Regulations: "If the granting of a perpetual conservation restriction has the effect of increasing the value of any other property owned by the donor or a related person, the amount of the deduction for the conservation contribution shall be reduced by the amount of the increase in the value of the other property, whether or not such property is contiguous."

Chapter 13
1. This section was adapted in part from Thomas L. Daniels, "Where Does Cluster Zoning Fit in Farmland Protection," *Journal of the American Planning Association* (Winter, 1997), 127–135.
2. Quoted in the *Lancaster New Era*, March 20, 1996.
3. *Baltimore Sun*, Jan. 29, 1995, p. 1L.
4. John Hart, *Farming on the Edge: Saving Family Farms in Marin County, California* (Berkeley: University of California Press, 1991), 145.
5. *Lancaster New Era*, Sept. 13, 1994, A-5.
6. *Lancaster New Era*, Sept. 12, 1994, A-6.

Chapter 14
1. *Hawaii Department of Budget & Finanace, Growth Management Issues in Hawaii* (Honolulu: State of Hawaii, 1977), 3.
2. *The Economist* (February 10, 1996): 30, and (December 2, 1995): 21.
3. *Washington Post*, March 17, 1996, 1.
4. See, for example, Thomas Hylton, *Save Our Land, Save Our Towns: A Plan for Pennsylvania* (York, Penn.: Strine Publishing, 1995).
5. Quoted in *Farmland Preservation Report* 5, no. 8 (June 1995): 6.
6. *Farmland Preservation Report* 5, no. 8 (June 1995): 7.
7. *Farmland Preservation Report* 4, no. 8 (June 1994): 4.
8. Steven A. Holmes, "Farm Count at Lowest Point Since 1850: Just 1.9 million," *New York Times*, Nov. 10, 1994.
9. For a discussion of how meat is produced in a farm factory setting, see Orville Schell, *Modern Meat* (New York: Vintage Books, 1985).

Glossary

The following words and phrases are common in farmland and open-space protection and are used throughout this handbook.

Agricultural district: An area in which farming is the preferred land use. A district is voluntarily created by landowners who receive a number of benefits (such as exemption from sewer and water assessments, greater protection against eminent domain, possibly use-value taxation, and eligibility to sell development rights) usually in return for not developing the land for a certain number of years.

Agricultural zoning: A zoning ordinance or zoning district designed to protect farmland from incompatible nonfarm uses. There are several types of agricultural zoning, which vary according to: 1) the uses allowed in the zone—exclusive or nonexclusive farm use; 2) the minimum farm size allowed, such as a 50-acre minimum lot size; 3) the number of nonfarm dwellings allowed, such as one building lot per 25 acres; and 4) the size of setbacks or buffer areas between farms and nonfarm properties.

Assessment: The process of determining the worth, or the market value, of land and buildings for taxation purposes. Most communities in the United States assess property at only a fraction of its real value; the percent of real value is called the assessment rate. Thus, multiplying the true value of all community property times the assessment rate yields the total assessed value of all real estate in the community.

Bargain sale: The sale of property or development rights for less than fair-market value. The seller may use the difference between the appraised fair-market value and the bargain sale price paid by a public agency or qualified nonprofit organization as a charitable contribution for an income tax deduction.

Basis: An accounting term that refers to the portion of a sale of property or development rights that is not subject to capital gains tax. The basis in a property is the original cost plus improvements minus depreciation.

Buffer area: A space between a farm and a nonfarm property that is kept undeveloped and often put in trees and bushes to minimize the spillover of noise, dust, and odors from the farm. The buffer is usually located on the nonfarm property.

Build-out scenario: When zoning farmland or open space at a certain density, it may be helpful to determine the maximum number of dwelling units that are actually allowed in that area. A build-out scenario can show on a map or series of maps whether the area would contain too much, too little, or the desired amount of development under the proposed zoning density.

Capital improvements program (CIP): A program of when, where, and how much a community or county plans to invest in public services over the next five to ten years. The program presents a capital budget each year, which is useful in drafting the community or county budget. A capital improvements program usually includes roads and bridges, school buildings, sewer and water lines and treatment plants, municipal buildings, solid-waste disposal sites, and police and fire equipment.

Cluster development: Grouping houses on part of a property while maintaining a large amount of open space on the remaining land. Cluster development should be seen as an open-space protection tool rather than a farmland protection tool.

Community-supported agriculture: Consumers buy shares in a farm's output and receive food directly from the farm. The farmer eliminates the intermediary and can earn a higher profit on what is grown.

Comprehensive plan: Also known as a master plan or a general plan. A comprehensive plan summarizes the current condition of a community or county, projects future needs, and develops general policy goals and objectives. The comprehensive plan acts as a legal basis for zoning and subdivision ordinances.

Concurrency: A government policy stating that a new development will be approved only when adequate public services, such as sewer and water, are in place.

Conditional use: A land use in a zoning district that is neither permitted outright nor prohibited outright. A conditional land-use permit may be granted after review by the planning commission and approval of the governing body. A conditional use usually has wide impacts, such as a telecommunications tower in an agricultural zone. The term *conditional use* may also refer to certain conditions placed in the permit that the holder of the permit must meet.

Conservation easement: A legal document that restricts the use of land to farming, open space, or wildlife habitat. A landowner may sell or donate an easement to a government agency or a private land trust.

Cost-of-services study: Estimating the cost of services demanded by residential development and by farmland for each dollar of property taxes generated by each land use.

Cumulative impact assessment: The evaluation of the impact of several recent and proposed development projects on community services and land-use patterns.

Density: The number of buildings or housing units on a particular area of land. High-density development leaves little open space; low-density development has few buildings or housing units per acre.

Development right: The right to develop land, which is one of several rights that come with landownership. The development right may be sold or given away separately from the other rights. If the development right is removed, the land is still private property, though the uses that are allowed are typically limited to farming and open space.

Differential assessment: Determining the value of farmland for property tax purposes, based on its use-value for farming rather than on its "highest and best" use for potential development. There are three types of differential assessment: preferential assessment, deferred taxation, and restrictive agreement.

Downzone: A change in a property's zone designation to a lower density or a less intensive use. For instance, a downzoning might change the allowed density on a piece of land from two houses per acre to only one house per 20 acres; or change the zoning on land from an industrial zone to an agricultural zone.

Easement: See Conservation easement.

Ex-urban: Refers to a region of countryside outside of cities and suburbs that is experiencing residential growth.

Farm: Land and buildings devoted to the production of crops and/or livestock. The U.S. Census of Agriculture defines a farm as producing at least $1,000 a year in crops and/or livestock. A commercial farm can be defined as producing at least $10,000 a year in gross sales.

Fee assessment: A charge to landowners when a sewer or water line is extended along their property, based on the linear footage of the line. These fees have driven metro-area farms out of business and speeded up the spread of development. Exemption from these fees is important for protecting farmland.

Fee interest and fee simple: Fee interest and fee simple refer to the ownership of all the rights to owning a property. Less than fee means that one or more rights to the property are not owned. For example, a development right held by a land trust is a less than fee interest in a certain property.

Geographic information system (GIS): A method of placing data into a computer to create a map or a series of maps. The data can also be used to predict what will happen under certain changes. Data might include: soils, parcels, roads, waterways, sewer and water lines, buildings, and zoning districts.

Greenbelt: An area of protected open space around a city or town, or an area that separates two built-up places.

Growth management: The use of regulations and incentives to influence the rate, timing, location, density, type, and style of development in the community.

Hobby farm: Produces less than $10,000 a year in gross sales.

Impermanence syndrome: A condition that occurs when development penetrates a farming area and farmers reduce investment in their farms as they anticipate the eventual sale of their land for development.

Land trust: A private nonprofit organization that qualifies as a charitable organization under Section 501(c)(3) of the Internal Revenue Code. A land trust may receive donations of property, development rights, or money. Donations may qualify as tax deductions. A land trust may also purchase property and development rights.

Land use: The function to which land is put or is classified for future uses: that is, for housing, agriculture, commercial, industrial, etc.

Land-use and development controls: Ordinances, resolutions, and codes enacted by communities, townships, and counties under the authority of state enabling legislation. Such controls are designed, and intended to be used, for the protection of the public health, safety, and welfare. Common land-use controls are: (1) zoning, which separates the planning area into zones and districts and regulates the use to which land can be put in those districts; (2) subdivision, which guides and controls the division of land for building purposes and the addition of new building areas to the community; and (3) the official map, which designates present and future community areas in which new development may or may not proceed.

Limited development: The development of a property with fewer houses than the local zoning ordinance allows. Limited development is often used to put some cash in the landowner's pocket while the remaining land is used for open space or farming.

Lot: A piece of land divided from a larger parcel.

Maximum lot size: The largest size lot allowed for a nonfarm dwelling in an agricultural zone, such as a 2-acre maximum lot size.

Minimum lot size: The smallest lot or parcel that can be built on in a particular zoning district. Also, the smallest lot that can be created by dividing a larger parcel. For example, an agricultural zone could require a landowner to have a minimum lot size of 40 acres to be able to build a farmhouse.

Nuisance: The use of land that brings harm or bother to adjacent property owners or the general public. Nuisances typically include noise, odors, visual clutter, and dangerous structures. A nuisance ordinance enacted by the governing body is a way to resolve nuisances. (See Right-to-farm law.)

Option: The right to buy land or development rights at a certain price within a certain amount of time. A potential seller sells an option to a potential buyer.

Parcel: A piece of land under single ownership or control.

Planning commission: An official panel appointed by the governing body of a city, township, or county that is responsible for drafting the comprehensive plan, the zoning ordinance, and subdivision regulations. The planning commission reviews proposed changes to the zoning and subdivision regulations and makes recommendations on the comprehensive plan, ordinances, and development proposals to the governing body.

Police power: The right of government to restrict an owner's use of property to protect the public health, safety, and welfare. Restrictions must be reasonable and be conducted according to due process.

Preacquisition: The purchase of land or development rights by one organization for eventual resale to another organization. For example, a government agency may ask a land trust to preacquire a conservation easement; then the agency will purchase the easement from the land trust. Preacquisitions are often done in urgent situations.

Prime agricultural land: Farmland that has a gentle slope and well-drained soils and requires a minimum of conservation practices. It is the easiest land to farm. Class I and II soils, as defined by the Natural Resources Conservation Service of the U.S. Department of Agriculture, are considered prime agricultural soils.

Private inurement: A problem that occurs when a landowner donates an easement on part of a property and then sells off another part of the property for development. The development portion may increase in value because of its proximity to the preserved land. As a result, the IRS may reduce the value of the donated easement.

Professional planner: A professional planner is a person who is qualified to make comprehensive plans for a community and to draft zoning ordinances and subdivision regulations. A professional planner usually will hold a master's degree in community planning and often will have obtained a professional registration.

Purchase of development rights (PDR): The voluntary sale of the rights to develop a piece of property by the landowner to a government agency or a land trust. The sale price is determined by an appraisal. The land is restricted to farming or open space.

Real estate transfer tax:	A state and/or local tax on the sale of real property. On occasion, transfer taxes have been used to finance land conservation programs. Transfer taxes normally do not apply to the sale or donation of a conservation easement.
Receiving areas:	Designated growth areas to which a developer may transfer development rights from a sending area in order to build more houses there than would normally be allowed.
Region:	An area including one or more counties that contain certain geographical, economic, and social characteristics in common.
Rezoning:	See Zone change.
Right-to-farm law:	A state law that denies nuisance suits against farmers who use standard farming practices. The law may also prevent local governments from enacting nuisance ordinances that prohibit normal farming practices.
Right-of-way:	The right to cross over property. Rights-of-way across private property are usually for utility lines or driveways.
Rural residential zone:	A zoning district in a rural area in which housing is the preferred land use. Densities may range from one house per acre to one house per 5 acres. On-site septic and well systems are commonly used.
Sending areas:	Areas in which landowners have transferable development rights. The rights may be sold to private developers or to a government TDR fund.
Setback:	The required distance of a building from a road, property line, or other building. This distance is specified in the zoning ordinance and may differ among zones.
Special exception:	Under unusual circumstances, an applicant may apply for a land use not normally allowed in a particular zone or district. A special use may be granted if the use is allowable as a special exception listed in the zoning ordinance. Special exceptions may involve decisions of the zoning board of adjustment, the planning commission, and the governing body. Special exceptions should be used sparingly.
Spot zoning:	The zoning of a particular parcel of land for a use that is different from the uses permitted in the surrounding zone. For example, a lot zoned for high-density residential use in the middle of an A-1 agricultural zone would be spot zoned. This practice should be avoided because of a potentially negative impact on neighborhoods and likely invalidation by the courts.
Sprawl:	Residential and commercial development that may take either of two forms: 1) a wave of urban or suburban expansion; or 2) scattered housing, offices, and stores throughout the countryside.
Subdivision:	The division of a parcel of land into lots for future sale and/or development. A subdivision occurs when three or four or more lots are created, depending on the particular state planning and zoning enabling act.
Taking:	A taking occurs when government takes private property without paying compensation to the landowner, in violation of the Fifth Amendment to the Constitution. A zoning ordinance that removes virtually all of the useful economic value of a property will be judged a taking by the courts.

TDR fund: A local government fund for purchasing and selling TDRs to keep the TDR market active. The fund may buy TDRs from landowners in sending areas and also sell TDRs to private developers for use in receiving areas.

Tract: A parcel of land under single ownership or control. A tract usually covers a large acreage and has the potential to be subdivided into lots.

Transfer of development rights (TDR): Property rights that may not be used on the land from which they come. TDRs may be sold to be used on a designated site in a receiving (growth) area. When TDRs are sold, the land they came from is then restricted to farming.

Urban growth boundary (UGB) or village growth boundary: A line agreed upon by a city and county, a village and county, or a village and township within which there is enough buildable land to accommodate development for up to twenty years. The governments agree not to extend urban-type services, especially public sewer and water, beyond the growth boundary. This encourages development inside the boundary.

Urban service area: An area where urban services, such as sewer, water, roads, and police and fire protection, will be provided to accommodate urban-type growth.

Use-value assessment: See Differential assessment.

Variance: Alteration of the provisions of a land-use ordinance, usually on a single piece of land. An area variance involves changing the zoning requirements for building height, lot coverage, setbacks, and yard size.

Zone: An area or areas of the community or county in which certain land uses are permitted and other uses are prohibited by the zoning ordinance. Common zones are residential, commercial, industrial, and agricultural.

Zone change: An action taken by the local governing body to change the type of zoning on one or more pieces of land. For example, a zone change or rezoning would be from A-1, agricultural, to C-2, medium-density commercial. A zone change for specific properties can happen in two ways. A property owner may ask for a zone change, which is a quasi-judicial action. Otherwise, either the planning commission or the governing body may seek a zone change through a legislative action. If a zone change is approved, the zoning map must also be amended. Some zone changes may require amending the comprehensive-plan map.

Zoning ordinance: A set of land-use regulations and a map adopted by the local governing body to create zoning districts that permit certain land uses and prohibit others, for example, an R-1 single-family residential district or a C-1 low-density commercial district. Land uses in each district are regulated according to type, density, height, and the coverage of buildings.

Bibliography

American Farmland Trust. *Fiscal Impacts of Major Land Uses in the Town of Hebron, Connecticut.* Washington, D.C.: American Farmland Trust, 1986.

———. *Planning and Zoning for Farmland Protection: A Community Based Approach.* Washington, D.C.: American Farmland Trust, 1987.

———. *Protecting Farmland through Purchase of Development Rights: The Farmers' Perspective.* Washington, D.C.: American Farmland Trust, 1988.

———. *Dutchess County: Cost of Community Services Study.* Washington, D.C.: American Farmland Trust, 1989.

———. *Does Farmland Protection Pay: The Cost of Community Services in Three Massachusetts Towns.* Washington, D.C.: American Farmland Trust, 1992.

———. *Florida's Growth Management Plans: Will Agriculture Survive?* Washington, D.C.: American Farmland Trust, 1992.

———. *Farming on the Edge.* Washington, D.C.: American Farmland Trust, 1993.

———. *Is Farmland Protection a Community Investment? How to Do a Cost of Community Services Study.* Washington, D.C.: American Farmland Trust, 1993.

———. *Alternatives for Future Urban Growth in California's Central Valley: The Bottom Line for Agriculture and Taxpayers.* Washington, D.C.: American Farmland Trust, 1995.

American Farmland Trust and Lake County, Ohio, Soil and Water Conservation District. *The Cost of Community Services in Madison Village and Township, Lake County, Ohio.* Washington, D.C.: American Farmland Trust, 1993.

Arendt, Randall. *Rural by Design.* Chicago: Planners Press, 1994.

Bowers, Deborah. *Farmland Preservation Report.* Street, Md.: Bowers Publishing Inc., 1990–1996.

Brenneman, Russell, and Sarah M. Bates, eds., *Land-Saving Action.* Covelo, Cal.: Island Press, 1984.

Conaway, James. *Napa.* New York: Avon Books, 1990.

Corser, Susan Ernst. *Preserving Rural Character through Cluster Development.* Chicago: American Planning Association, Planners Advisory Service Memo, July 1994.

Coughlin, Robert E. "Formulating and Evaluating Agricultural Zoning Programs." *Journal of the American Planning Association* 57, no. 2 (1991): 183–192.

———. *The Adoption and Stability of Agricultural Zoning in Lancaster County, Pennsylvania.* Research Report Series #15, Department of City and Regional Planning, Graduate School of Fine Arts, University of Pennsylvania, 1992.

Coughlin, Robert E., and John C. Keene, eds. *National Agricultural Lands Study, The Protection of Farmland: A Reference Guidebook for State and Local Governments.* Washington, D.C.: U.S. Government Printing Office, 1981.

Coughlin, Robert E., and Stanford M. Lembeck. *Agricultural Protection Zoning: A Key Element in an Effective Agricultural Protection Program for Pennsylvania Municipalities.* Harrisburg, Penn.: Center for Rural Pennsylvania, 1995.

Crosson, Pierre R., ed. *The Cropland Crisis: Myth or Reality.* Baltimore, Md.: Johns Hopkins University Press, 1982.

Daniels, Thomas L. "Hobby Farming in America: Rural Development or Threat to Commercial Agriculture." *Journal of Rural Studies* 2, no. 1 (1986): 31–40.

———. "Evaluating Farmland Retention in Oregon: A Comparison of Agricultural Zoning Experience." Paper presented at the American Collegiate Schools of Planning Conference, Portland, Oregon, October 6, 1989.

———. "Policies to Preserve Prime Farmland in the U.S.A.: A Comment." *Journal of Rural Studies* 6, no. 3 (1990): 331–336.

———. "The Purchase of Development Rights: Preserving Agricultural Land and Open Space." *Journal of the American Planning Association* 57, no. 4 (1991): 421–431.

———. "Do Tax Breaks on Farmland Help Protect It from Conversion?" *Farmland Preservation Report*, Special Report, January 1991.

———. "Limited Development of Farmland: Solution or Smokescreen?" American Planning Association National Conference, Washington, D.C., May 1992.

———. "Preserving Farmland: Matching Techniques to Local Circumstances," *Farmland Preservation Report*, Special Report, June 1993.

———. "Agricultural Zoning: Managing Growth, Protecting Farms." *Zoning News* (American Planning Association, August 1993).

———. "Where Does Cluster Zoning Fit in Farmland Protection?" *Journal of the American Planning Association* 63, no. 1 (1977): 127–135.

Daniels, Thomas L., John W. Keller, and Mark B. Lapping. *The Small Town Planning Handbook*, 2nd edition. Chicago: Planners Press, 1995.

Daniels, Thomas L., and Arthur C. Nelson. "Is Oregon's Farmland Preservation Program Working?" *Journal of the American Planning Association* 52, no. 1 (1986): 22–32.

Diamond, Henry L., and Patrick F. Noonan, eds. *Land Use in America.* Covelo, Cal.: Island Press, 1996.

Diehl, Janet, and Thomas S. Barrett, eds. *The Conservation Easement Handbook.* Alexandria, Va.: Land Trust Exchange and Trust for Public Land, 1988.

Doherty, Joseph C. *Growth Management in Countrified Cities.* Alexandria, Va.: Vert Milon Press, 1984.

Dunford, Richard. *The Development and Current Status of Federal Farmland Retention Policy.* Washington, D.C.: Congressional Research Service, Library of Congress, Report No. 85-21 ENR., Nov. 21, 1984.

Easley, V. Gail. *Staying Inside the Lines: Urban Growth Boundaries.* Planners Advisory Service Report Number 440. Chicago, Ill.: American Planning Association, 1992.

Endicott, Eve, ed. *Land Conservation through Public-Private Partnerships.* Washington, D.C.: Island Press, 1994.

Gottsegen, Amanda Jones. *Planning for Transfer of Development Rights: A Handbook for New Jersey Municipalities.* Mount Holly, N.J.: Burlington County Board of Chosen Freeholders, 1992.

Guttenplan, Charles L. "Townships Responsible for Preserving Natural Resources," *Pennsylvania Township News* (November 1993): 17–20.

Harl, Neil E. *Farm Estate and Business Planning.* Ames, Iowa: Century Communications, 1994.

Hart, John. *Farming on the Edge: Saving Family Farms in Marin County, California.* Berkeley: University of California Press, 1991.

Hawaii Department of Budget & Finance. *Growth Management Issues in Hawaii.* Honolulu: State of Hawaii, 1977.

Healy, Robert, and James L. Short. *The Market for Rural Land.* Washington, D.C.: Conservation Foundation, 1981.

Heimlich, Ralph E. "Metropolitan Agriculture: Farming in the City's Shadow." *Journal of the American Planning Association* 55, no. 4 (1989): 457–466.

Herbers, John. *The New Heartland.* New York: Times Books, 1986.

Hiemstra, Hal, and Nancy Bushwick, eds. *Plowing the Urban Fringe.* Fort Lauderdale, Fla.: Florida Atlantic University and Florida International University, 1989.

Hoagland, Glenn. "Limited Development: Packaging Projects to Benefit Farmers and Their Neighbors." In American Farmland Trust, ed., *Saving the Land That Feeds America: Conservation in the Nineties.* Washington, D.C.: American Farmland Trust, 1991.

Howard, Robert West. *The Vanishing Land.* New York: Ballantine Books, 1985.

Jonovic, Donald J., and Wayne D. Messick. *Passing Down the Farm: The Other Farm Crisis.* Cleveland, Ohio: Jamieson Press, 1994.

Land Trust Alliance. *National Directory of Conservation Land Trusts.* Washington, D.C.: Land Trust Alliance, 1992.

———. *The Standards and Practices Guidebook: An Operating Manual for Land Trusts.* Washington, D.C.: Land Trust Alliance, 1994.

Land Trust Alliance and the National Trust for Historic Preservation. *Appraising Easements: Guidelines for Valuation of Historic Preservation and Land Conservation Easements,* 2nd edition. Washington, D.C.: Land Trust Alliance and the National Trust for Historic Preservation, 1990.

Lapping, Mark B. "Agricultural Land Retention: Responses, American and Foreign." In A. M. Woodruff, ed., *The Farm and the City: Rivals or Allies?* Englewood Cliffs, N.J.: Prentice-Hall, 1980.

Lapping, Mark B., Thomas L. Daniels, and John W. Keller. *Rural Planning and Development in the United States.* New York: Guilford Publications, 1989.

Lapping, Mark B., George E. Penfold, and Susan Macpherson. "Right-to-Farm Laws: Do They Resolve Land Use Conflicts?" *Journal of Soil and Water Conservation* 26, no. 6 (1983): 465–467.

Little, Charles. *Greenways for America.* Baltimore, Md.: Johns Hopkins University Press, 1990.

Lockeretz, William, ed. *Sustaining Agriculture Near Cities.* Ankeny, Iowa: Soil and Water Conservation Society, 1987.

Looker, Dan. *Farmers for the Future.* Ames: Iowa State University Press, 1996.

Louv, Richard. *America II.* New York: Viking/Penguin, 1983.

Montgomery County Office of Economic Development. *Farmland Preservation Almanac.* Montgomery County, Md.: Office of Economic Development, 1990.

National Wildlife Federation. *1996 Conservation Directory.* Washington, D.C.: National Wildlife Federation, 1996.

New Jersey Conservation Foundation. *Farmland Forum.* Morristown, N.J.: New Jersey Conservation Foundation, 1990–96.

Oregon Department of Land Conservation and Development. *Development Activity in Exclusive Farm Use Zones.* Salem, Ore.: Department of Land Conservation and Development, 1991.

Peters, James E. "Saving Farmland: How Well Have We Done?" *Planning* 56, no. 9 (1990): 12–17.

Real Estate Research Corporation. *The Cost of Sprawl: Environmental and Economic Costs of Alternative Development Patterns at the Urban Fringe.* Washington, D.C.: U.S. Government Printing Office, 1974.

Roddewig, Richard J., and Cheryl A. Inghram. *Transferable Development Rights Programs.* Planners Advisory Service Report. Chicago, Ill.: American Planning Association, 1987.

Rusk, David. *Cities Without Suburbs.* Washington, D.C.: Woodrow Wilson Center Press, 1993.

Schiffman, Irving. *Alternative Techniques for Managing Growth.* Berkeley: Institute of Governmental Studies, University of California at Berkeley, 1989.

Small, Stephen J. *Preserving Family Lands,* 2nd edition. Boston: Landowner Planning Center, 1992.

Solnit, Albert. *The Job of the Planning Commissioner.* Chicago, Ill.: Planners Press, 1987.

Steiner, Frederick, James R. Pease, and Robert E. Coughlin, eds. *A Decade with LESA: The Evolution of Land Evaluation and Site Assessment.* Ankeny, Iowa: Soil and Water Conservation Society, 1994.

Steiner, Frederick, and John Theilacker, eds. *Protecting Farmland.* Westport, Conn.: Avi Publishing, 1984.

Stokes, S., A.E. Watson, Genevieve P. Keller, and J. Tomothy Keller. *Saving America's Countryside.* Baltimore, Md.: Johns Hopkins University Press, 1989.

Sutro, Suzanne. *Reinventing the Village.* Planners Advisory Service Report Number 430. Chicago, Ill.: American Planning Association, 1990.

Toner, William. "Ag Zoning Gets Serious." *Planning* 50, no. 12 (1984): 19–24.

Trust for Public Land. *Doing Deals: A Guide to Buying Land for Conservation.* Washington, D.C.: Land Trust Alliance and the Trust for Public Land, 1995.

Tustian, Richard. "Preserving Farmland through Transferable Development Rights." *American Land Forum* 3, no. 3 (1983): 63–76.

Urban Land Institute. *The Costs of Alternative Development Patterns: A Review of the Literature.* Washington, D.C.: Urban Land Institute, 1992.

U.S. Department of Agriculture, Soil Conservation Service. *National Agricultural Land Evaluation and Site Assessment Handbook.* Washington, D.C.: U.S. Government Printing Office, 1983.

U.S. Department of Commerce, Bureau of the Census. *1992 Census of Agriculture.* Washington, D.C.: U.S. Government Printing Office, 1994.

U.S. House of Representatives, Committee on Banking, Finance and Urban Affairs, Subcommittee on the City. *Compact Cities: Energy Saving Strategies for the Eighties.* Washington, D.C.: U.S. Government Printing Office, 1980.

Vaughan, Gerald F. *Land Use Planning in the Rural-Urban Fringe.* Extension Bulletin Number 157. Newark: University of Delaware, 1994.

Wiebe, Keith, Abedayehu Tegene, and Betsey Kuhn. *Partial Interests in Land: Policy Tools for Resource Use and Conservation.* Economic Research Service Report Number 744. Washington, D.C.: U.S. Department of Agriculture, 1996.

Wolf, Peter M. *Land in America: Its Value, Use, and Control.* New York: Pantheon Books, 1981.

Yaro, Robert D., Randall G. Arendt, Harry L. Dodson, and Elizabeth A. Brabec. *Dealing with Change in the Connecticut River Valley: A Design Manual for Conservation and Development.* Amherst: University of Massachusetts, Center for Rural Massachusetts, 1988.

Contacts

American Farmland Trust
1920 N Street NW, Suite 400
Washington, DC 20036
(202) 659-5170
American Farmland Trust has field offices in California, Illinois, Massachusetts, Michigan, New York, and Ohio.

American Farmland Trust also has set up a Farmland Information Library on the Internet. The Information Library contains a bibliography and abstracts of farmland protection topics and state statutes relating to farmland protection. American Farmland Trust operates the Farmland Information Library as part of the Center for Agriculture in the Environment with Northern Illinois University. To reach the Farmland Information Library Home Page from the World Wide Web:
http://farm.fic.niu.edu/fic/home.html.

To reach AFT's Home Page, from the World Wide Web, type: http://www.farmland.org

American Planning Association
Small Town and Rural Division
c/o Dr. John Keller
Department of Regional and Community Planning
Seaton Hall
Kansas State University
Manhattan, KS 66506
(913) 532-5958

Farmland Preservation Report
900 La Grange Road
Street, MD 21154
(410) 692-2708
email: dbowers@harford.campus.mci.net
A national newsletter for professionals in land use, conservation, and agriculture, published since 1990.

The Land Trust Alliance
1319 F St. NW, Suite 501
Washington, DC 20004
(202) 638-4725
Homepage: http://www.lta.org
The Land Trust Alliance sponsors an annual Land Trust Rally, which features a wide range of presentations on land trust operations, land protection, and land stewardship topics. It's highly recommended.

The Land Trust Alliance together with the Land Conservation Law Institute publishes The Back Forty, a newsletter featuring legal aspects of tax regulations, estate planning, appraisals, and charitable giving. There is also a telephone research service, at a favorable cost, for answers to legal questions.

The National Trust for Historic Preservation
1785 Massachusetts Avenue NW
Washington, DC 20036

The Nature Conservancy
1815 North Lynn
Arlington, VA 22209

The Trust for Public Land
116 New Montgomery Street, 4th Floor
San Francisco, CA 94105
For an introduction to the tax and estate planning issues of land preservation, see Stephen Small, Preserving Family Lands, *available from Preserving Family Lands for $8.95 plus $3 for postage and handling at P.O. Box 2242, Boston, MA 02107.*

For information on the installment purchase agreement with tax-exempt interest payments, contact:

Daniel P. "Pat" O'Connell, President
Evergreen Capital Advisors
14 East 76th Street, Box 3057
Harvey Cedars, NJ 08008-0306

Index

Aggie bond programs, 66
Agins vs. City of Tiburon (zoning and land value), 69
Agribusiness, 11, 60–61, 258
Agricultural District Approach, The (Bills & Boisvert), 100
Agricultural districts, voluntary, 90
    advantages/disadvantages, 236
    characteristics of, 98–99
    Delaware, 99–100
    Lancaster County (PA), 239
    laws in various states, 299
    Pennsylvania, 100–101
    purchase of development rights, 168–69
    purposes of, 264
    transfer of development rights, 190
Agricultural section of comprehensive plan, 37–40
Agricultural use notice, 270
Agriculture, U.S. Department of (USDA), 24, 65–66, 75, 84, 217
Air pollution, 62
American Dream, 6–7, 133–34
American Farmland Trust (DC):
    cost-of-community studies, 15–16
    dual purpose of, 194–95
    metropolitan areas, 10
    population pressures on farming areas, 24
    sprawling suburban development, 3
    taxes, 55
Animal confinement units, 258
Appraisals and conservation easements/development rights, 155–57, 202–6
Area-based allocation, 116, 118–19
Area variance, 46
Arendt, Randall, 123
Ariyoshi, George R., 251
Assessing farmland conversion, 77–81, 154–55

Attitudes toward the land, 7–8
Auto-dependent sprawl, 1, 135

Baltimore County (MD), 143
Bank of America, 3
Bankruptcy, 81
Bargain sale in land trust process, 200, 230
Barley, John, 69
Beale, Calvin, 257
Beginning farmer loan programs, 66–67
Berner, Bob, 199
Beyond Sprawl, 3
Bidwell, Dennis, 193, 214
Biosecurity, 62
Black Hawk County (IA), 113
Boissoneault, Jeff, 146–47
Bonds for financing purchase of development rights programs, 163–64, 165
Bonneville County (ID), 270
Boulder (CO), 143, 212
Boyd, Darvin, 70, 152
Brandywine Conservancy, 195, 211, 212
Buckingham Township (PA), 185
Bucks County (PA), 17
Buelt, Ken, 31
Buffer zones, agricultural, 116, 119–21
Bureau of Land Management, 209
Bureau of the Census, U.S., 251–52
Burton, John, 242
Business of farming in America:
    agribusiness exerting power over farmers, 60–61
    challenges to, 62
    changing role of farmland, 67–70
    federal farm programs, 65–67
    ownership of farmlands, 70–73
    romantic view of farming, 59
    structure of farming, 63–65
    summary, 73–74

California:
    agricultural districts, voluntary, 99
    buffers, agricultural, 120
    farmland acres out of production, 1–3
    financing purchase of development rights programs, 163, 166–67
    geographic information systems, 52
    in-area transfers, 187
    Marin County and package of farmland protection techniques, 243–45
    population increases, 24, 252
    purchase of development rights, 147, 153
    restrictive agreements, 95, 96
    rezoning, 127
    San Mateo and transfer of development rights, 185–86
    subdivision regulations, 113
    urban growth boundaries, 143
    zoning, agricultural, 106
California Coastal Conservancy, 196, 197
California Farm Bureau, 53
Calthorpe, Peter, 136
Calvert County (MD), 123, 179–80, 252
Cananan, Dennis, 177
Capital gains taxes, 160
Capital improvements program, 50–51
Carroll County (MD), 241–42
Case-by-case determination in purchase of development rights programs, 189–90
Casey, Robert, 279
Center for Urban Policy Studies, 16
Central Valley (California), 2–3, 9, 24
Champlain Valley (VT), 210
Charette, planning, 40–41

Chester County (PA), 164, 239
Chicago region (IL), 24, 252, 253
Choices, making land-use policy, 27–29
Circuit breaker property tax relief, 97–98
Cities Without Suburbs (Rusk), 254
Citizen groups, 26–27, 34–40
Civic-industrial complex, 6
Clarke County (VA), 118
Cluster developments, 116
Cluster subdivisions, 48
Cluster zoning, 121–23
Colorado:
  farmland acres out of production, 1
  municipal land trust, 212
  population increases, 252
  state farmland protection programs, 87
  subdivision regulations, 113
  urban growth boundaries, 143
Colorado Cattlemen's Association, 196
Commitment to protection programs, 247
Community involvement, 12
  attitude toward the land, 7
  citizen groups, 26–27, 34–40
  land trust process, 214
  public-private partnerships, 209–12, 293–94
  transfer of development rights, 176–78
  zoning issues, 124–25
  see also Package of farmland protection techniques
Compact development/housing, 134, 136
Comparable sales approach for determining after-easement value, 205
Compensable zoning, 180
Compensation and property rights, 33
Competition for land, 10
Comprehensive plan, 35–40, 236, 245
Concurrency, 51
Conditional land uses, 46–47, 111, 112

Connecticut:
  deferred taxation, 94
  financing purchase of development rights programs, 164
  property tax abatement, 98
Conservancies, 195
Conservation Easement Handbook (Diehl & Barrett), 202
Conservation easements, 82–83, 120
  advantages/disadvantages, 236
  appraisals, 202–6
  assignment of, 297–98
  document form, 199–203, 272–78
  estate and transfer planning, 228–34
  IRS form 8283 (noncash charitable contributions), 229, 300–302
  ranking system used for sale applications, 289–92
  reimbursable acquisition agreement, 295–96
  taxes, 202
Conservation Fund, 193
Conservation Reserve Program, 83–84
Constitution of the United States:
  property rights, 32–33
  zoning, agricultural, 107, 108
Conventional subdivision, 48
Conversion *vs.* preservation, 168
Cooperation around zoning issues, 124–25
Cooperation between political jurisdictions, 26
Corporate farming, 11, 60–61, 258
Cost-of-community-services study, 15–16, 55–57
County-state approach to purchase of development rights, 153, 163
Cycle of farmland conversion, 6

Dairy farms, 65
Debt-reduction-for-easements program, 81–82
Deed of easement, 149–50
Deferred taxation, 94–95

Delaware:
  agricultural districts, voluntary, 99–100
  buffers, agricultural, 120
  land evaluation and site assessment system, 80–81
Detroit (MI), 134
Developer exactions to finance purchase of development rights programs, 166–67
Development, two types of, 4–5
Development rights, buying/transferring, 194
  see also Conservation easements; Purchase of development rights (PDR); Transfer of development rights (TDR)
Differential assessment:
  advantages/disadvantages, 236
  deferred taxation, 94–95
  improving the process, 96
  Lancaster County (PA), 239
  preferential assessment, 93–94
  purchase of development rights, 152, 169
  restrictive agreements, 95
  tax burden, shifting the, 95–96
  use-value assessment, 92–93
Dispersed development, 4
Downzoning, 47, 175, 178, 190
Due diligence in land trust process, 198

Economic development, agricultural, 12
  cost-of-community-services studies, 15–16
  crops/livestock, prices received for, 60–61, 62
  incentives for, 12, 177, 190
  Lancaster County (PA), 240
  local diverse economy, 17–18
  profits, importance of, 248–49
  residential and farmland development, comparing, 55–56
  state farmland protection programs, 101–2
Edge cities, 10
Educating the public:
  active farmland protection, 259–60
  nonfarm neighbors, 62

package of farmland protection techniques, 247
transfer of development rights, 178
zoning issues, 124
Elected officials:
attitudes toward the land, 7
lack of political will, 26–27
protecting farms and farmland, 12
rezoning, 126
tax base concerns, 254
TDR programs, support for, 176–78
Emergency Wetlands Reserve Program of 1993, 84–85
Eminent domain, power of, 152
Environmental laws/issues, 21, 83–85, 258
Environmental Protection Agency, U.S. (EPA), 83
Epitaph for a Peach (Masumoto), 217
Equipment and supply dealers, farm, 18
Equity issues, 67–69
Esseks, Dixon, 69
Estate and transfer planning:
combination of sale and gift, 224–25
conservation easements, 228–34
financial situation of landowners, 220–21
gifts to family members, 224
life insurance, 226–27
living trust, 227
S corporations, 227–28
selling the property, 221–22
summary, 234
taxes, 218, 220
will, transfer by, 225–26
European countries and urban growth boundaries, 139
Exclusive farm use (EFU) zones, 115–16, 128–29
Expectations and goals, realistic, 5–7, 21–26, 54

Factory farming, 11, 60–61, 258
Family corporation, land held by a, 229–30

Farm Bills of 1985/90/96, 81–84
Farm Credit System, 66
Farmers Home Administration, 82
Farming on the Edge, 24
Farmland Conversion Rating Forms, 77
Farmland Protection Policy Act of 1981, 76–77, 90, 279
Farmlands:
acres out of production, 1–4
changing role of, 67–70
fluctuations in the land market, 189
main farming regions, 9
size of, 63–65, 112–13
variety of, 9
see also Lot size; Value of land
Farm Service Agency, 66, 77, 81–82
Farms for the Future Act, 82
Federal Highway Administration, 77
Federal programs promoting farmland protection:
debt-reduction-for-easements program, 81–82
environmental laws, 83–85
Farm Bills of 1985/90/96, 81–84
Farmland Protection Act of 1981, 76–77
Farms for the Future Act, 82
lack of, 75–76
land evaluation and site assessment system, 77–81
lending programs, 66–67
subsidies, 65
TDR banks, 173, 177, 190
Federal Reserve Board, 81
Fee simple ownership, 32, 200
Fifth Amendment to Constitution of the United States, 32, 108
Financing land trust process, 197, 207
Financing purchase of development rights programs, 162–67
First-time farm buyers, 66–67, 224

Fixed area-based allocation, 116, 118–19
Florida:
concurrency, 51
farmland acres out of production, 1
population increases, 24, 252
urban growth boundaries, 144
Fluctuations in the land market, 189
Food-processing plants, 19
Food shortages, worldwide, 25
Forest Service, U.S., 209
Forsyth County (NC), 165
Foundations, 195
Fourteenth Amendment to Constitution of the United States, 32
Fragmentation of the land base, 62
France, land banking in, 247–48
Fremont County (ID), 270
Fresno County (CA), 3, 24, 113

Garber, Gene, 199
Garden Cities, 256
Geographic Information Systems (GIS), 39–40, 52
Gift of land and estate planning, 224, 233
Global perspective on population growth and agriculture, 257
Goals and expectations, realistic, 5–7, 21–26, 54
Golden Gate National Recreation Area (CA), 154
Golf courses, 134–35
Government, see Federal programs promoting farmland protection; State farmland protection programs
Governor's executive order on farmland preservation policy, 279–81
Grossi, Ralph, 243, 245
Groundwater, 121–22
Growth management/pressure and transfer of development rights, 177

Harford County (MD), 159, 165
Hawaii, agricultural zoning in, 106, 115
Heuer, Robert, 6, 34–35
Hildebrand, John, 219–20
Hood River Valley (OR), 71
Hopkins, Steve, 3
Hopper, Leslie, 120
Howar, Robert W., 1
Howard, Ebeneezer, 256
Howard, Robert W., 60
Howard County (MD), 122, 159, 164, 165
Hudson Valley (NY), 24

Idaho:
    buffers, agricultural, 120
    resource management easements, 270
Identifying land to protect for purchase of development rights, 153–57
Illinois:
    minimum-lot-size zoning, 117
    population increases, 252
    urban decline and suburban sprawl linked, 253
Impermanence syndrome, 72–73
In-area transfers, 187
Incentives for agricultural economic development, 12, 177, 190
Income approach for determining after-easement value, 205
Income tax credit, state, 97–98
Incompatible uses in zoning ordinance, 267
Installment payments for purchase of development rights, 157–60
Installment sale of farms, 223–24
Interest rates, 81
Interior, U.S. Department of the, 209
Internal Revenue Service (IRS), 160, 162, 194
    see also Taxes
Inventory of land resources, 39–40
Investing for the long-term, 62
Iowa:

Iowa Natural Heritage Foundation, 195
    subdivision regulations, 113
IRS form 8283 (noncash charitable contributions), 229, 300–302
Isabella County (MI), 188

Jackson, Scott, 142
Jackson Hole Land Trust, 196

Kaii-Ziegler, Steve, 123
Kennedy, John F., 60
Kentucky:
    deferred taxation, 95
    urban growth boundaries, 139
King County (WA), 23, 149, 163
Krupa, Ken, 75

Lancaster County (PA):
    buffer zones, agricultural, 121
    fixed-area allocation, 119
    land sales information, 52
    package of farmland protection techniques, 237–40
Lancaster Farmland Trust (PA), 193–94
Land, attitudes toward, 7–8
Land banking, 247–48
Land evaluation and site assessment system (LESA), 77–81, 154–55
Land fragmentation, 134–35
Land sales information, 52–53
    see also Value of land
Land tenure, 70–73
Land Trust Alliance, 81, 193
Land trusts:
    advantages/disadvantages, 236
    community planning, 214
    future of, 215
    limited development, 212–14
    managing and monitoring procedures, 203, 208–9
    operating, 197–99
    origin of, 195–96
    practices implemented by, 199–207
    public-private partnerships, 209–12
    structure of, 196–97
    summary, 215–16

Land-use planning:
    choices, making land-use policy, 27–29
    comprehensive plan, 36–40
    land trusts and community, 214
    planning commission, 35–36
    political process, 34–35
    protecting farmland promotes good, 20
    state farmland protection programs, 87, 90
Large minimum lot size (agricultural zoning), 116, 117
Latin America, debt-for-nature-protection swaps with, 82
Laws:
    agricultural district, state, 299
    environmental, 83–85
    see also Right-to-farm laws
Lease signing in land trust process, 200
Legal aspects of agricultural zoning, 107–8
Legal foundations of managing growth, 31–34
Lending programs, federal farm, 66–67, 81–82
Libby, Jim, 82
Life insurance and estate planning, 226–27
Like-kind exchange and purchase of development rights, 160–62, 232
Limited development on farmland, 195, 231–34
Livestock operations, 65
Living trust, 227
Loans, 66–67, 70, 81–82, 152
Local Agency Formation Commissions (LAFCOs), 143
Local diverse economy, 18
Local public-private partnership between land trusts/government, 211
Local supply of fresh food, 20–21
Loeben, Arthur, 246
Lot size:
    buffer zones, agricultural, 116
    large minimum, 117

legal aspects of agricultural zoning, 108
minimum number of acres and farm related dwellings, 112–13, 114
misperceptions in managing growth of community, 53–54
Loudoun County (VA), 252
Lump-sum payment for purchase of development rights, 157
Lump-sum sale of farms, 222

Maine:
  buffers, agricultural, 121
  deferred taxation, 95
  purchase of development rights, 88
Managing and monitoring land trusts, 203, 208–9
Managing community growth:
  capital improvements program, 50–51
  citizen participation, 34–40
  legal foundations of, 31–34
  monitoring changes in the land base, 51–53
  problems and misperceptions, 53–56
  subdivision regulations, 47–50
  summary, 56–57
  vision for the community, creating a, 40–41
  zoning, agricultural, 41–47
Manheim Township (PA), 189, 238
Manure runoff, 258
Mapping the Farm (Hildebrand), 217
Maps, agricultural zone, 39, 52, 109–10
Marin Agricultural Land Trust (CA), 193, 194, 196, 197
Marin County (CA):
  comprehensive planning process, 154
  package of farmland protection techniques, 243–45
  special taxation districts, 166
  successful practices in, 188
  zoning, agricultural, 106
Maryland:

Calvert County and history of TDRs, 179–80
Carroll County and package of farmland protection techniques, 241–42
cluster zoning, 122–23
deferred taxation, 95
financing purchase of development rights programs, 164–67
fixed area-based allocation, 118–19
Howard County and securitized installment procedure agreement, 159
parklands buffering protected farmlands, 21
preservation vs. conversion, 168
property tax abatement, 98
purchase of development rights, 147–49, 153, 154
transfer of development rights, 188
urban growth boundaries, 143
zoning, agricultural, 106
see also Montgomery County (MD)
Maryland Environmental Trust (MET), 208–9, 214
Massachusetts:
  land trusts, 195
  purchase of development rights, 148
  zoning, agricultural, 108
Massachusetts Farmland and Conservation Trust, 194
Masumoto, Mas, 217
Meikle, John, 3
Mennitto, Donna, 122
Mesa County Land Conservancy (CO), 194
Metropolitan counties/areas, 10, 24
  see also Sprawling suburban development
Michigan:
  income tax credits, state, 97
  population pressures in farming areas, 24
  purchase of development rights, 147, 160

right-to-farm laws, 91
sprawling suburban development, 134
state farmland protection programs, 87
transfer of development rights, 188
Milk-price support program, 65
Mineral rights in land trust process, 207
Minimum-lot-size zoning, 116, 117, 119, 121
Minneapolis–St. Paul (MN):
  population pressures in farming areas, 24
  regional land-use control, 28
  tax-base sharing, 254–55
Minnesota, farmland acres out of production in, 1
Mixed-use zoning, 54
Moderate-strength farming communities, maintaining, 240–45
Monitoring and managing land trusts, 203, 208–9
Monitoring changes in the land base, 51–53
Monitoring development proposals/changes to the zones, 125–27
Montana Land Reliance, 194
Montgomery County (MD):
  downzoning, 175
  family lot exemption, 188
  fluctuations in the land market, 189
  points-based appraisal formulas, 156–57
  prices of TDR acres, 176
  transfer of development rights, 180–82
Montgomery County (PA), 246
Moseley, James, 75
Municipal land trust, 212

Napa County Land Trust (CA), 194
National Agricultural Land Evaluation and Site Assessment Handbook, 79
National Agricultural Lands Study (NALS), 76

National Audubon Society, 193
Natural Resources Conservation Service (NRCS), 8, 39, 77
Nebraska, deferred taxation in, 95
Neotraditional development, 256
Newcomers to the countryside, 5
New England states and purchase of development rights, 153
New Hampshire:
  purchase of development rights, 88
  restrictive agreements, 95
New Jersey:
  agricultural districts, voluntary, 101
  farmland protection vs. development, 16–17
  financing purchase of development rights programs, 164
  purchase of development rights, 147
  transfer of development rights, 176, 182–85
  urban growth boundaries, 144
New Urbanism, 136
New York:
  agricultural districts, voluntary, 99
  deferred taxation, 94
  environmental issues, 21
  population pressures in farming areas, 24
  purchase of development rights, 147
  state farmland protection programs, 87
Niche marketing of specialty crops, 19
Nock, Denis, 212
Nonexclusive farm zoning, 115–21
Nonfarm dwellings, locating, 121
Nonfarm neighbors:
  agricultural use notice, 270
  conflicts with, farmland protection minimizing, 19–20
  educating, 62
  right-to-farm laws, 90–91
  zoning, agricultural, 105
North Carolina:
  financing purchase development rights programs, 165
  population pressures in farming areas, 24
  purchase of development rights, 88
  right-to-farm laws, 91
  rushing into farmland protection programs, 25
  zoning, agricultural, 108
Northeastern Illinois Planning Commission, 134
Northup, Jim, 55
Nuisance doctrine, 90–91, 113–14

Office parks, 134
1000 Friends of Oregon, 141
On-site systems, 108
Open-space protection, 12–13
Option to purchase in land trust process, 200
Orange County (NC), 25, 108
Oregon:
  concurrency, 51
  deferred taxation, 95
  land tenure, 71
  population pressures in farming areas, 24
  state farmland protection programs, 87
  urban development, attractive, 28
  urban growth boundaries, 139–41
  zoning, agricultural, 106, 128–30
Oregon Farm Bureau, 4, 18
Orfield, Myron, 254–55
Ownership of farmlands, 70–73

Package of farmland protection techniques, 235, 237
  moderate-strength farming communities, 240–45
  pitfalls to avoid, 247–49
  strong farming communities, 237–40
  summary, 249
  weak farming communities, 245–46
Parklands buffering protected farmlands, 21
Partnerships between local government/private landowners/citizen organizations, 26–27
Pay-as-you-go financing for purchase of development rights programs, 165
Payment options for purchase of development rights:
  like-kind exchange, 160–62
  securitized installment purchase agreement, 158–60
  tax considerations, 157
Peninsula Township (MI), 160
Pennsylvania:
  agricultural districts, voluntary, 99, 100–101
  buffers, agricultural, 121
  comprehensive plan, 38
  deferred taxation, 94
  farmland protection vs. development, 17
  financing purchase of development rights programs, 164
  groundwater, 121–22
  limited development on farmland, 212
  minimum-lot-size zoning, 117, 119
  Montgomery County and package of farmland protection techniques, 246
  population pressures in farming areas, 24
  public-private partnership between land trusts/government, 211, 293–94
  purchase of development rights, 147, 149, 153, 154, 162
  restrictive agreements, 95
  transfer of development rights, 185
  see also Lancaster County (PA)
Permitted land uses in zoning ordinance, 111, 112
Pesticides, 258
Pinelands, New Jersey, 176, 182–85
Planned agricultural district (PAD), 186
Planning commission, 35–36, 114–15, 125
  see also Land-use planning
Plat books, 39

Points-based appraisal formulas, 156–57
Politics of land protection:
  choices, making land-use policy, 27–29
  governor's executive order on farmland preservation policy, 279–81
  land-use planning, 34–35
  partnerships between local government/private landowners/citizen organizations, 26–27
  transfer of development rights, 177
Pollution, 62, 83, 258
Population pressures on farming areas, 24, 257
  see also Sprawling suburban development
Population Profile of the United States, 257
Portland Metropolitan Service District (MSD), 140–41
Preferential assessment, 93–94
Preliminary plat application, 50
Preservation *vs.* conversion, 168
Prices of land, see Value of land
Prices received for crops/livestock, 60–61, 62
Prime farmland, 8–9, 10
Prince Edward Island (Canada), 248
Private land in America, 4, 8
Private responsibility for land conservation, 12, 27–28
  see also Land trusts
Production costs, 62
Profit for farmers and farmland protection, see Economic development, agricultural
Property rights, 31–33, 43
Property taxes, 18
  abatement, 98
  cost to public purse, 215
  differential assessment, 92–96
  like-kind exchange and purchase of development rights, 160–62
  purchase of development rights, 152, 157
  state income tax providing relief for, 97–98

Proposition 70 (CA), 243
Protecting farms and farmland:
  advantages/disadvantages, 236
  economic development policy, good, 15–18
  effective methods, 11–12
  future scenarios, 247–48
  goals, achievable, 5–7, 21–26
  local economy, promoting a diverse, 18
  local supply of fresh food, 20–21
  nonfarm neighbors, conflicts with, 19–20
  open-space, 12–13
  politics, understanding, 26–29
  problems created by development, 4–5
  summary, 258–60
  uniting citizens/farmers/politicians, 259–60
  see also Federal programs promoting farmland protection; Package of farmland protection techniques; State farmland protection programs
Public good aspects of land, 8
Public interest in land, defining, 33–34
Public participation in the comprehensive planning process, 34–40
  see also Community involvement
Public-private partnerships between land trusts/government agencies, 209–12, 293–94
Purchase of development rights (PDR), 145
  advantages/disadvantages, 236
  buying back development rights, 151
  Farm Bill of 1996, 82–83
  farmers benefited by, 19
  financing programs, 162–67
  how it works, 148–50
  identifying what land to protect, 153–57
  combination with other growth management techniques, 167–69
  landowner's questions, 150–52

  local implementation, 88
  local programs, leading, 147
  Montgomery County (MD), 130
  payment options, 157–62
  setting up the program, 153
  states involved, 146
  summary, 169
  weak farming communities, 245

Queen Anne's County (MD), 122–23

Radburn (NJ), 256
Ramsey, Carolyn, 188
Ranking system for conservation easement sale applications, 289–92
Real estate market/development proposals, tracking, 52–53
Real estate transfer taxes for financing purchase of development rights programs, 165–66
Realistic goals and expectations, 5–7, 21–26, 54
Receiving areas in TDR programs, 173–76, 177
Recessions, 189
Record-keeping in transfer of development rights programs, 188
Records of changes to the agricultural zone, 127–28
Regional land-use control, 28–29
Registry in land trust process, 200
Regulations, government:
  challenges to farming business, 62
  land-use, 33–34
  subdivisions, 47–50, 112–13
  see also Federal programs promoting farmland protection
Reimbursable conservation easement acquisition agreement, 295–96
Renting farmland, 71–72
Resource management easements, 120, 270–71
Resources Conservation Act Appraisal of 1987, 83

Restrictive agreements, 95, 96
Reyes National Seashore (CA), 154
Rezoning land from one use to another, 47, 52, 126–27
Rhode Island, deferred taxation in, 94
Right of first refusal in land trust process, 200
Right-to-farm laws, 20
  advantages/disadvantages, 236
  Carroll County (MD), 242
  Lancaster County (PA), 239
  nuisance doctrine, 90–91
Romantic view of farming, 59
Rural-urban fringe regions, 4
Rusk, David, 253–54
Russell, Joel, 54

Sacred lands, long-term protection for, 28
Sale taxes to finance purchase of development rights programs, 166
San Joaquin County (CA), 166
San Luis Obispo County (CA), 187
San Mateo (CA), 185–86
S corporations, 227–28
Securitized installment purchase agreement, 158–60
Selling farms vs. sticking it out, 3–4, 72–73
Selling property, estate planning and, 221–24
Sending areas in TDR programs, 173–76
Septic systems, on-site, 108
Setbacks from property boundaries, 113
Shopping centers, 134
Shrewsbury Township (PA), 38
Sietsema Farms, 52
Single-use zoning, rigid, 54
Sliding scale agricultural zoning, 116, 118
Small, Stephen J., 282
Soil and Water Resources Conservation Act of 1977, 83
Soil(s):
  classes of, 8, 78–79
  erosion, 258

  pollution, 83
  quality, 62
  surveys, 39, 78–79
Sokolow, Alvin, 2
Solano County (CA), 166
Sonoma County (CA), 143, 166
South Windsor (CT), 164
Special exceptions to zoning ordinance, 46, 112, 265–67
Special taxation districts for financing purchase of development rights programs, 166–67
Spirit of Community, The (Etzioni), 251
Sprawling suburban development:
  causes of, 135
  compact development, 136
  costs of, 135
  farmlands succumbing to, 1–4
  government aided, 10
  negative aspects of, 1
  population increases and shifts, 251–52
  rise of, 133–34
  urban decline connected to, 253–55
  urban growth boundaries, 136–44
Staff in transfer of development rights programs, 188
Stanislaus County (CA), 120
State farmland protection programs:
  agricultural districts, 98–101
  differential assessment, 92–96
  economic development, agricultural, 101–2
  governor's executive order on farmland preservation policy, 279–81
  income tax, state, 97–98
  powers to influence farmland protection, four, 87–88
  property tax abatement, 98
  public-private partnerships around land trusts, 210
  purchase of development rights, 163
  right-to-farm laws, 90–91
  summary, 88–89, 103
  zoning, agricultural, 88

Stauber, Dick, 97
Stormwater management ordinance, 49
Strong farming communities, maintaining, 237–40
Structure of American farming, 63–65
Subdivision regulations, 47–50, 112–13
Subsidies for major crops, 65
Suburbs, *see* Sprawling suburban development
Suffolk County (NY), 147
Supply-and-demand imbalance in transfer of development rights programs, 189
Supreme Court, Illinois, 117
Supreme Court, U.S.:
  developer exactions, 166
  zoning, agricultural, 42, 69
Surveys in land trust process, 206
Sussex County (DE), 120
Sustainable development and growth, 255–58
Swampbuster/sodbuster provisions of the 1985 Farm Bill, 84

Tax-base sharing, 28–29, 254–55
Taxes:
  appraisals of conservation easements, 204–6
  capital gains, 160
  conservation easements, 229, 300–302
  deferred taxation, 94
  estate and transfer planning, 218, 220
  land trusts, 194
  real estate transfer, 165–66
  sales, 166
  services for taxes collected, 55–56
  special taxation districts, 166–67
  will, transferring the farm by, 225–26
  see also Property taxes
TDR, *see* Transfer of development rights (TDR)
Technology and increasing crop yields, 10

Tenant farmers, 71–72
Tenth Amendment to Constitution of the United States, 33, 42, 107
Testamentary easement, 229, 282–83
Texas, population increases in, 252
Thompson, Ed, Jr., 15
Thurston County (WA), 186
Timing and farmland protection programs, 247
Title search of the property in land trust process, 206, 207
Tourist industry, 18
Town resolutions, 271
Transfer of development rights (TDR), 171
 advantages/disadvantages, 236
 basic ideas behind, 172–73
 history of, 179–80
 combination with other farmland protection tools, 190
 innovations, workable, 186–87
 Montgomery County (MD), 130, 180–82
 New Jersey Pinelands, 182–85
 Pennsylvania, 185
 San Mateo (CA), 185–86
 shortcomings of, 187–90
 successful programs, four elements in, 174–78
 summary, 190–91
 Thurston County (WA), 186
 weak farming communities, 245
Transferring the farm, see Estate and transfer planning
Transit-oriented development (TOD), 136
Trimble, Harold, 145
Trustees of Reservations (MA), 195
Tulare County (CA), 113, 127

Undivided interest in land trust process, 200
Union of Concerned Scientists, 257
United Nations, 25
Urban decline and outward expansion, 253–55

Urban development, attractive, 28
Urban growth boundaries (UGBs):
 advantages/disadvantages, 236
 agreement form, 284–88
 Lancaster County (PA), 141–42
 Oregon, 128, 139–41
 purchase of development rights, 169
 purpose of, 136–37
 research prior to implementing, 138
 spread of, 142–44
 track record, 139
 weak farming communities, 245
Urban Land Institute, 134
Use-value assessment, 5, 92–93, 96
Use variance, 46
Utah, agricultural zoning in, 138

Value of land:
 changing role of farmland, 67–69
 increased through conservation, 54
 land trusts, 202–6
 like-kind exchange with development rights payment, 161
 purchase of development rights, 152
 transfer of development rights, 176
Variances from the zoning text or map, 46
Variety of farms, 9
Vaughan, Gerald, 72
Ventura County Agricultural Land Trust and Conservancy (CA), 194
Vermont:
 Farms for the Future Act, 82
 restrictive agreements, 95
 right-to-farm laws, 91
 state farmland protection programs, 87
 urban growth boundaries, 144
Vermont Conservation and Housing Board (VCHB), 210

Vermont Land Trust, 193, 194, 210
Vermont Natural Resources Council, 55
Village of Euclid Ohio vs. Ambler Realty Co. (zoning), 42
Virginia:
 population increases, 24, 252
 purchase of development rights, 88, 147, 159–60
 sliding-scale agricultural zoning, 118
 urban growth boundaries, 144
 use-value assessment, 96
Virginia Beach (VA), 144, 159–60

Wagner, Bob, 122, 237–38
Washington:
 financing purchase of development rights programs, 163
 purchase of development rights, 149
 Thurston County and TDRs, 186
 urban growth boundaries, 144
Washington Post, 252
Water supply/quality, 62, 121–22, 258
Weak farming communities, maintaining rural character in, 245–46
Weather, 62
Weaver, Larry, 246
Well systems, on-site, 108
Wetlands, 82, 84
Will, transferring the farm by, 225–26
Will County (IL), 253
Williamson Act, 95, 99
Wisconsin:
 income tax credits, state, 97
 population pressures in farming areas, 24
 purchase of development rights, 147
 zoning, agricultural, 106, 115–16
Wisconsin Farmland Conservancy, 194
Wolf, Peter, 8, 87, 133
Wright, Lloyd, 75

Yolo County (CA), 143

Zoning, agricultural:
    administration, 46–47
    advantages/disadvantages, 236
    cluster, 121–23
    comprehensive plan, 38
    conservation easements, 120
    cost to public purse, 215
    creating and supporting, 123–28
    differential assessment, 96
    difficult tool to use, 41–42
    downzoning, 175, 178
    farmer's perspective on, 109
    guidelines, 42–44
    Lancaster County (PA), 238–39
    legal aspects of, 107–8
    opposition to, 18, 42
    ordinance, 44–46, 109–15, 263–69
    Oregon, 128–30
    purposes of, 105–6, 110–11
    single-use, rigid, 54
    state farmland protection programs, 88
    summary, 130–31
    types of, 115–21
    unrealistic, 54
    value of land, 68–69
    weak farming communities, 2This is the third column.

About the Authors

Tom Daniels is director of the Agricultural Preserve Board of Lancaster, Pennsylvania, where he has been involved in the preservation of 140 farms and over 13,000 acres. He holds a Ph.D. in agricultural economics and has taught rural and small town planning at Iowa State and Kansas State universities. He has written several articles on farmland protection and is a coauthor of *The Small Town Planning Handbook* (Chicago: Planners Press, 1995) and *Rural Planning and Development in the United States* (New York: Guilford Publications, 1989).

Deborah Bowers is the editor and publisher of *Farmland Preservation Report*, based in Street, Maryland. She has worked as a staff writer for the *Baltimore News American* and for the *Potomac News* (Woodbridge, Virginia), where she wrote on planning and land use issues. She holds a Bachelor of Fine Arts degree in writing from Emerson College in Boston.

Island Press Board of Directors

CHAIR   Susan E. Sechler

Executive Director, Pew Global Stewardship Initiative

VICE-CHAIR   Henry Reath

President, Collector's Reprints, Inc.

SECRETARY   Drummond Pike

President, The Tides Foundation

TREASURER   Robert E. Baensch
Senior Consultant, Baensch International Group Ltd.

Peter R. Borrelli
Executive Director, Center for Coastal Studies

Catherine M. Conover

Lindy Hess
Director, Radcliffe Publishing Program

Gene E. Likens
Director, The Institute of Ecosystem Studies

Jean Richardson
Director, Environmental Programs in Communities (EPIC), University of Vermont

Charles C. Savitt
President, Center for Resource Economics/Island Press

Victor M. Sher
President, Sierra Club Legal Defense Fund

Peter R. Stein
Managing Partner, Lyme Timber Company

Richard Trudell
Executive Director, American Indian Resources Institute